T0350773

NUMERICAL CALCULATION OF ELASTOHYDRODYNAMIC LUBRICATION

NUMERICAL CALCULATION OF ELASTOHYDRODYNAMIC LUBRICATION

METHODS AND PROGRAMS

Ping Huang

School of Mechanical and Automotive Engineering
South China University of Technology
People's Republic of China

This edition first published 2015

Published by John Wiley & Sons Singapore Pte Ltd, 1 Fusionopolis Walk, #07-01 Solaris South Tower, Singapore 138628, under exclusive license granted by Tsinghua University Press for all media and languages excluding Chinese and throughout the world excluding Mainland China, and with non-exclusive license for electronic versions in Mainland China.

Registered Office
John Wiley & Sons Singapore Pte Ltd, 1 Fusionopolis Walk, #07-01 Solaris South Tower, Singapore 138628

For details of our global editorial offices, for customer services and for information about how to apply for permission to reuse the copyright material in this book please see our website at www.wiley.com.

Library of Congress Cataloging-in-Publication Data

Huang, Ping, 1957–
Numerical calculation of elastohydrodynamic lubrication : methods and programs / Ping Huang.
pages cm
Includes bibliographical references and index.
ISBN 978-1-118-92096-1 (cloth)
1. Elastohydrodynamic lubrication. 2. Lubrication and lubricants–Mathematical models. 3. Tribology. I. Title.
TJ1075.H68 2015
621.8′9–dc23

2015005240

A catalogue record for this book is available from the British Library.

Set in 11/13pt Times by SPi Global, Pondicherry, India
Printed and bound in Singapore by Markono Print Media Pte Ltd

1 2015

Contents

Preface

The book is a companion book of *Numerical Calculation of Elastohydrodynamic Lubrication: Methods and Programs* [2] written by the author and his graduate students. Since its Chinese publication, it has been widely approved, so we felt necessary to write a book similar to that in elastohydrodynamic lubrication (EHL).

The present book is based on numerical methods and programs of EHL and it describes the calculation of isothermal and thermal EHL problems in line contact, point contact, and elliptical contact in detail. In addition, numerical calculations of some special problems of EHL, such as velocities in different directions, grease lubrication, electric double layer effect, unsteady state, rough surface, and micro-polar fluid are introduced as well.

EHL calculation has a very close relationship with practical engineering. For example, EHL exists in rolling contact bearings and gears. However, compared with calculation of hydrodynamic lubrication, EHL calculation is more difficult. This is because the Reynolds equation, elastic deformation equation, viscosity pressure equation, and energy equation should be solved simultaneously. As EHL solutions are always obtained through iterations, if the iteration method is inappropriately chosen, it often causes the calculation process to be divergent. In this book, we have chosen the more reliable iteration methods and have checked and confirmed that all the programs and examples provided give satisfactory results.

However, it must be noted that because of the complexities of EHL problems, an EHL program is commonly not universal, that is, when the working conditions change, the program may not be able to get a converged or satisfied solution when used directly. In such cases, the reader should note the following conditions.

1. **Divergence for given accuracy:** There always exist some numerical errors in numerical calculation. If the accuracy given in the program cannot be met for some EHL problems, for example, double precision is not used, the calculation process may repeated at a lower level of accuracy. However, it is not really divergent. In such cases, although the accuracy does not meet the requirement, the result can be still considered as a converged solution so that the solution is useful.
2. **No convergence due to limitation of iteration number:** In any iteration program, in addition to level of accuracy, in order not to calculate infinitely, a number must be set to limit the iteration loops. Therefore, although the iteration may have begun to converge, but the limitation of the iteration number has been reached, the calculation will be stopped. In this case, although the result is not convergent, the solution is available. If readers need a more accurate solution, he or she can increase the limitation of number of iterations.
3. **Divergence for small calculation region:** The calculation region of EHL is generally set to be six times of the half width of the contact zone, where the inlet zone is about four times the half-width of the contact zone and the exit zone is about 1.5–2 times. Such a width generally meets the requirement to solve typical EHL problems. However, for some EHL problems, such as light-load, high-speed, high-viscosity, or non-Newtonian fluids, the length may be not large enough to bring the convergent result. For such a situation, please increase the calculation region so as to get a converged solution. However, this may usually bring loss of accuracy and the loss cannot be compensated by the approach to increase the node number.
4. **Divergence for unsuitable problems:** The maximum contact pressure of EHL commonly is between 0.2 and 0.8 GPa in the line or point contacts. If the maximum pressure is very low or very high, the lubrication state may tend to be a hydrodynamic lubrication or the film thickness may be too thin. In such situations, the present EHL calculation programs are not suitable anymore. In such cases, if the reader cannot obtain a convergent solution, he or she must find an appropriate program suitable for his own problem.

Readers must note that most EHL calculation programs are not universal. It is necessary for a user to be familiar with the EHL properties, and modify the corresponding statements in the program accordingly to satisfy the calculation requirements. However, this does not affect the role of the book as it shares the same purpose of the book *Numerical Calculation of Elastohydrodynamic Lubrication: Methods and Programs*. That is, the authors hope to establish an EHL calculation platform so as to avoid repeated EHL calculations and help users to complete EHL calculations and obtain solutions more quickly.

This book is also written by Lai Tianmao (Section 2.3, Chapters 8 and 10), Yang Qianqian (Chapter 4), Wang Yazhen (Chapters 6 and 9), Chen Yingjun (Chapter 11), Bai Shaoxian and Zuo Qiyang (Chapter 12) and Zhang Wei (Chapter 13). Ping Huang is responsible for the remaining chapters and has edited the whole book. Due to time and scope limitations, errors and problems cannot be avoided entirely in the book. We hope that readers' invaluable comments will help to further improve the subsequent editions of this book.

Ping Huang

Introduction

In this book, numerical methods and programs of elastohydrodynamic lubrication (EHL) are described. The contents of the book are divided into 15 chapters. In Chapters 1–3, basic numerical methods of EHL, numerical methods of elastic deformation, and numerical methods of energy equation are given, based on which the subsequent chapters are described. In Chapters 4–10, isothermal EHL problems in line contact, point contact, and elliptical contact are discussed, which are the basic and the most important contents in EHL calculation. Moreover, in Chapter 9, the EHL problem in the elliptical contact with different direction velocities is discussed. In Chapters 11–15, grease lubrication, electric double layer effect, unsteady EHL, rough surface and micro-polar fluid EHL are introduced, which are modified from the former and basic programs.

This book can serve as a textbook or a reference book for teachers and postgraduate students in mechanical engineering. It can be used as reference material for professional techniques in EHL computational analysis and engineering research as well.

Nomenclature

a	is the half-width in the point or ellipse contact region in the x direction (the velocity direction). For the point contact, $a = b = (3wR_x/2E)^{1/3}$
a	is the characteristic length in the x direction
b	is the half-width in the line contact region, $b = \sqrt{8wR/\pi E}$
b	is the radius of the elliptical contact region in the y direction (cross the velocity direction)
b	is the characteristic length in the y direction
c_p	is the specific heat capacity of lubricant
c	is the pressure viscosity coefficient of Cameron equation, which is approximate to $\alpha/15$
c_1 and c_2	are the specific heats of the materials on the up and down surfaces
D	is the density–temperature coefficient, $D = -0.00065\ \mathrm{K}^{-1}$
D_{ij}^{kl}	are the two-dimensional elastic deformation stiffnesses
e_k	is the ellipse rate, $e_k = R_x/R_y$
E	is the equivalent elastic modulus of the two contact surface materials, $1/E = 1/2\left(1-\nu_1^2/E_1 + 1-\nu_2^2/E_2\right)$
E_1 and E_2	are the elastic modulus corresponding to the up and down surfaces of the materials
G^*	is a material parameter, $G^* = \alpha E$
h	is the lubricant film thickness
h_1	is the minimum lubricant thickness film

h_0	is the central film thickness to be determined based on the load balancing condition
H	is the nondimensional film thickness, for the line contact $H = hR/b^2$, for the point or elliptical contact $H = hR_x/a^2$
H_0	is the nondimensional central film thickness
J	is the mechanical equivalent of heat
J	is the inertia factor of micropolar fluid
k	is the thermal conductivity of lubricant
k_1 and k_2	are the thermal conductivities of the up and down surfaces
K_{ij}	is the one-dimensional elastic deformation stiffness
l	is the characteristic length of micropolar fluid
L	is the nondimensional characteristic length of micro-polar fluid
m	is the number of nodes in the y direction
n	is the number of nodes in the x direction
n	is the rheological index, ≤ 1
N	is the coupling coefficient of micropolar fluid
p	is the pressure
p_0	is the pressure–viscosity coefficient, $p_0 = 1.96 \times 10^8$
p_H	is Hertz contact pressure, for the line contact $p_H = 2w/\pi b$, for the point contact $p_H = 3w/2\pi a^2$, and for the elliptic contact $p_H = 3w/2\pi ab$
P	is the nondimensional pressure, $P = p/p_H$
P_{tr}	the deformation coefficient of elliptical contact,$P_{tr} = 3wR_x/\pi^2 a^2 bE$
R	is the equivalent radius of curvature, $1/R = 1/R_1 \pm 1/R_2$; for the outer contact, take + and for the inner contact, take –
R_1 and R_2	are the integrated radii of curvature of the up and down surfaces in the line or point contact
R_x and R_y	are the equivalent radii of curvature in the x and y direction of both surfaces
s	is the sliding–rolling ratio, $s = (u_1 - u_2)/u_s$
t	is the time
T	is the temperature
T	is one of the nondimensional time, $T = Ut/b$
T_0	is the initial temperature, here $T_0 = 303$ K
T^*	is the nondimensional temperature, $T^* = T/T_0$
$u_s = u_1 + u_2/2$	is the average velocity of the up and down surfaces along the x direction
u	is the fluid velocity component in the x direction
u_1 and u_2	are the tangential velocities respectively to the up and down surface in the x direction
U^*	is the velocity parameters, for the line contact $U^* = \eta_0 u_s/ER$, for the point or ellipse contact $U^* = \eta_0 u_s/ER_x$
v	is the fluid velocity component in the y direction
v_s	is the average speed in y direction along the two surfaces, $v_s = (v_1 + v_2)/2$

v_1 and v_2 are the tangential velocities respectively to the up and down surface in the y direction

$v(x)$ is the displacement of the elastic deformation generated by pressure

V^* is the velocity parameter, $V^* = \eta_0 v_s / ER_x$

w is the load, for line contact w is the load per unit length of the contact, and for the point or ellipse contact w is the total load

w is the fluid velocity component in the z direction

W is the nondimensional load, for the line contact $W = \pi/2$, for the point contact $W = 2\pi/3$, and for the elliptical contact $W = 2\pi b/3a$

W^* is the load parameters, for the line contact $W^* = w/ER$, for the point or elliptical contact $W^* = w/ER_x^2$

x is the coordinate which is in the same direction of the main speed

x_0 is the inlet coordinate

x_e is the outlet coordinate

X is nondimensional coordinate of x, for the line contact $X = x/b$, for the point contact $X = x/a$

X_0 and X_e are the nondimensional coordinates of the inlet and outlet

y is the coordinate which is vertical to the main speed

Y is the nondimensional coordinate in the y direction, $Y = y/a$

z is the coordinate in the film thickness direction

z is the coefficient of viscosity pressure formula, for a mineral oil it is generally $z = 0.68$

Z is the nondimensional coordinate of z, for the line contact $Z = zR/b^2$, for the point or ellipse contact $Z = zR_x/a^2$

α is the pressure coefficient of oil or base oil of grease in Barus viscosity–pressure formula

α is the length proportional factor of the contact ellipse, $\alpha = a/b$

α_T is the coefficient in the density–temperature equation, its unit is °C^{-1}

β is the viscosity–temperature coefficient in the Barus formula, for oil it usually is 0.03°C^{-1}

ΔX is the equally divided nondimensional increment between the nodes of the mesh, $\Delta X = X_i - X_{i-1}$

ε is the Reynolds coefficient, for oil $\varepsilon = \rho^* H^3 / \eta^* \lambda$, for grease

$$\varepsilon = \lambda\left(H^{(2+1/n)} / \phi^{*1/n}\right)$$

$$\varepsilon_0 = \varepsilon_{i-1/2,j} + \varepsilon_{i+1/2,j} + \varepsilon_{i,j-1/2} + \varepsilon_{i,j+1/2}$$

$$\varepsilon_{i\pm 1/2} = \frac{1}{2}(\varepsilon_i + \varepsilon_{i\pm 1})$$

$$\varepsilon_{i\pm 1/2,j} = \frac{1}{2}\left(\varepsilon_{i,j} + \varepsilon_{i\pm 1,j}\right)$$

ϕ is the plastic viscosity

ϕ_0	is the plastic viscosity of grease at the normal pressure and at the room temperature
ϕ^*	is the nondimensional viscosity, $\phi^* = \phi/\phi_0$
γ	is the material constant of micropolar fluid
λ	is the parameter of the coefficients ε, for the line contact $\lambda = 12\eta_0 u_s R^2 / b^3 p_H$, for the point or elliptical contact $\lambda = 12\eta_0 u_s R_x^2 / a^3 p_H$, for grease $\lambda = {p_H}^{1/n} b^{2+1/n} / 2u_s(2+1/n) R^{(1+1/n)} 2^{1/n} {\phi_0}^{1/n}$
η	is the viscosity of the lubricant
η_0	is the viscosity of the lubricant at $p = 0$ and $T = T_0$
η^*	is the nondimensional viscosity of the lubricant, $\eta^* = \eta/\eta_0$
μ	is the viscosity of Newtonian fluid of micropolar fluid
ρ	is the lubricant density
ρ_0	is the density of lubricant at $p = 0$ and $T = T_0$
ρ^*	is the nondimensional density of lubricant, $\rho^* = \rho/\rho_0$
ρ_1 and ρ_2	are the material densities of the up and down surfaces
χ	is the rotary viscosity of micropolar fluid
ω_1, ω_2, and ω_3	are the rotational angular velocities of micropolar fluid respectively in the x, y, and z directions
ν_1 and ν_2	are the Poisson ratios of the two surface materials

Note: If the above symbols are stated in the text otherwise, the content described here is no longer valid.

1

Basic equations of elastohydrodynamic lubrication

1.1 Basic equations

1.1.1 One-dimensional Reynolds equation of elastohydrodynamic lubrication

One-dimensional isothermal Reynolds equation of elastohydrodynamic lubrication (EHL) is given as follows [1, 2]:

$$\frac{d}{dx}\left(\frac{\rho h^3}{\eta}\frac{dp}{dx}\right)=12u_s\frac{d(\rho h)}{dx} \tag{1.1}$$

Let us assume that the surfaces do not stretch and the density of lubricant does not change with time, the general form of the one-dimensional Reynolds equation of hydrodynamic lubrication can be written as follows:

$$\frac{\partial}{\partial x}\left(\frac{\rho h^3}{\eta}\frac{\partial p}{\partial x}\right)=6(u_0-u_h)\frac{\partial(\rho h)}{\partial x}+12\rho(w_h-w_0) \tag{1.2}$$

where, the last term is obtained from $\partial(\rho h)/\partial t=\rho(w_h-w_0)$ as we assume that the density of lubricant does not change with time.

Numerical Calculation of Elastohydrodynamic Lubrication: Methods and Programs, First Edition.
Ping Huang.

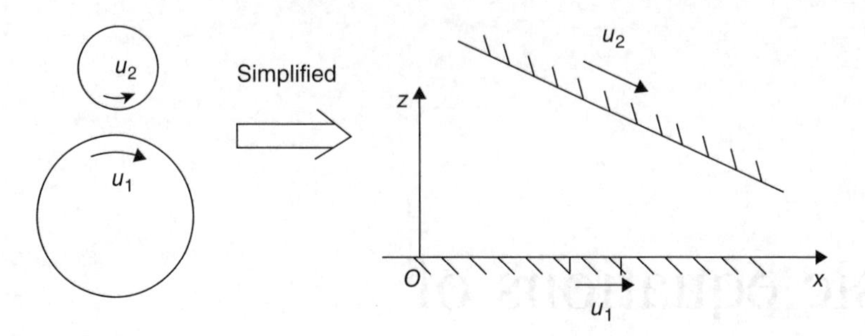

Figure 1.1 Velocity analysis of rolling problem: (a) two rolling cylinders and (b) simplified model

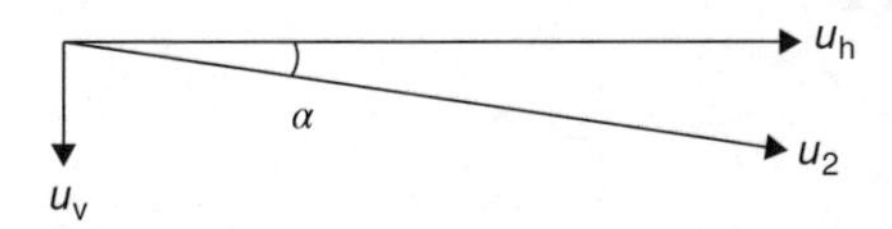

Figure 1.2 Velocity decomposition of rolling problem

In order to transform the general form of Reynolds equation (1.2) into the EHL Reynolds equation (1.1), let us analyze the problem of two surfaces with the rolling speed, see Figure 1.1a. First, expand the surfaces as shown in Figure 1.1b, where we set the down surface to be horizontal and the up surface to be declined.

In order to substitute the velocities into the Reynolds equation (1.2), let us analyze the velocities given in Figure 1.1b. If set u is the velocity in the x direction and w in the z direction, we can see that on the down surface, $u_0 = u_1$ and $w_0 = 0$. The velocity on the up surface can be decomposed into the horizontal and the vertical two components as shown in Figure 1.2.

After decomposition, the two velocity components can be written as follows:

$$\begin{aligned} u_\mathrm{h} &= u_2 \cos\alpha \\ u_\mathrm{v} &= u_2 \sin\alpha = w_\mathrm{h} \end{aligned} \tag{1.3}$$

where, α is the declined angle of the up surface.

Because the declined angle α is very small, we can approximately take $\cos\alpha \approx 1$ and $\sin\alpha \approx (\partial h/\partial x)$. Therefore, we have

$$\begin{aligned} u_\mathrm{h} &= u_2 \cos\alpha \approx u_2 \\ w_\mathrm{h} &= u_2 \sin\alpha \approx u_2 \frac{\partial h}{\partial x} \end{aligned} \tag{1.4}$$

Substituting u_0, u_h, w_0, and w_h into the Reynolds equation (1.2), we have:

$$\frac{d}{dx}\left(\frac{\rho h^3}{\eta}\frac{dp}{dx}\right)=6(u_1-u_2)\frac{\partial(\rho h)}{\partial x}+12\rho u_2\frac{\partial h}{\partial x} \tag{1.5}$$

Let $u_1+u_2=2u_s$ and because $u_0=u_1$ and $u_h=u_2$, we have:

$$\frac{d}{dx}\left(\frac{\rho h^3}{\eta}\frac{dp}{dx}\right)=6(u_1+u_2)\frac{\partial(\rho h)}{\partial x}=12u_s\frac{\partial(\rho h)}{\partial x} \tag{1.6}$$

Therefore, Equation 1.1 is obtained same as Equation 1.6.

It should be noted that when simplifying Equation 1.5 into Equation 1.6, $\rho(\partial h/\partial x)=(\partial(\rho h)/\partial x)$ is used in the second term of the equation. However, in an EHL problem, the lubricant density usually varies with pressure and hence it varies with coordinate x too. Therefore, the result of Equation 1.6 or 1.1 is an approximation, but the error caused by this approximation is negligible.

1.1.2 Two-dimensional Reynolds equation of EHL

Similar to the one-dimensional isothermal EHL problem, if the direction of the coordinate x is set to be consistent with the surface velocity direction, the Reynolds equation of the two-dimensional isothermal EHL can be written as follows:

$$\frac{\partial}{\partial x}\left(\frac{\rho h^3}{\eta}\frac{\partial p}{\partial x}\right)+\frac{\partial}{\partial y}\left(\frac{\rho h^3}{\eta}\frac{\partial p}{\partial y}\right)=12u_s\frac{\partial(\rho h)}{\partial x} \tag{1.7}$$

The EHL problems in the point and elliptical contacts all belong to the two-dimensional problems so that they have the same Reynolds equation as (1.7). Only their nondimensional treatment processes are different because the sizes in the different directions are different.

1.1.3 EHL Reynolds equation with two direction velocities

Some EHL problems cannot set the x coordinate to be the same as the direction of the main velocity, such as with rotational speed, or the velocity is not collinear with the main axis of the contact ellipse. Then, we have to consider the problem with two velocity components in x and y directions. In such a problem, the EHL Reynolds equation will become:

$$\frac{\partial}{\partial x}\left(\frac{\rho h^3}{\eta}\frac{\partial p}{\partial x}\right)+\frac{\partial}{\partial y}\left(\frac{\rho h^3}{\eta}\frac{\partial p}{\partial y}\right)=12u_s\frac{\partial(\rho h)}{\partial x}+12v_s\frac{\partial(\rho h)}{\partial y} \tag{1.8}$$

1.1.4 Time-dependent EHL Reynolds equation

Adding the squeezing item, the EHL Reynolds equation becomes time dependent. Then, we should use the time-dependent EHL Reynolds equation as follows.

$$\frac{\partial}{\partial x}\left(\frac{\rho h^3}{\eta}\frac{\partial p}{\partial x}\right) = 12u_s\frac{\partial(\rho h)}{\partial x} + 12\rho\frac{\partial h}{\partial t} \tag{1.9}$$

As the squeezing effect caused by rolling has been already considered in Equation 1.9 (the first term on the right side of the equation), it should be noted that when we derive EHL Reynolds equation, we need not consider the squeezing effect again. However, what we need to consider is the velocity moves vertically to the center, which is also time-dependent.

1.2 Film thickness equation without elastic deformation

1.2.1 Film thickness equation in line contact

Here, first we derive the film thickness equation in the line contact. The same method can also be applied to derive the film thickness equation in the point contact.

An EHL problem in the line contact with two rolling cylinders is shown in Figure 1.3.

Assume that the radii of the two cylinders are R_1 and R_2, respectively, and set the origin of the coordinates at the center connecting the line of the two cylinders; the x-axis is perpendicular to this connecting line and the undeformed

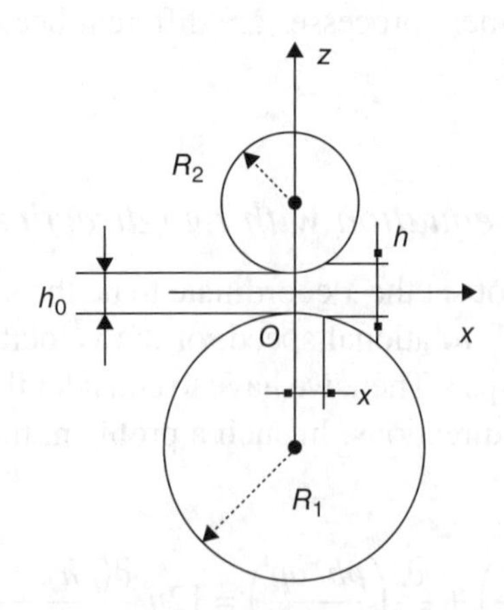

Figure 1.3 Two cylinder EHL in line contact

distance of the two cylinders is h_0, then the film thickness h at the point x can be written as:

$$h = h_0 + R_1 - \sqrt{R_1^2 - x^2} + R_2 - \sqrt{R_2^2 - x^2} \tag{1.10}$$

Expand Equation 1.10 and consider that x is much small compared to the two radii so that the higher order terms can be omitted, then we obtain:

$$h = h_0 + \frac{x^2}{2}\left(\frac{1}{R_1} + \frac{1}{R_2}\right) \tag{1.11}$$

Let R be equal to $(1/R) = (1/R_1) + (1/R_2)$, which is called as the equivalent radius, and the film thickness Equation 1.11 can be written as:

$$h = h_0 + \frac{x^2}{2R} \tag{1.12}$$

Therefore, the film thickness of the two cylinders in the line contact is similar to that of a flat plane that contacts a cylinder with radius R.

1.2.2 Film thickness equation in point contact

Similar to the line contact situation, we can derive the film thickness equation in the point contact with no deformation as follows:

$$h = h_0 + \frac{x^2 + y^2}{2R} \tag{1.13}$$

It should be noted that in the derivation of the film thickness equation in the point contact, we assume that the curvature radii of the two surfaces in contact are respectively R_1 and R_2 and the equivalent radius of expression $(1/R) = (1/R_1) + (1/R_2)$ is similar to that in the line contact.

1.2.3 Film thickness equation in ellipse contact

If two objects are arbitrary ellipsoids, the contact area is an ellipse. We can choose two equivalent curvature radii to be R_x and R_y at h_0, and their tangential directions are the contact bodies of the main axes, x and y. Then, the equation of the film thickness can be written as:

$$h = h_0 + \frac{x^2}{2R_x} + \frac{y^2}{2R_y} \tag{1.14}$$

1.3 Surface elastic deformation

1.3.1 One-dimensional elastic deformation equation

In EHL calculation, the elastic deformation of the surface must be considered and it should be added to the undeformed film thickness. Therefore, the final film thickness equation in the line contact, that is Equation 1.12, will be written as:

$$h = h_0 + \frac{x^2}{2R} + v(x) \tag{1.15}$$

The last term in Equation 1.15 is the elastic deformation. It is obtained from the deformation formula of elasticity theory. If a load that acts on a semi-infinite plane causes the plane to deform, as shown in Figure 1.4, then the elastic deformation equation can be written as:

$$v(x) = -\frac{2}{\pi E}\int_{x_0}^{x_e} p(s)\ln(s-x)^2 ds + c \tag{1.16}$$

where, c is a constant to be determined.

In calculation, c is merged in the rigid film thickness h_0 in Equation 1.15, which can be determined by balancing the load.

1.3.2 Two-dimensional elastic deformation equation

Similar to the line contact situation of EHL, the film thickness in the point contact can also be obtained by adding the elastic deformation to the film thickness without

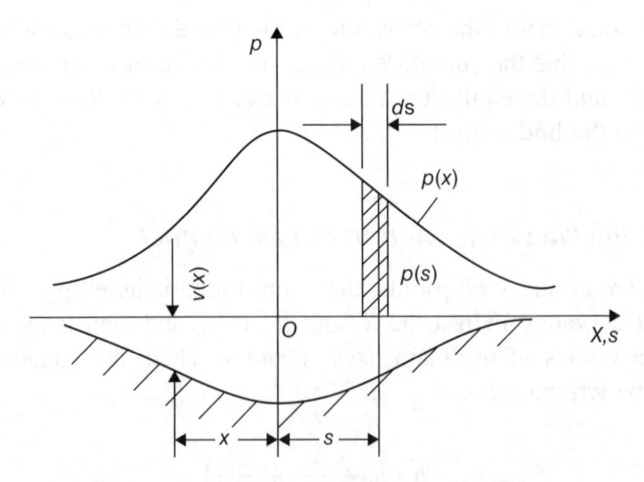

Figure 1.4 Elastic deformation in line contact

deformation, that is, Equation 1.13. The deformed film thickness in the point contact can be written as follows:

$$h = h_0 + \frac{x^2 + y^2}{2R} + v(x,y) \tag{1.17}$$

The last term in Equation 1.17 is the elastic deformation. It is obtained from elasticity theory, in which a normal distributing pressure $p(s, t)$ is applied on a semi-infinite plane. The elastic deformation can be written as follows:

$$v(x,y) = \frac{2}{\pi E}\iint_{\Omega} \frac{p(s,t)}{\sqrt{(x-s)^2 + (y-t)^2}} dsdt \tag{1.18}$$

Similar to the point contact problem, the film thickness equation with elastic deformation in the ellipse contact can be written as follows:

$$h = h_0 + \frac{x^2}{2R_x} + \frac{y^2}{2R_y} + v(x,y) \tag{1.19}$$

The elastic deformation in the elliptical contact is the same as that in the point contact. Therefore, Equation 1.18 can be used.

1.4 Viscosity and density equations varying with pressure and temperature

1.4.1 Viscosity equations

1.4.1.1 Viscosity–pressure equation

When a liquid or gas is subject to high pressure, the distance between its molecules will be reduced but the intermolecular force increases so that its viscosity increases. For example, when pressure in a mineral oil is higher than 0.02 GPa, the variation of its viscosity with pressure is very significant. The rate of viscosity varying with pressure increases too. If pressure is higher than 1 GPa, the viscosity of lubricant can be several orders higher than the usual. While pressure is high enough, the liquid feature of a mineral oil will be lost and it will become a waxy solid. Therefore, under a heavy load, the viscosity–pressure characteristic is important and essential to EHL.

The common viscosity–pressure equations to describe the relationship between viscosity and pressure are as follows:

$$\text{Barus} \quad \eta = \eta_0 e^{\alpha p} \tag{1.20}$$

$$\text{Roelands} \quad \eta = \eta_0 \exp\left\{ (\ln \eta_0 + 9.67) \left[-1 + \left(1 + \frac{p}{p_0} \right)^z \right] \right\} \tag{1.21}$$

$$\text{Cameron} \quad \eta = \eta_0 (1 + cp)^{16} \tag{1.22}$$

When pressure is higher than 1 GPa, the viscosity calculated by Barus equation will be too large but that by Roelands equation is more reasonable.

The viscosity–pressure coefficient α of a mineral oil is generally equal to $2.2 \times 10^{-8}\ \mathrm{m^2 \cdot N^{-1}}$. The viscosity–pressure coefficients of some lubricants are given in Tables 1.1 and 1.2.

1.4.1.2 Viscosity–temperature equations

When considering the effect of temperature on viscosity, Reynolds viscosity-temperature equation is usually used. It is as follows.

$$\eta = \eta_0 \exp[-\beta(T - T_0)] \tag{1.23}$$

Although Rolelands viscosity–temperature equation is slightly more complicated, it is more realistic and hence usually more useful.

$$\eta = \eta_0 \exp\left\{ (\ln \eta_0 + 9.67) \left[\left(\frac{T - 138}{T_0 - 138} \right)^{-1.1} - 1 \right] \right\} \tag{1.24}$$

Table 1.1 Viscosity–pressure coefficient of some mineral oil α ('$\times 10^{-8}\ \mathrm{m^2 \cdot N^{-1}}$)

	Cycloalkyl			Paraffinic		
Temperature, °C	Spindle oil	Light oil	Heavy oil	Light oil	Heavy oil	Cylinder oil
30	2.1	2.6	2.8	2.2	2.4	3.4
60	1.6	2.0	2.3	1.9	2.1	2.8
90	1.3	1.6	1.8	1.4	1.6	2.2

Table 1.2 Viscosity–pressure coefficient α of some base oil viscosity at 25°C ($\times 10^{-8}\ \mathrm{m^2 \cdot N^{-1}}$)

Type of lubricating oil	α	Type of lubricating oil	α
Paraffinic	1.5–2.4	Alkyl silicone	1.4–1.8
Cycloalkyl	2.5–3.6	Polyether	1.1–1.7
Aromatic	4–8	Fragrance silicone	3–5
Polyolefin	1.5–2.0	Chlorinated paraffins	0.7–5
Diester	1.5–2.5		

1.4.1.3 Viscosity–pressure–temperature equations

When taking into account the effects of temperature and pressure on the viscosity at the same time, the following equations of viscosity–pressure–temperature are commonly used.

1. Barus and Reynolds equation

$$\eta=\eta_0 \exp[\alpha p-\beta(T-T_0)] \tag{1.25}$$

2. Roelands equation

$$\eta=\eta_0 \exp\left\{(\ln\eta_0+9.67)\left[\left(1+\frac{p}{p_0}\right)^z\times\left(\frac{T-138}{T_0-138}\right)^{-1.1}-1\right]\right\} \tag{1.26}$$

Equation 1.25 is relatively simple and is easily used, but Equation 1.26 is more accurate so that it is more used in practice.

1.4.2 *Density equation*

1.4.2.1 Density-pressure equation

The density of a lubricant increases with increasing pressure. In order to simplify the calculation, the variation of the density with pressure is often described by the following equation:

$$\rho=\rho_0\left(1+\frac{0.6p}{1+1.7p}\right) \tag{1.27}$$

where, p is the pressure, whose unit is GPa.

1.4.2.2 Density–temperature equation

Influence of temperature on density of a lubricant is not very significant. The volume of lubricant increases due to thermal expansion and hence the density decreases with increase in temperature. Under the usual lubrication conditions, influence of temperature on the density is usually very small and hence can be ignored. When the influence of temperature on the density of lubricant should be considered, the following density–temperature equation can be used.

$$\rho=\rho_0[1-\alpha_T(T-T_0)] \tag{1.28}$$

where, α_T is the thermal expansion coefficient of lubricant and it is usually calculated with one of the following two equations.

If the unit of the viscosity is mPa · s, when the viscosity is less than 3000 mPa · s, and $\log \eta \leq 3$, then, α_T is equal to

$$\alpha_T = \left(10 - \frac{9}{5}\log \eta\right) \times 10^{-4} \tag{1.29}$$

When the viscosity is larger than 3000 mPa · s, and $\log \eta > 3$, then, α_T is equal to

$$\alpha_T = \left(5 - \frac{3}{8}\log \eta\right) \times 10^{-4} \tag{1.30}$$

1.4.2.3 Density–pressure–temperature equation

When influences of pressure and temperature on the density together are taken into account, we can combine the density–temperature and density–pressure equations. A commonly used equation is as follows:

$$\rho = \rho_0 \left[1 + \frac{0.6p}{1 + 1.7p} + D(T - T_0)\right] \tag{1.31}$$

Because the aforementioned formulas are analytical, they can be directly used in numerical calculations.

1.5 Load balancing condition

1.5.1 Load balancing equation

Because an EHL problem must be numerically solved under a certain given load, pressure must satisfy the equilibrium condition or equation. An EHL solution is said to be correct only when both the pressure distribution and the load balancing condition are satisfied. Generally, the load balancing condition is called the load balancing equation. It is more appropriate to call the load balancing condition because the load is already given before the pressure is obtained.

1.5.1.1 Load balancing equation in line contact

$$w - \int_{x_0}^{x_e} p(s)ds = 0 \tag{1.32}$$

The nondimensional form of the load balancing equation

$$\frac{\pi}{2}-\int_{X_0}^{X_e} P(X)dX=0 \tag{1.33}$$

Its discrete form

$$\frac{\pi}{2}-\sum_{i=1}^{n} P_i \Delta X_i=0 \tag{1.34}$$

1.5.1.2 Load balancing equation in point contact

$$w-\int_{y_a}^{y_b}\int_{x_0}^{x_e} p(x,y)dxdy=0 \tag{1.35}$$

The nondimensional form of the load balancing equation

$$\frac{2\pi}{3}-\int_{x_0}^{x_e}\int_{y_0}^{y_e} P(X,Y)dXdY=0 \tag{1.36}$$

Its discrete form

$$\frac{2\pi}{3}-\sum_{i=1}^{n}\sum_{j=1}^{m} P_{ij}\Delta X_i \Delta Y_j=0 \tag{1.37}$$

1.5.2 Numerical calculation of load balancing

It must be noted that even if the pressure has converged, the real solution of the problem may not be obtained. The reason is that unlike the film thickness of an ordinary hydrodynamic lubrication problem is fixed, the film thickness of an EHL problem must be adjusted by the given load so that the pressure distribution will change. This means that the load is not calculated but is pre-given and the sum of the pressure must be equal to it. Therefore, readers must pay full attention to the condition of the load balancing.

When a pressure distribution has been obtained from the Reynolds equation and the iterative convergence condition is satisfied, we must check whether the load balancing equation, that is, Equation 1.34 or 1.37, is satisfied or not. If satisfied or approximately satisfied, the pressure can be considered to be the correct one. For the one-dimensional problem, the convergence criterion is as follows:

$$\left|\frac{\pi}{2}-\sum_{i=1}^{n} P_i \Delta X_i\right| \le \varepsilon_w \tag{1.38}$$

where, ε_w is the load balancing tolerance.

During calculation, especially at the beginning, the load balancing condition is not satisfied. Let us consider the line contact EHL problem to illustrate the load balancing method.

We record the difference between the load and the pressure sum as follows:

$$\Delta W = \frac{\pi}{2}-\sum_{i=1}^{n} P_i \Delta X_i \tag{1.39}$$

If Equation 1.38 is not satisfied, we have

$$|\Delta W| = \left|\frac{\pi}{2}-\sum_{i=1}^{n} P_i \Delta X_i\right| > \varepsilon_w \tag{1.40}$$

Therefore, we must adjust the rigid film thickness h_0 in the film thickness equation. The nondimensional form of the film thickness equation (1.12) can be written as follows:

$$H(X) = H_0 + \frac{X^2}{2} + v(X) \tag{1.41}$$

According to the relationship between the load and the film thickness, it is known that:

1. If $\Delta W > 0$, the pressure sum is smaller than the load. Therefore, in order to increase the pressure, it is necessary to reduce the film thickness, that is, to decrease H_0.
2. If $\Delta W < 0$, the pressure sum is larger than the load. Therefore, in order to decrease the pressure, it is necessary to increase the film thickness, that is, to increase H_0.

The key problem is that H_0 may dramatically change the pressure distribution. Therefore, if ΔH_0 large, it easily causes the solving process to be divergent. However, if ΔH_0 is too small, it will increase the computational workload. For different EHL problems, it is difficult to give ΔH_0 a same value.

Here, we introduce two methods to balance the load in EHL numerical calculation.

1.5.2.1 Bisection method

Because a single suitable ΔH_0 is difficult to find, we can gradually reduce ΔH_0 according to the number of iterations. This approach is valid to most EHL problems, that is:

1. By using the available empirical formula of the relationship between the film thickness and the load, calculate the initial H_0. For an example, one of the nondimensional minimum film thickness formula is:

$$H_{\min} = aG^{*\alpha}U^{*\beta}W^{*\gamma} \tag{1.42}$$

where, a is a constant and α, β, and γ are the indicates in the empirical formula.
Then, the initial H_0 can be calculated by the following formula:

$$\min\left\{H_0 + \frac{X_i^2}{2} + v_i\right\} = H_{\min} \tag{1.43}$$

where, X_i is the nondimensional coordinate of Node i, v_i is the elastic deformation, and $H_{\min}$ is obtained from Equation 1.42.

2. Choose ΔH_0

$$\Delta H_0 = (0.01 \sim 0.005)H_{\min} \tag{1.44}$$

If $H_{\min}$ in Equation 1.44 cannot be accurately given, ΔH_0 can be determined to try.

3. Correction H_0
(1) If $\Delta W > 0$, $H_0 - \Delta H_0 \rightarrow H_0$.
(2) If $\Delta W < 0$, $H_0 + \Delta H_0 \rightarrow H_0$.

4. Adjust ΔH_0
After k loops of iterations, the difference of the load and the pressure sum will continue to reduce. Therefore, we can half ΔH_0.

$$\Delta H_0^{Mk+1} \underset{\Delta}{\Leftarrow} \frac{\Delta H_0^{Mk}}{2} \tag{1.45}$$

5. Ending condition

After k loops, if ΔH_0 is less than the given error ε_{H_0}, then the calculation is finished. If the pressure satisfies the convergence criterion, the solution can be thought as the real EHL solution. If the pressure does not satisfy the convergence criterion, the result can also be used as a reference solution.

Now consider a program used to calculate an isothermal EHL problem in the point contact (PROGRAM POINTEHL) as an example. The following statements are cited from the source codes of the program.

(1) In the main program, the Dowson–Hamrock empirical formula [1] of film thickness is used to calculate the minimum film thickness H_{min}.

```
......
HM0=3.63*(RX/B)**2*G**0.49*U**0.68*W0**(-0.073)
......
```

(2) In Subroutine EHL, we use the following statements according to Equation 1.45 to adjust ΔH_0 after several loops (here $k = MK = 20$):

```
......
IF(MK.GE.20)THEN
MK=1
DH=0.5*DH
ENDIF
......
```

(3) In Subroutine HREE, we use the following statements according to Equation 1.43 to calculate the initial rigid film thickness H_0 and ΔH_0, and then use ΔH_0 to modify H_0:

```
   ......
   IF(KK.EQ.0)THEN
   KK=1
   DH=0.01*HM0
   H00=-HMIN+HM0
   ENDIF
   W1=0.0
   DO 32 I=1,N
   DO 32 J=1,N
32 W1=W1+P(I,J)
   W1=DX*DX*W1/G0
   DW=1.-W1
   IF(DW.LT.0.0)H00=H00+DH
   IF(DW.GT.0.0)H00=H00-DH
   ......
```

With calculation, it is shown that the method is effective.

1.5.2.2 Derivative method

If the load is unbalanced, the difference ΔW of the load and the pressure sum can be considered to be proportional to the film thickness increment ΔH_{min} corresponding to Equation 1.42. Therefore, we can derivate the equation to obtain:

$$\Delta H_{min} = \gamma G^{*\alpha} U^{*\beta} W^{*\gamma-1} \Delta W \tag{1.46}$$

Because $\Delta W = \pi/2 - \sum_{i=1}^{N} P_i \Delta X_i$, which has been obtained from Equation 1.39, the rigid displacement H_0 can be corrected as follows:

$$\Delta H_0 = \Delta H_{min} = \gamma G^{*\alpha} U^{*\beta} W^{*\gamma-1} \Delta W \tag{1.47}$$

This method has been applied into the program with multigrid method to solve the line contact EHL problems in this book. The following statements are cited from Subroutine HREE of program M1.

1. In the main program, we use the Dowson–Higginson formula [1] to estimate a nondimensional EHL film thickness in the line contact. We can obtain the derivative CW of the nondimensional load W as follows.

```
......
C3=1.6*(R/B)**2*G**0.6*U**0.7*W1**(-0.13)
......
CW=-1.13*C3
......
```

2. In Subroutine HREE, modify DW and correct H_0 according to the difference of the load and the pressure sum. In the program, in order to ensure that the thickness modified is not negative, we use a conditional sentence to adjust H_0.

```
   ......
   IF(H0+CW*DW.GT.0.0)HM0=H0+CW*DW
   IF(H0+CW*DW.LE.0.0)HM0=HM0*C3
48 H00=HM0-HMIN
   ......
```

A similar approach can also be used in the point contact EHL problem.

1.6 Finite difference method of Reynolds equation

If the boundary conditions are used to solve a differential equation, it is known as the boundary value problem. In EHL calculations, the finite difference method is commonly used to solve the Reynolds equation. The major steps of the finite difference method are as follows.

1.6.1 Discretization of equation

First, change the partial differential equations into nondimensional forms. This is accomplished by expressing variables in the universal form.

Then, divide the solution region into a mesh with uniform or nonuniform grids. In Figure 1.5, a uniform mesh is given; the x direction consists of nodes m and y direction of nodes n and hence the total nodes are equal to $m \times n$. The division of a mesh is determined by calculation accuracy. For a common EHL problem, $m \times n = 50–100 \times 50–100$ will usually meet the requirement of accuracy. Sometimes, in order to improve accuracy, if the unknown variables undergo a rapid change in the region, the grid needs to be refined by using two or more different subgrids.

Let us consider pressure p as an example. The distribution of pressure in the whole region can be expressed by each node p_{ij}. According to the differential regularities, the

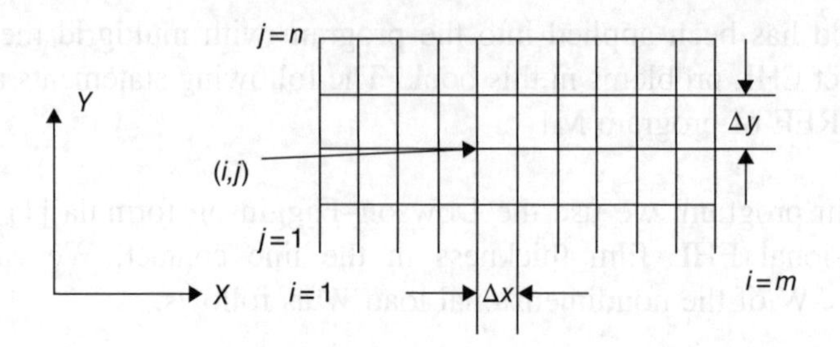

Figure 1.5 Uniform mesh

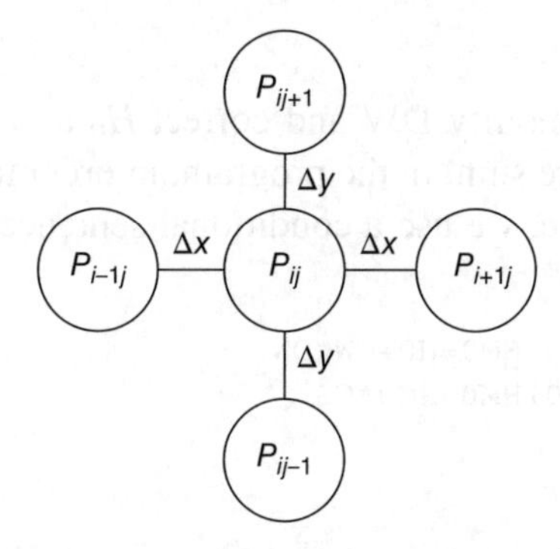

Figure 1.6 Relationship of difference

partial derivatives at the node (i,j) can be represented by the surrounding node variables.

As shown in Figure 1.6, the expressions of partial derivatives of the intermediate difference at node (i, j) can be written in the following forms:

$$\left(\frac{\partial p}{\partial x}\right)_{i,j} = \frac{p_{i+1,j} - p_{i-1,j}}{2\Delta x}$$
$$\left(\frac{\partial p}{\partial y}\right)_{i,j} = \frac{p_{i,j+1} - p_{i,j-1}}{2\Delta y} \tag{1.48}$$

The second-order partial derivatives of the intermediate difference are as follows:

$$\left(\frac{\partial^2 p}{\partial x^2}\right)_{i,j} = \frac{p_{i+1,j} + p_{i-1,j} - 2p_{i,j}}{(\Delta x)^2}$$
$$\left(\frac{\partial^2 p}{\partial y^2}\right)_{i,j} = \frac{p_{i,j+1} + p_{i,j-1} - 2p_{i,j}}{(\Delta y)^2} \tag{1.49}$$

In order to obtain the unknown variables near the border, the forward or backward difference formulas are used as follows:

$$\left(\frac{\partial p}{\partial x}\right)_{i,j}=\frac{p_{i+1,j}-p_{i,j}}{\Delta x}$$
$$\left(\frac{\partial p}{\partial y}\right)_{i,j}=\frac{p_{i,j+1}-p_{i,j}}{\Delta y} \tag{1.50}$$

$$\left(\frac{\partial p}{\partial x}\right)_{i,j}=\frac{p_{i,j}-p_{i-1,j}}{\Delta x}$$
$$\left(\frac{\partial p}{\partial y}\right)_{i,j}=\frac{p_{i,j}-p_{i,j-1}}{\Delta y} \tag{1.51}$$

Usually, the accuracy of the intermediate difference is high. The following intermediate difference formula can also be used in calculation:

$$\left(\frac{\partial p}{\partial x}\right)_{i,j}=\frac{p_{i+1/2,j}-p_{i-1/2,j}}{\Delta x} \tag{1.52}$$

1.6.2 Different forms of Reynolds equation

Based on the above formulas, the two-dimensional Reynolds equation can be written in a standard form of the second-order partial differential equation.

$$A\frac{\partial^2 p}{\partial x^2}+B\frac{\partial^2 p}{\partial y^2}+C\frac{\partial p}{\partial x}+D\frac{\partial p}{\partial y}=E \tag{1.53}$$

where, *A*, *B*, *C*, *D*, and *E* are the known parameters.

Equation 1.53 can be applied to each node. According to Equations 1.48 and 1.49, the relationship of pressure $p_{i,j}$ at Node (i, j) with the adjacent pressures can be written as follows:

$$\tilde{p}^k_{i,j}=C_N p^k_{i,j+1}+C_S p^{k+1}_{i,j-1}+C_E p^k_{i+1,j}+C_W p^{k+1}_{i-1,j}+G \tag{1.54}$$

where, $C_N=\left[\frac{B}{\Delta y^2}+\frac{D}{2\Delta y}\right]/K$; $C_S=\left[\frac{B}{\Delta y^2}-\frac{D}{2\Delta y}\right]/K$; $C_E=\left[\frac{A}{\Delta x^2}+\frac{C}{2\Delta x}\right]/K$; $C_W=\left[\frac{A}{\Delta x^2}-\frac{C}{2\Delta x}\right]/K$; $G=-\frac{E}{K}$; $K=2\left[\frac{A}{\Delta x^2}+\frac{B}{\Delta y^2}\right]$.

By using Equation 1.54, we can write out the finite different forms of equations of each node. The variables of border nodes should satisfy the boundary conditions because their values are pre-given. Therefore, a set of linear algebraic equations can be obtained. The number of the equations obtained is equal to the unknown variables so that it is a definite problem to be solved. By using an elimination method or iterative method, we can solve the algebraic equations. If the convergent result meets with the given precision, the solution of each node has been found.

The following sections will describe how to solve lubrication problems by an iteration method.

1.6.3 *Iteration of differential equation*

During iterating, in order to guarantee the convergence criterion, a relaxation or super relaxation iteration method is often used, that is, weightily to add the old pressure to the iterated pressure to obtain a new pressure as shown in Equation 1.55.

$$p_{ij}^{k+1} = (1-\alpha)p_{ij}^{k} + \alpha\tilde{p}_{ij}^{k} \tag{1.55}$$

where, p_{ij}^{k} is the old pressure, p_{ij}^{k+1} is the new pressure, $\tilde{p}_{ij}^{k}$ is the iterated pressure by Equation 1.54, and α is a positive number, usually larger than 0 and less than 1.

Then, use the new pressure to check whether the convergency is satisfied. If not, carry out the next iteration.

1.6.4 *Iteration convergence condition*

By using the differential Equation 1.54, the next iterative pressure is obtained to combine the new and old pressures of a node, and the correction of all the nodes in the lubrication area will be completed only when a whole iteration is finished. Usually, Equation 1.54 is not satisfied. However, through iteration, the pressure often converges to the real solution.

There are two general methods to be used to make sure whether the iterative process meets the accuracy requirement or not. They are the absolute and relative accuracy criteria.

1.6.4.1 Absolute accuracy criterion

When a new iteration is finished, Equation 1.27 can be rewritten as follows:

$$r_{ij}^{k+1} = p_{i,j}^{k+1} - C_N p_{i,j+1}^{k+1} - C_S p_{i,j-1}^{k+1} - C_E p_{i+1,j}^{k+1} - C_W p_{i-1,j}^{k+1} - G \tag{1.56}$$

where, r_{ij}^{k+1} is the residual of the differential equation at Node (i, j).

According to the accuracy criterion, the residual of all the nodes should be less than a very small positive number ε_1, that is:

$$\left| r_{ij}^{k+1} \right| \leq \varepsilon_1 \tag{1.57}$$

Sometimes, in order to facilitate the use of two iterations, the pressure difference can be used to determine the convergence, that is:

$$\left| p_{ij}^{k+1} - p_{ij}^{k} \right| \leq \varepsilon_2 \tag{1.58}$$

where, ε_2 is a very small positive number.

1.6.4.2 Relative accuracy criteria

For most lubrication problems, because the residuals of the equation may vary, it is quite difficult to accurately give the absolutely accuracies, whether ε_1 or ε_2. Therefore, the relative accuracy criteria are often used in practice.

There are two commonly used criteria of the relative accuracy. The stricter one is to have the pressure on each node to meet the relative accuracy, that is:

$$\left| \frac{p_{i,j}^{k+1} - p_{i,j}^{k}}{p_{i,j}^{k+1}} \right| \leq \varepsilon_3 \tag{1.59}$$

A less strict criterion of relative accuracy is that the relative accuracy of the two iteration loads meets the following requirement:

$$\frac{\sum\sum \left| p_{i,j}^{k+1} - p_{i,j}^{k} \right|}{\sum\sum p_{i,j}^{k+1}} \leq \varepsilon_4 \tag{1.60}$$

In the above two equations, ε_3 and ε_4 are the relative accuracies. According to the convergence difficulty of the problem, they are often chosen between 0.01 and 10^{-6}, that is, the relative precision of convergence is from one percent to one per million.

Generally speaking, if the working conditions are not too bad, the relative accuracy is convenient and practical. Therefore, it is used as a primary convergence criterion. However, the reader must be aware that in the EHL numerical calculation by using the relative accuracy, although the pressure is converged, it may not be the real solution. A simple example is that as the two error pressure solutions are close to each other, they meet the relative convergence criterion but they are not the real solutions.

2

Numerical calculation method and program of elastic deformation

2.1 Numerical method and program of elastic deformation in line contact

2.1.1 Equations of elastic deformation

In EHL calculation, we need to superimpose the elastic deformation on the original film thickness. An equivalent problem that arises when an elastic cylinder contacts with a rigid plane is shown in Figure 2.1. The expression of the deformed film thickness can be written as follows:

$$h(x) = h_0 + \frac{x^2}{2R} + v(x) \tag{2.1}$$

For the line contact, as the length and the radius of the surfaces are much larger than the contact width, the problem can be considered in the plane strain state, that is, its elastic deformation is equivalent to that is produced by a pressure distribution acting on an infinite-wide elastic plane as shown in Figure 2.2.

According to the theory of elasticity, the elastic deformation in the vertical direction can be written as follows:

$$v(x) = -\frac{2}{\pi E}\int_{s_1}^{s_2} p(s)\ln(s-x)^2 ds + c \tag{2.2}$$

Numerical Calculation of Elastohydrodynamic Lubrication: Methods and Programs, First Edition.
Ping Huang.

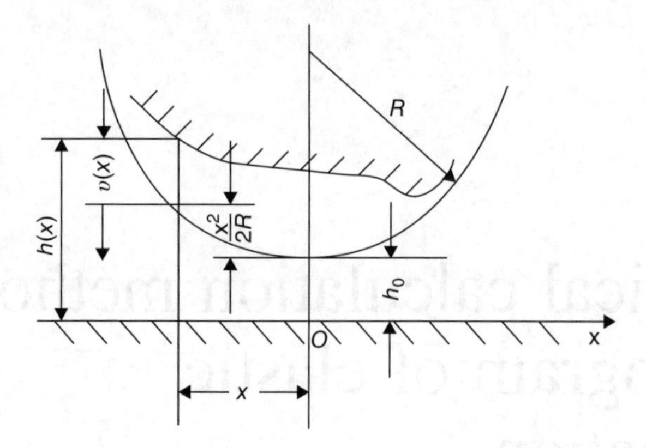

Figure 2.1 Shape of film thickness in line contact

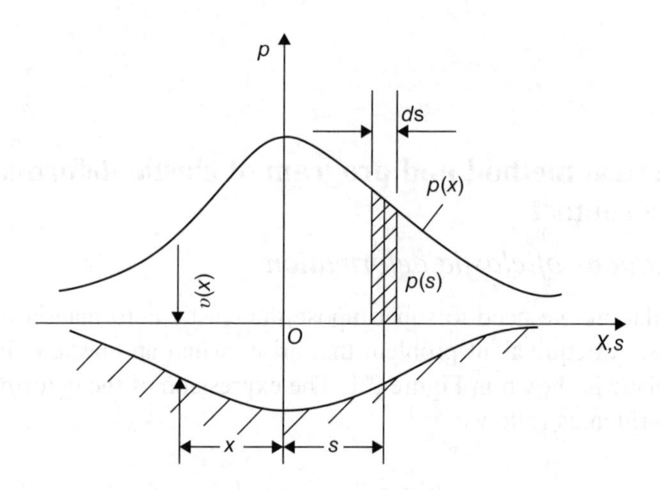

Figure 2.2 Elastic deformation in line contact

where, s is an additional coordinate of the x axis, which indicates the distance between the load $p(s)ds$ and the origin of the coordinate; $p(s)$ is the pressure at s; s_1 is the start point of the pressure $p(s)$ and s_2 is the end point; and c is a constant to be determined. Because the central film thickness h_c in Equation 2.1 can be determined based on the load balance condition, c can be usually mergered in it. Therefore, c or h_c is no longer considered separately.

2.1.2 *Numerical method of elastic deformation*

Because the pressure $p(x)$ can be obtained by solving Reynolds equation, the elastic deformation $v(x)$ can be solved numerically by integrating the pressure. The integral part of the deformation Equation 2.2 is as follows:

$$v = \int_{s_1}^{s_2} p(s)\ln(s-x)^2 ds \tag{2.3}$$

Although a singular point may exist as x is equal to s in Equation 2.3, the point can be removed in the numerical integration. If an integration interval contains the singular point, the mean value theorem can be used, that is, the two nodes near the midpoint are used so as to avoid the singular point. Therefore,

$$\Delta v = \int_{x}^{x+\Delta x} p(s)\ln(s-x)^2 ds \approx p\left(x+\frac{\Delta x}{2}\right)\ln\left(\frac{\Delta x}{2}\right)^2 \Delta x \tag{2.4}$$

Thus, the singular point can be removed out of the integration interval without any significant difference.

For the equant grid, the numerical integration of Equation 2.4 can be simplified as follows:

$$v_i = \Delta x \sum_{j=1}^{n} (a_{i-j} + \ln \Delta x) p_j \tag{2.5}$$

where,
$a_{i-j} = (i-j+0.5)(\ln(i-j+0.5)-1) - (i-j-0.5)(\ln|i-j-0.5|-1),\ i-j=0,\ldots,N.$

Nondimensionlize Equation 2.5, we have

$$V_i = -\frac{\Delta X}{\pi} \sum_{j=1}^{n} \left[K_{ij} + \ln(\Delta X)\right] P_j \tag{2.6}$$

where, $K_{ij} = (|i-j|+0.5)[\ln(|i-j|+0.5)-1] - (|i-j|-0.5)(\ln||i-j|-0.5|-1),\ i,j=0,$ $1,\ldots,n.$

2.1.3 Calculation diagram and program

2.1.3.1 Flowchart (Figure 2.3)

2.1.3.2 Program

Elastic deformation calculation subroutine VI

In the program, the input variables are the node number N, the equidistant distance DX between two neighbor nodes, and the pressure P(I, J) of each node. The output variable is the elastic deformation array V(I).

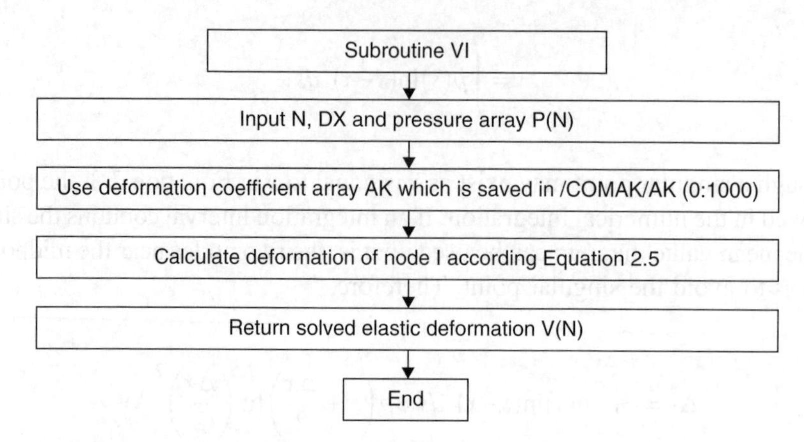

Figure 2.3 Calculation diagram of elastic deformation in line contact

Coefficient calculation subroutine SUMAK

As the coefficient K_{ij} is independent of pressure, it is calculated only once, but it needs to be used many times in elastic deformation calculation. In order to save time, K_{ij} will be calculated once in advance by Subroutine SUBAK.

In the subroutine, MM is substituted, which is equal to the number of nodes. The equation in the subroutine to calculate the coefficients of elastic deformation $K_{ij} = (|i-j|+0.5)[\ln(|i-j|+0.5)-1] - (|i-j|-0.5)(\ln||i-j|-0.5|-1)$ is $\mathrm{AK(I)} = (\mathrm{I}+0.5)*(\mathrm{ALOG(ABS(I}+0.5))-1.)-(\mathrm{I}-0.5)*(\mathrm{ALOG(ABS(}I-0.5))-1.)$.

Subroutine SUBAK is called only once at the very beginning. The elastic deformation coefficient AK(I) is put in a common region COMAK so that it is used by other subroutines.

It should be noted that in the array AK(IJ), IJ is the number between deformation node v_i to the pressure node p_j, that is, IJ = |I – J|.

```
   SUBROUTINE VI(N,DX,P,V)
   DIMENSION P(N),V(N)
   COMMON /COMAK/AK(0:1100)
   PAI1=0.318309886
   C=ALOG(DX)
   DO 10 I=1,N
   V(I)=0.0
   DO 10 J=1,N
   IJ=IABS(I-J)
10 V(I)=V(I)+(AK(IJ)+C)*DX*P(J)
   DO I=1,N
   V(I)=-PAI1*V(I)
   ENDDO
   RETURN
   END
   SUBROUTINE SUBAK(MM)
   COMMON /COMAK/AK(0:1100)
    DO 10 I=0,MM
```

```
10 AK(I)=(I+0.5)*(ALOG(ABS(I+0.5))-1.)-(I-0.5)*(ALOG(ABS(I-0.5))-1.)
   RETURN
   END
```

2.1.3.3 Calculation diagram

Figure 2.4 shows the diagram used to calculate the elastic deformation V(I) and it is superimposed on the initial film thickness to obtain the deformed film thickness. Note that before calculating the elastic deformation, call SUBAK to calculate the elastic deformation coefficients first.

In this diagram, N is the node number, X0 is the inlet coordinate, X1 is the outlet coordinate, AK is the coefficient array of elastic deformation, V is the elastic deformation vector, X is the coordinate vector, P is the pressure vector, and H is the film thickness vector.

2.1.4 Example

Because the curve of a cylinder is close to $h=x^2/2$ in the contact area, the pressure distribution will be a parabola, that is,

$$p_i=0 \qquad |x_i|>b$$
$$p_i=p_H\sqrt{1-x_i^2} \qquad |x_i|\le b$$

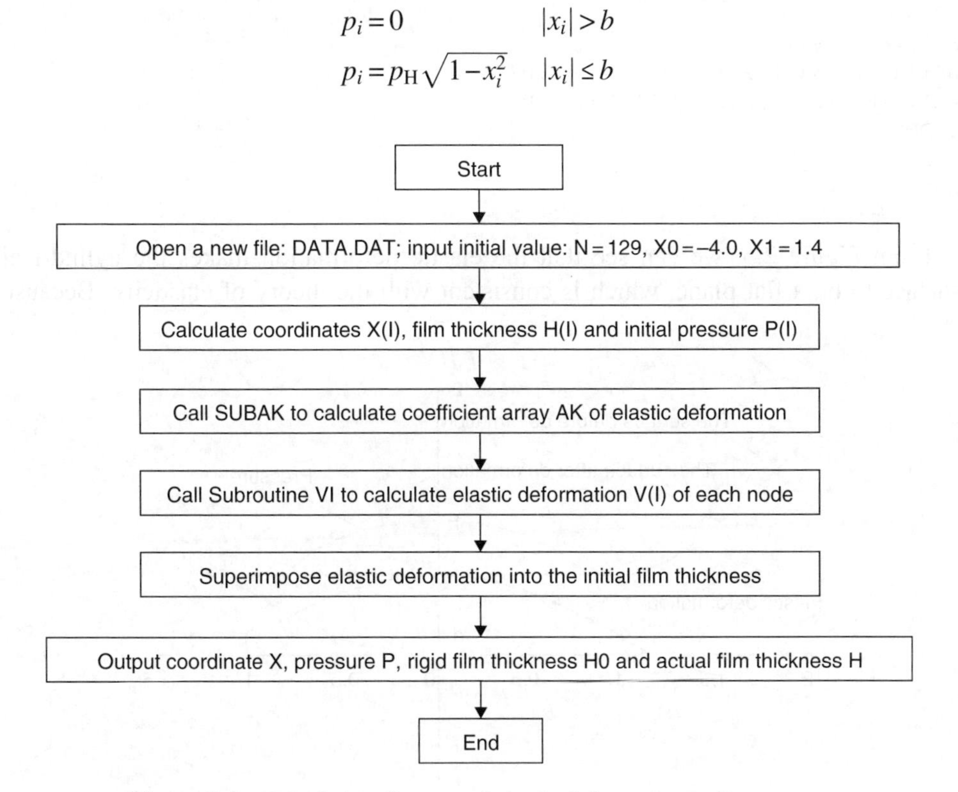

Figure 2.4 Calculation diagram of elastic deformation in line contact

The nondimensional film thickness can be expressed as $H(X) = H_c + X^2/2 + V(X)$ and the nondimensional Hertz contact pressure is $P = \sqrt{1-X^2}$ $(|X| \leq 1)$. In the program, the total nodes are 129. If the nondimensional coordinate X is from –4.0 to 1.4, the EHL calculation program in the line contact is as follows:

```
DIMENSION P(1000),H0(1000),H(1000),V(1000),X(1000)
OPEN(8,FILE='DATA.DAT',STATUS='UNKNOWN')
N=129
X1=1.4
X0=-4.0
DX=(X1-X0)/(N-1.0)
DO I=1,N
X(I)=-4.0+(I-1)*DX
H0(I)=0.5*X(I)*X(I)
H(I)=H0(I)
P(I)=0.0
IF(X(I).GE.-1.0.AND.X(I).LE.1.0)THEN
P(I)=SQRT(1-X(I)*X(I))
ENDIF
ENDDO
CALL SUBAK(N)
CALL VI(N,DX,P,V)
DO I=1,N
H(I)=H(I)+V(I)
WRITE(*,*)X(I),P(I),V(I),H0(I),H(I)
WRITE(8,*)X(I),P(I),V(I),H0(I),H(I)
ENDDO
STOP
END
```

From Figure 2.5, we can see that the elastic deformation makes the cylindrical surface to be a flat plane, which is consistent with the theory of elasticity. Because

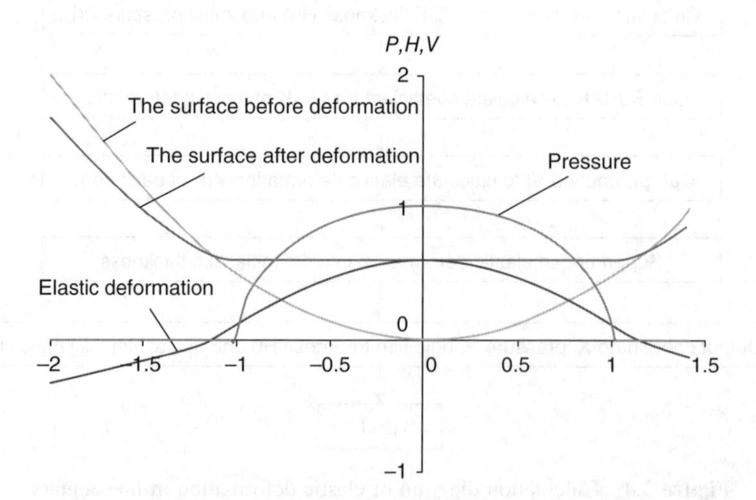

Figure 2.5 Calculated results of elastic deformation of line contact

the load balance condition is not used, there is a gap between two surfaces in the figure. For a real contact, the actual gap is zero.

2.2 Numerical method and program for elastic deformation in point contact

2.2.1 *Equation of elastic deformation*

For the point contact, the solution region is generally a rectangular area as shown in Figure 2.6, where *AB* is the inlet border, *CD* is the outlet border, and *AD* and *BC* are the leakage sides. In the figure, α, β, and γ can be used to determine the solution region.

Furthermore, a point contact problem can usually be simplified into an equivalent elastic ball contacting with a rigid plane. When considering the elastic deformation, the film thickness can be expressed as follows:

$$h = h_0 + \frac{x^2 + y^2}{2R} + v(x,y) \tag{2.7}$$

According to the theory of elasticity, the relationship between the pressure $p(x,y)$ and the deformation $v(x,y)$ can be expressed as follow:

$$v(x,y) = \frac{2}{\pi E} \iint_{\Omega} \frac{p(s,t)}{\sqrt{(x-s)^2 + (y-t)^2}} dsdt \tag{2.8}$$

where, s and t are the additional coordinates in the x and y directions and Ω is the solving domain as shown in Figure 2.7.

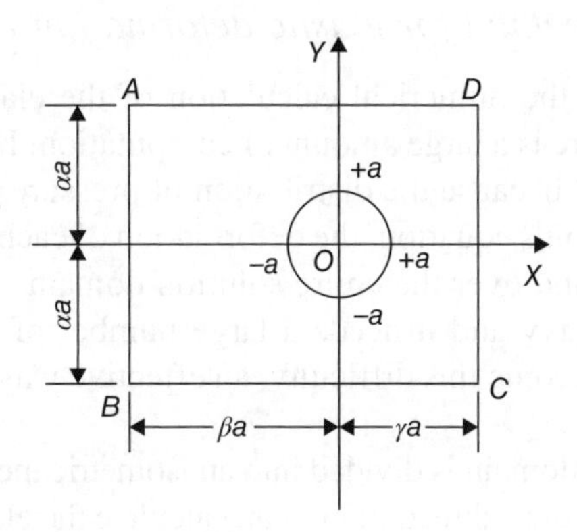

Figure 2.6 Solution region in point contact

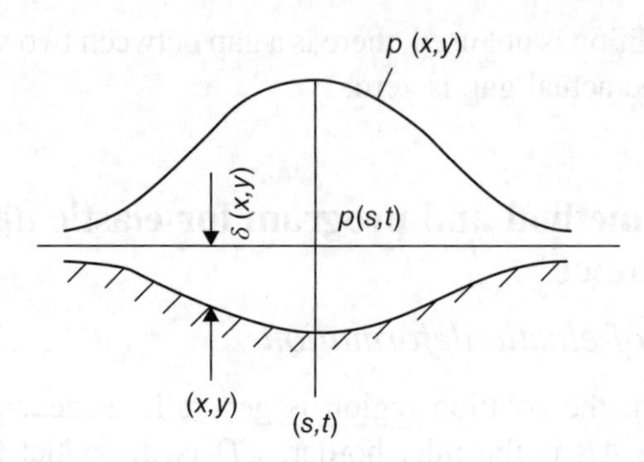

Figure 2.7 Elastic deformation in point contact

The denominator of the integral part in Equation 2.8 is the distance between the pressure point (s, t) and the deformed point (x, y). If $x = s$ and $y = t$, the integral is singular. However, it can be removed with the variable substitution method. By transforming the coordinate origin to (s, t), that is, $x' = x - s$ and $y' = y - t$, Equation 2.8 becomes

$$v(x,y) = \frac{2}{\pi E}\iint_{\Omega} \frac{p'(x,y)}{\sqrt{x^2 + y^2}} dxdy \tag{2.9}$$

Then, by using the polar coordinate, that is, $x' = r\cos\theta$ and $y' = r\sin\theta$, we will obtain the following integral equation:

$$v(x,y) = \frac{2}{\pi E}\iint_{\Omega} p''(r,\theta) drd\theta \tag{2.10}$$

The above integral equation has no singular point anymore.

2.2.2 *Numerical method for elastic deformation in point contact*

The main difficulty in the numerical calculation of the elastic deformation in the point contact is that there is a large amount of computation. If a usual numerical integration method is used, because the distribution of pressure $p(x, y)$ is different from that generated by Reynolds equation, the deformation of each node will be calculated in each iteration loop and over the entire solution domain. This makes the computational work much heavy and it needs a large number of computer storage units as well. In order to overcome this difficulty, an effective way is to use a deformation matrix.

Suppose the solution domain is divided into an isometric mesh with m nodes in the x direction and n nodes in the y direction, we can calculate the elastic deformation $a_{i-k,j-l}$

at Node (k, l) when there is a unit pressure acting on Node (i, j) while the rest nodes have no pressure anymore. Then, the discrete form of the elastic deformation equation becomes:

$$v_{ij} = \frac{2}{\pi E}\sum_{k=1}^{n}\sum_{l=1}^{m} a_{i-k,j-l}p_{kl} \tag{2.11}$$

where, v_{ij} is the elastic deformation at Node (i, j), $a_{i-k,j-l}$ is the elastic deformation coefficient between Node (k, l) and Node (i, j), and p_{kl} is the pressure at Node (k, l).

Nondimensionlize Equation 2.11, and we have

$$V_{ij} = \frac{2\Delta X}{\pi^2}\sum_{k=1}^{n}\sum_{l=1}^{m} D_{ij}^{kl}P_{kl} \tag{2.12}$$

where, V_{ij} is the elastic deformation at Node (i, j), $\Delta X = \Delta Y$ is the nondimensional distance of the two neighbor nodes, P_{lk} is the nondimensional pressure at Node (k, l), and D_{ij}^{kl} is the nondimensional elastic coefficient.

Similar to the line contact, we need to calculate D_{ij}^{kl} once and store it for later calculation. Substitute D_{ij}^{kl} in Equation 2.12 to calculate the deformation repeatedly in the iteration process, and therefore the amount of computation can be significantly reduced with no loss of accuracy.

As we use the isometric grid mesh, we have

$$\begin{aligned} D_{ij}^{kl} &= D_{kj}^{il} \\ D_{ij}^{kl} &= D_{il}^{kj} \end{aligned} \tag{2.13}$$

Finally, the calculated point number decreases to $m \times n$. The formula used to calculate D_{ij}^{kl} is as follows:

$$\begin{aligned} D_{ij}^{kl} = {} & (|j-l|+0.5)\ln[f(|i-k|+0.5,|j-l|+0.5)/f(|i-k|-0.5,|j-l|+0.5)] \\ & +(|i-k|-0.5)\ln[f(|j-l|-0.5,|i-k|-0.5)/f(|j-l|+0.5,|i-k|-0.5)] \\ & +(|j-l|-0.5)\ln[f(|i-k|-0.5,|j-l|-0.5)/f(|i-k|+0.5,|j-l|-0.5)] \\ & +(|i-k|+0.5)\ln[f(|j-l|+0.5,|i-k|+0.5)/f(|j-l|-0.5,|i-k|+0.5)] \end{aligned} \tag{2.14}$$

where, $f(x,y) = x + \sqrt{x^2 + y^2}$.

2.2.3 Calculation diagram

The diagram used to calculate the elastic deformation in the point contact is shown in Figure 2.8. The deformation coefficient AK should be calculated in advance.

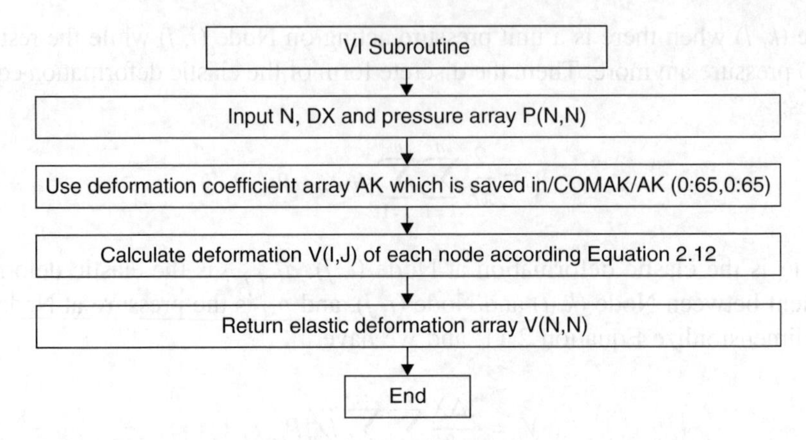

Figure 2.8 Calculation diagram of elastic deformation in point contact

2.2.4 *Program*

```
      SUBROUTINE VI(N,DX,P,V)
      DIMENSION P(N,N),V(N,N)
      COMMON /COMAK/AK(0:65,0:65)
      PAI1=0.2026423
      DO 40 I=1,N
      DO 40 J=1,N
      H0=0.0
      DO 30 K=1,N
      IK=IABS(I-K)
      DO 30 L=1,N
      JL=IABS(J-L)
   30 H0=H0+AK(IK,JL)*P(K,L)
   40 V(I,J)=H0*DX*PAI1
      RETURN
      END
      SUBROUTINE SUBAK(MM)
      COMMON /COMAK/AK(0:65,0:65)
      S(X,Y)=X+SQRT(X**2+Y**2)
      DO 10 I=0,MM
      XP=I+0.5
      XM=I-0.5
      DO 10 J=0,I
      YP=J+0.5
      YM=J-0.5
      A1=S(YP,XP)/S(YM,XP)
      A2=S(XM,YM)/S(XP,YM)
      A3=S(YM,XM)/S(YP,XM)
      A4=S(XP,YP)/S(XM,YP)
      AK(I,J)=XP*ALOG(A1)+YM*ALOG(A2)+XM*ALOG(A3)+YP*ALOG(A4)
   10 AK(J,I)=AK(I,J)
      RETURN
      END
```

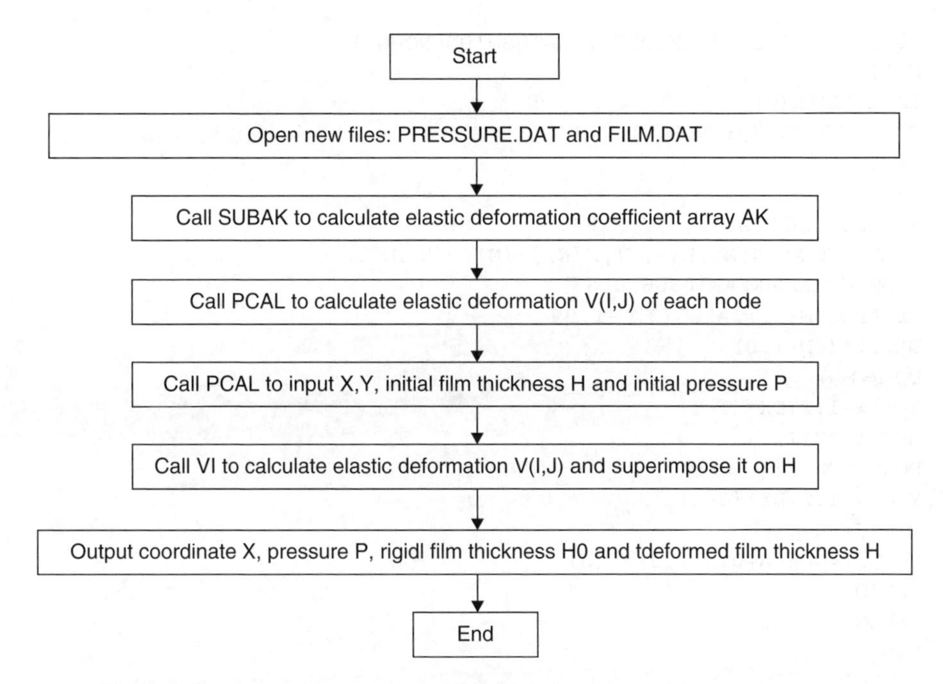

Figure 2.9 Main calculation diagram of elastic deformation in point contact

It should be noted that in the program, the input node number N is only in one direction. Therefore, the total nodes number is equal to $N \times N$. DX = DY is the nondimensional spacing of the isometric mesh. P(I, J) is the nondimensional pressure array, which will be obtained by solving the Reynolds equation. V(I, J) is the nondimensional elastic deformation array, which will be the output after being calculated. In addition, because $a_{i-k,j-l}$ is not related to the pressure, it can be calculated in advance and be ready to be used to calculate the elastic deformation. Therefore, in order to save computing time, Subroutine SUBAK is used to calculate $a_{i-k,j-l}$.

2.2.5 *Example*

The elastic deformation of a ball with a radius R contacting with a rigid plane is calculated by using the Hertzian contact stress distribution. The solving domain is $X \times Y = (-1.2, 1.2) \times (-1.2, 1.2)$. The region is divided into 33×33 isometric nodes. The calculation diagram is shown in Figure 2.9.

2.2.5.1 Main program diagram

2.2.5.2 Program

The codes of the main program of the elastic deformation in the point contact are as follows:

```
DIMENSION P(4500),H(4500),V(4500),X(65),Y(65)
OPEN (8,FILE='PRESS.DAT',STATUS='UNKNOWN')
```

```
      OPEN (10,FILE='FILM.DAT',STATUS='UNKNOWN')
      N=33
      CALL SUBAK(N)
      CALL PCAL(N,X,Y,P,H,V)
      STOP
      END
      SUBROUTINE PCAL(N,X,Y,P,H,V)
      DIMENSION P(N,N),H(N,N),X(N),Y(N),V(N,N)
      COMMON /COMAK/AK(0:65,0:65)
      KL=ALOG(N-1.)/ALOG(2.)-1.99
      DX=2.4/(N-1.0)
      DO I=1,N
      X(I)=-1.2+DX*(I-1)
      A=X(I)*X(I)
      DO J=1,N
      Y(J)=-1.2+DX*(J-1)
      P(I,J)=0.0
      H(I,J)=0.5*A+0.5*Y(J)*Y(J)
      ENDDO
      ENDDO
      M=0
      DO I=1,N
      DO J=1,N
      A=1.0-X(I)*X(I)-Y(J)*Y(J)
      IF(A.GE.0.0) P(I,J)=SQRT(A)
      ENDDO
      ENDDO
      CALL VI(N,DX,P,V)
      DO 10 I=1,N
      DO 10 J=1,N
      H(I,J)=H(I,J)+V(I,J)
   10 CONTINUE
      XP=1.0
      WRITE(8,20)XP,(Y(I),I=1,N)
      WRITE(10,20)XP,(Y(I),I=1,N)
      DO I=1,N
      WRITE(8,20)X(I),(P(I,J),J=1,N)
      WRITE(10,20)X(I),(H(I,J),J=1,N)
      ENDDO
   20 FORMAT(1X,34(F6.3,1X))
      STOP
      END
```

2.2.5.3 Calculation results

The pressure distribution and the deformed film thickness are shown in Figure 2.10. In the calculation of EHL, the rigid film thickness h_0 in Equation 2.7 or H0 in the program should be adjusted by the load balance condition.

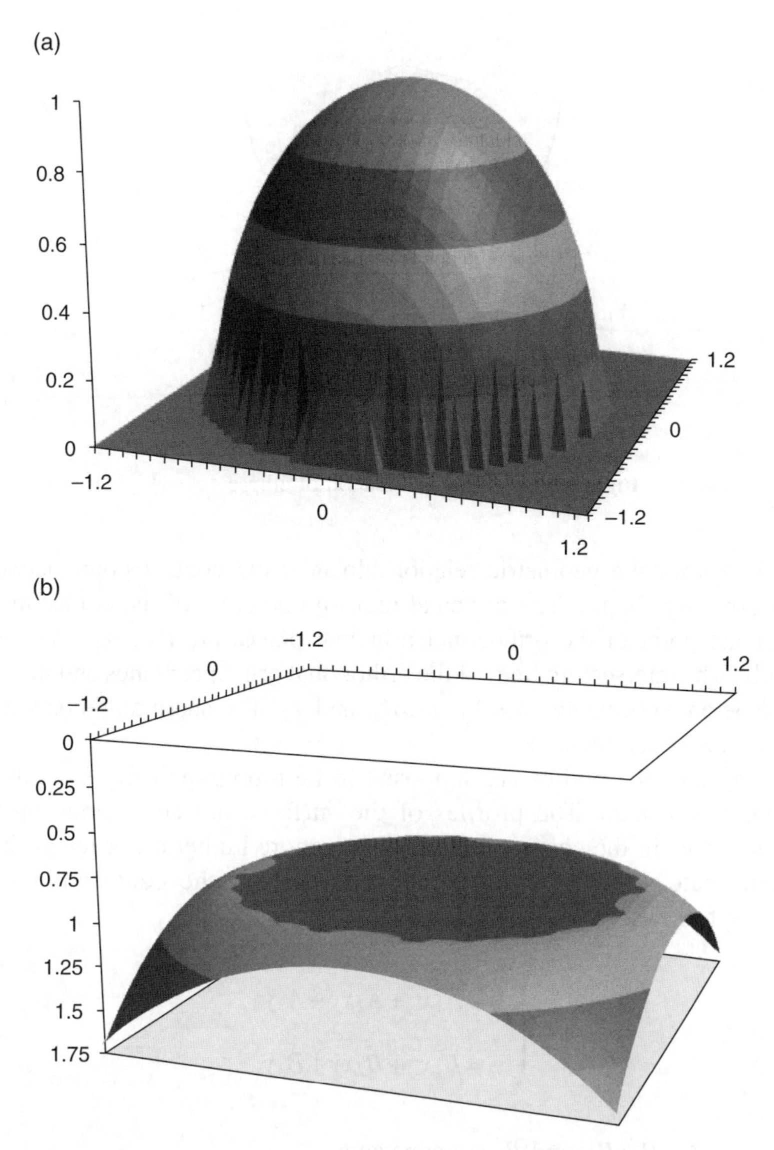

Figure 2.10 Calculation results of elastic deformation in point contact. (a) Pressure distribution and (b) Film thickness after deformation

2.3 Numerical method and program of elastic deformation in ellipse contact

2.3.1 Contact geometry

The contact between two bodies with arbitrary shapes can be expressed by the contact of ellipsoids constructed by the principal radii of curvature of the contact points.

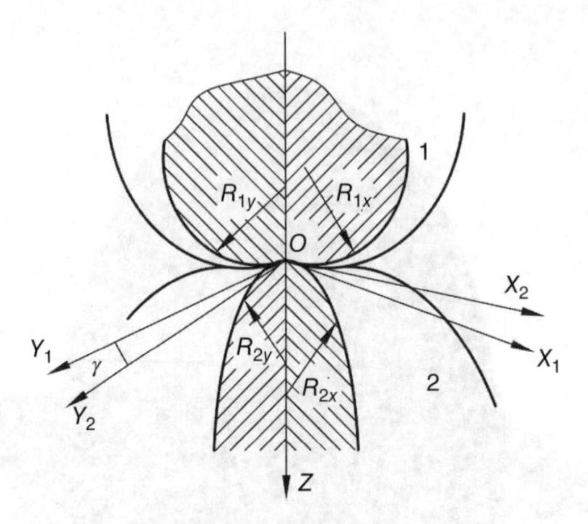

Figure 2.11 General situation of ellipse contact

Figure 2.11 shows the geometric relationship near the contact point between two bodies of arbitrary shapes. The principal radii of curvature of the contacting bodies on the contact point of the orthogonal principal planes are R_{1x}, R_{1y}, R_{2x}, and R_{2y}, respectively. The intersecting lines of the orthogonal principal planes and the common tangent plane are coordinate axes X_1, Y_1, X_2, and Y_2. The sharp angle between these coordinate axes is defined as γ.

The surfaces of the bodies are supposed to be topographically smooth in both micro- and macroscale. The profiles of the surfaces are continuous up to their second derivative in the contact region. If we ignore higher order terms, in a Cartesian coordinate system (x, y, z), the surfaces near the contact point can be expressed as follows:

$$\begin{cases} z_1 = A_1x^2 + A_2xy + A_3y^2 \\ z_2 = B_1x^2 + B_2xy + B_3y^2 \end{cases} \tag{2.15}$$

where, A_1, A_2, A_3, B_1, B_2, and B_3 are constants.

The distance between these surfaces along the z direction is then given by $s = z_2 - z_1 = (B_1 - A_1)x^2 + (B_2 - A_2)xy + (B_2 - A_2)y^2$. By a suitable choice of axes x and y, Equation 2.15 will not contain the xy item, hence

$$s = Ax^2 + By^2 \tag{2.16}$$

where, A and B are constants, which are related to the geometrical shapes of the bodies, and can be expressed as follows:

$$\begin{cases} A+B=\dfrac{1}{2}\left(\dfrac{1}{R_{1x}}+\dfrac{1}{R_{2x}}+\dfrac{1}{R_{1y}}+\dfrac{1}{R_{2y}}\right) \\ |B-A|=\dfrac{1}{2}\left[\left(\dfrac{1}{R_{1x}}-\dfrac{1}{R_{1y}}\right)^2+\left(\dfrac{1}{R_{2x}}-\dfrac{1}{R_{2y}}\right)^2+2\left(\dfrac{1}{R_{1x}}-\dfrac{1}{R_{1y}}\right)\left(\dfrac{1}{R_{2x}}-\dfrac{1}{R_{2y}}\right)\cos 2\gamma\right]^{1/2} \end{cases} \tag{2.17}$$

For engineering problems, the principal planes of the bodies always overlap with each other. That is, $\gamma = 0°$ holds. Since it is relatively simple and universal, in this chapter, we will only discuss this kind of contact. As shown in Figure 2.12, *XOY* plane is the common tangent plane. In the *XOZ* plane, the principal radii of curvature of the contacting bodies on the contact point are R_{1x} and R_{2x}. In the *YOZ* plane, the principal radii of curvature of the contacting bodies on the contact point are R_{1y} and R_{2y}. The relationships of the constants A and B can be expressed as $A+B=\frac{1}{2}\left(\frac{1}{R_x}+\frac{1}{R_y}\right)$ and $|B-A|=\frac{1}{2}\left|\frac{1}{R_x}-\frac{1}{R_y}\right|$, where $\frac{1}{R_x}=\frac{1}{R_{1x}}+\frac{1}{R_{2x}}$ and $\frac{1}{R_y}=\frac{1}{R_{1y}}+\frac{1}{R_{2y}}$. The distance between these surfaces along the z direction is then given by the following:

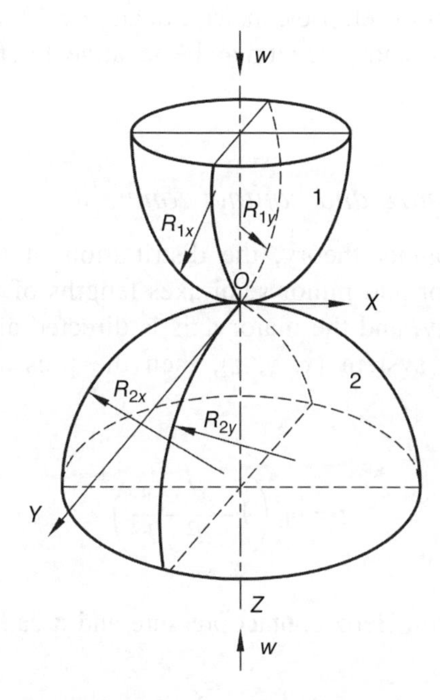

Figure 2.12 Ellipse contact where principal planes of bodies overlap with each other

$$s = \frac{x^2}{2R_x} + \frac{y^2}{2R_y} \tag{2.18}$$

From this equation, we see that the elliptical contact between two elastic bodies can be transformed into the contact between an elastic ellipsoid and a rigid flat surface. When they are loaded together along the z axis, the contact area is an ellipse. The equivalent radii of curvature of the ellipsoid is R_x and R_y, and equivalent elastic modulus of the ellipsoid is E.

As shown in Figure 2.12, if a lubricating oil film exists between these surfaces, h_c is the film thickness in the contact center, and $v(x, y)$ is the elastic deformation, then the film thickness can be written as follows:

$$h(x,y) = h_c + s(x,y) + v(x,y) - v(0,0) \tag{2.19}$$

If the center film thickness $h_c = h_0 + v(0,0)$, where h_0 is the rigid central film thickness, then the film thickness can also be expressed as follows:

$$h(x,y) = h_0 + \frac{x^2}{2R_x} + \frac{y^2}{2R_y} + v(x,y) \tag{2.20}$$

For EHL calculation of the ellipse contact, Equation 2.20 will be used. Note that h_0 is not a true film thickness and hence it may be negative, but the film thickness $h(x, y)$ cannot be negative.

2.3.2 Contact pressure and contact zone

According to Hertz contact theory, the distribution of the contact pressure is an ellipsoid. If the major and minor semi-axes lengths of the ellipse contact area are a and b, respectively, and the major axis is directed along x-axis direction in a Cartesian coordinate system (x, y, z), then the pressure distribution can be written as:

$$p = p_H\left(1 - \frac{x^2}{a^2} - \frac{y^2}{b^2}\right)^{\frac{1}{2}} \tag{2.21}$$

where, p_H is the maximum Hertz contact pressure and it can be expressed as:

$$p_H = \frac{3w}{2\pi ab} \tag{2.22}$$

The major and minor semi-axes lengths of the contacting ellipse can be expressed as:

$$a = k_a \left[\frac{3w}{2E(A+B)}\right]^{1/3} \tag{2.23}$$

$$b = k_b \left[\frac{3w}{2E(A+B)}\right]^{1/3} \tag{2.24}$$

where, k_a and k_b are the parameters for calculating a and b. They are given by:

$$k_a = \left[\frac{2E(e)}{\pi(1-e^2)}\right]^{1/3} \tag{2.25}$$

$$k_b = k_a\sqrt{1-e^2} \tag{2.26}$$

where, e is the ellipticity, $e = \sqrt{1-(b/a)^2}$, $E(e)$ is the complete ellipse integrals of second kind, $E(e) = \int_0^{\pi/2} \sqrt{1-e^2\sin^2\alpha d\alpha}$, and α is a variable of integration. The equation is usually used to find the ellipticity in the ellipse contact. It is as follows:

$$\frac{|A-B|}{A+B} = \frac{2(1-e^2)}{e^2}\frac{E(e)-K(e)}{K(e)} + 1 = \cos\theta \tag{2.27}$$

where, $K(e)$ is the complete ellipse integrals of the first kind, that is, $K(e) = \int_0^{\pi/2} d\alpha/\sqrt{1-e^2\sin^2\alpha}$ and θ is the parameter for calculation.

Generally, for EHL in the ellipse contact, we should obtain the major and minor semi-axes lengths a and b accurately as possible. a and b are computed by the parameters k_a and k_b. k_a and k_b are computed by $E(e)$ and the ellipticity e. In order to obtain e, first we should obtain Equation 2.27. The numerical method of this problem is as follows.

One difficulty in solving the problem is that $K(e)$ is a singular integral equation. A differential quadrature method is always introduced to solve the integral equation. Here, we use the Simpson's method with variable step length to obtain $K(e)$. When e tends to be 1, the denominator in the integral is very small, so the values of some nodes may tend to be infinite. To handle this problem, we give a large set value, say 10^{35}, in the program. If a node value is larger than this, we just let the node value to be equal to the set value. In this way, the integral value will be finite.

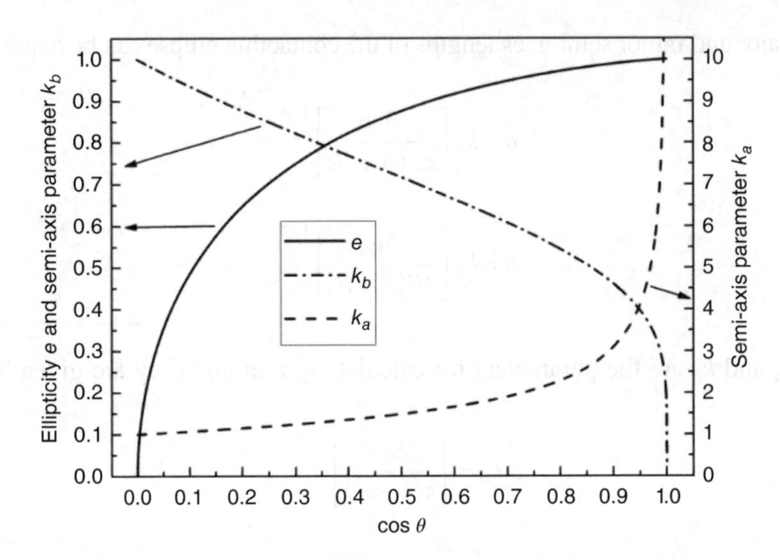

Figure 2.13 Relationships of parameters k_a, k_b, and ellipticity e with cosθ.

After getting $K(e)$ and $E(e)$, we should solve Equation 2.27. It is a second-order transcendental equation with complete ellipse integrals. We can rewrite it as a function and is as follows:

$$f(e)=2\left(1-e^2\right)(E(e)-K(e))+(1-\cos\theta)e^2K(e)=0 \tag{2.28}$$

Solving this equation is another difficult point in the problem. We locate the roots of the equation by using the bisection approach. Bisection the approach is a simple method to approximate one root in an interval where the function is continuous. If $0<e<1$ and if we set $e_0=10^{-30}$ and $e_2=1$ at the end points, we will find out that $f(e_0)>0$ and $f(e_2)>0$. Then we find a midpoint e_1 by searching method, such that $f(e_1)<0$. Since one of the roots of Equation 2.28 is very close to 0 and the other one changes as cosθ increases, the step length should be chosen carefully. In this program, we choose 0.0001 as the searching step length. Then, we can approximate the two roots in the intervals $[e_0, e_1]$ and $[e_1, e_2]$ by the bisection approach. The bigger one is the required root. As shown in Figure 2.13, as cosθ increases from 0 to 1, e also increases from 0 to 1, k_a from 1 to infinite, and k_b decreases from 1 to infinitesimal.

2.3.3 Calculation program

In this program, the elastic deformation of an elastic ellipsoid and a rigid flat surface will be calculated by using the theory of Hertz contact. Here, $R_x=0.3$ m, $R_y=0.5$ m, the loading force $w=200$ N, the equivalent elastic modulus $E=2.21\times10^{11}$ Pa. The nondimensional regions in the x and y directions $X\times Y$ are $(-2, 2)\times(-2, 2)$. The number of nodes is 65×65 (Figure 2.14).

2.3.3.1 Calculation diagram

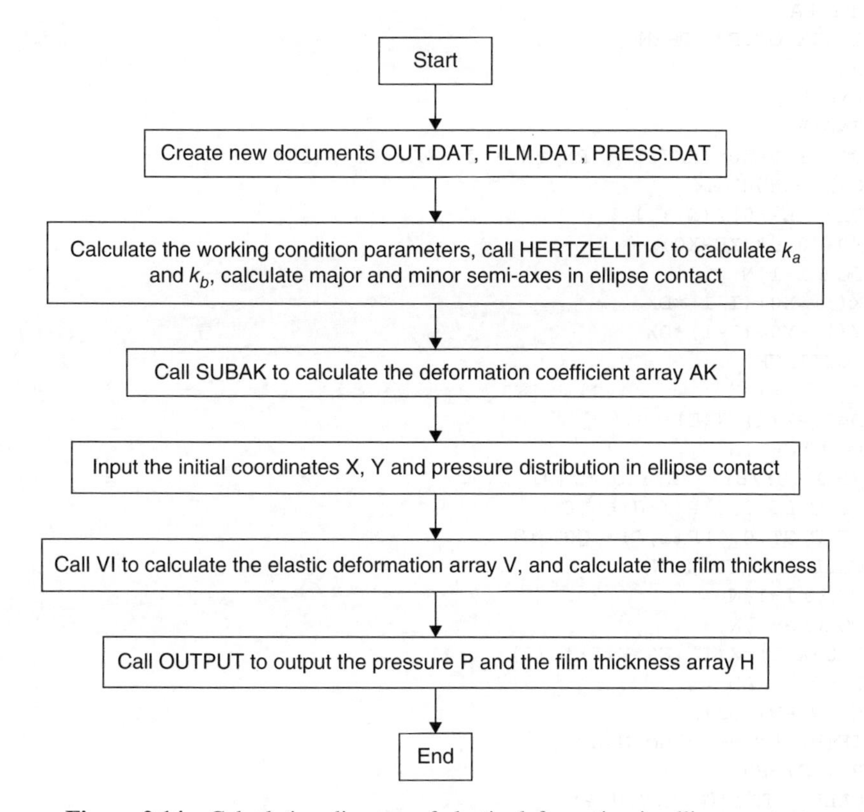

Figure 2.14 Calculation diagram of elastic deformation in ellipse contact

2.3.3.2 Calculation program

```
DIMENSION X(65),Y(65),H(65,65),P(65,65),W(65,65)
 REAL*8 RX,RY,KA,KB
COMMON /COMEK/EK,BX,BY,PTR
DATA N,E1,RX,RY,X0,XE,W0,PAI,PAI1/65,2.21E11,0.3,0.5,
-2.0,2.0,200.,3.14159265,0.2026423/
OPEN(4,FILE='OUT.DAT',STATUS='UNKNOWN')
OPEN(8,FILE='FILM.DAT',STATUS='UNKNOWN')
OPEN(10,FILE='PRESS.DAT',STATUS='UNKNOWN')
EK=RX/RY
AA=0.5*(1./RX+1./RY)
BB=0.5*ABS(1./RX-1./RY)
 CALL HERTZELLIPTIC(RX,RY,KA,KB)
EA=KA*(1.5*W0/AA/E1)**(1./3.0)
EB=KB*(1.5*W0/AA/E1)**(1./3.0)
PH=1.5*W0/(EA*EB*PAI)
WRITE(4,*)N,X0,XE,W0,PH,E1,RX,RY
```

```
      MM=N-1
      BX=EB
       BY=EA
      IF(RX.GT.RY) THEN
      BX=EA
      BY=EB
      ENDIF
      PTR=3.0*W0*(RX/BY)*(RX/BX)**2/(PAI**2)/(E1*RX**2)
      CALL SUBAK(MM)
      DX=(XE-X0)/(N-1.)
      Y0=-0.5*(XE-X0)
      DO 5 I=1,N
      X(I)=X0+(I-1)*DX
      Y(I)=Y0+(I-1)*DX
5     CONTINUE
      DO 10 I=1,N
      D=1.-X(I)*X(I)
      DO 10 J=1,N
       C=D-(BX/BY)**2*Y(J)*Y(J)
      IF(C.LE.0.0)P(I,J)=0.0
10    IF(C.GT.0.0)P(I,J)=SQRT(C)
      CALL VI(N,DX,P,W)
      DO 30 I=1,N
      DO 30 J=1,N
      RAD=X(I)*X(I)+EK*Y(J)*Y(J)
      W1=0.5*RAD
      H0=W1+W(I,J)
      IF(H0.LT.HMIN)HMIN=H0
30    H(I,J)=H0
      CALL OUTPUT(N,X,Y,H,P)
      STOP
      END
      SUBROUTINE VI(N,DX,P,V)
      DIMENSION P(N,N),V(N,N)
       COMMON /COMEK/EK,BX,BY,PTR
      COMMON /COMAK/AK(0:65,0:65)
      DO 40 I=1,N
      DO 40 J=1,N
      H0=0.0
       DO 30 K=1,N
      IK=IABS(I-K)
      DO 30 L=1,N
      JL=IABS(J-L)
30    H0=H0+AK(IK,JL)*P(K,L)
40    V(I,J)=H0*DX*PTR
      RETURN
      END
      SUBROUTINE SUBAK(MM)
      COMMON /COMAK/AK(0:65,0:65)
      S(X,Y)=X+SQRT(X**2+Y**2)
      DO 10 I=0,MM
```

```
      XP=I+0.5
      XM=I-0.5
      DO 10 J=0,I
      YP=J+0.5
      YM=J-0.5
      A1=S(YP,XP)/S(YM,XP)
      A2=S(XM,YM)/S(XP,YM)
      A3=S(YM,XM)/S(YP,XM)
      A4=S(XP,YP)/S(XM,YP)
      AK(I,J)=XP*ALOG(A1)+YM*ALOG(A2)+XM*ALOG(A3)+YP*ALOG(A4)
10    AK(J,I)=AK(I,J)
      RETURN
      END
      SUBROUTINE OUTPUT(N,X,Y,H,P)
      DIMENSION X(N),Y(N),H(N,N),P(N,N)
      A=0.0
      WRITE(8,110)A,(Y(I),I=1,N)
      DO I=1,N
      WRITE(8,110)X(I),(H(I,J),J=1,N)
      ENDDO
      WRITE(10,110)A,(Y(I),I=1,N)
      DO I=1,N
      WRITE(10,110)X(I),(P(I,J),J=1,N)
      ENDDO
110   FORMAT(66(E12.6,1X))
      RETURN
      END
SUBROUTINE HERTZELLIPTIC(RX,RY,KA,KB)
      IMPLICIT NONE
      REAL*8, EXTERNAL :: EE,KE
       REAL*8 RX,RY,BPA,BMA,CTH,THT,PAI,E1,KA,KB
      DATA PAI/3.1415926/
      BPA=0.5*(1./RX+1./RY)
      BMA=0.5*ABS(1./RX-1./RY)
      CTH=BMA/BPA
       THT=ACOS(CTH)*180.0/PAI
       CALL CACUE(CTH,E1)
      KA=(2.*EE(E1)/(PAI*(1-E1**2)))**(1/3.)
      KB=KA*(1.-E1**2)**(1/2.)
      RETURN
      END
       SUBROUTINE CACUE(CTH,E1)
      IMPLICIT NONE
      REAL*8, EXTERNAL :: EE,KE
        REAL*8, EXTERNAL :: FAB
      INTEGER FLG,I
      REAL*8 PAI,CTH,E1,E11,E12,DX,A,B,A1,A2,A3,A4,A5,T1,T2,T3,T4,T5,ER0
      DATA PAI,DX,FLG,I,T1,T5,ER0/3.1415926,0.0001,1,1,1.E-30,1.,1.E-12/
       IF(CTH.LT.1.E-6)THEN
    WRITE(*,*)"!NOTE:COS(THETA) IS TOO SAMLL TO CALCULATE,E1 IS SETTED TO 0."
      E1=0.
```

```
RETURN
ENDIF
IF(CTH.GT.0.9999999999)THEN
E1=1.
RETURN
ENDIF
 A1=FAB(T1,CTH)
A5=FAB(T5,CTH)
DO WHILE(FLG.EQ.1)
T3=T1+I*DX
A3=FAB(T3,CTH)
I=I+1
IF((A1*A3.LT.0.).AND.(A3*A5.LT.0.)) THEN
FLG=0
END IF
END DO
DO WHILE((T3-T1).GT.ER0)
T2=(T1+T3)/2.
 A2=FAB(T2,CTH)
IF(A2.GT.0.) T1=T2
IF(A2.LT.0.) T3=T2
IF(A2.EQ.0.)THEN
E11=T2
EXIT
END IF
END DO
E11=T2
DO WHILE((T5-T3).GT.ER0)
T4=(T3+T5)/2.
 A4=FAB(T4,CTH)
IF(A4.GT.0.) T5=T4
IF(A4.LT.0.) T3=T4
IF(A4.EQ.0.)THEN
E12=T2
EXIT
END IF
END DO
E12=T4
E1=E11
IF(E11.LT.E12) E1=E12
 RETURN
END
REAL*8 FUNCTION FAB(E1,CTH)
IMPLICIT NONE
REAL*8 E1,CTH,T1,T2
REAL*8, EXTERNAL :: EE,KE
 T1=EE(E1)
T2=KE(E1)
FAB=2*(1-E1**2)*(T1-T2)+(1.-CTH)*E1**2*T1
RETURN
END
```

```
REAL*8 FUNCTION KE(E1)
IMPLICIT NONE
INTEGER N,I,FLG
REAL*8 E1,PAI,H,T,T1,T2,S1,S2,P,Q
PAI=3.1415926
IF(E1.EQ.1) THEN
KE=1.E10
RETURN
ENDIF
IF(E1.LT.1.E-20) THEN
KE=PAI/2.
RETURN
ENDIF
 N=1
H=PAI/2.
 Q=SQRT(1.-E1*E1*SIN(H)*SIN(H))
IF(Q.LT.1.E-35) Q=1.E35
Q=1./Q
T1=.5*H*(1+Q)
S1=T1
FLG=1
DO WHILE(FLG.EQ.1)
 P=0.
DO I=0,N-1
T=(I+0.5)*H
 Q=SQRT(1.-E1*E1*SIN(T)*SIN(T))
IF(Q.LT.1.E-35) Q=1.E35
Q=1./Q
P=P+Q
END DO
 T2=(T1+H*P)/2.
S2=(4.*T2-T1)/3.
 IF(ABS(S2-S1).LT.ABS(S2)*1.E-7) FLG=0
T1=T2
S1=S2
N=N+N
H=.5*H
END DO
 KE=S2
RETURN
END
 REAL*8 FUNCTION EE(E1)
IMPLICIT NONE
INTEGER N,I,FLG
REAL*8 E1,PAI,H,T,T1,T2,S1,S2,P,Q
PAI=3.1415926
 N=1
H=PAI/2.
IF(E1.EQ.1) THEN
EE=1.
RETURN
```

```
ENDIF
IF(E1.LT.1.E-20) THEN
EE=PAI/2.
RETURN
ENDIF
 Q=SQRT(1.-E1*E1*SIN(H)*SIN(H))
T1=.5*H*(1+Q)
S1=T1
FLG=1
DO WHILE(FLG==1)
 P=0.
DO I=0,N-1
T=(I+0.5)*H
 Q=SQRT(1.-E1*E1*SIN(T)*SIN(T))
P=P+Q
END DO
 T2=(T1+H*P)/2.
S2=(4.*T2-T1)/3.
IF(ABS(S2-S1).LT.ABS(S2)*1.E-7) FLG=0
T1=T2
S1=S2
N=N+N
H=.5*H
END DO
 EE=S2
RETURN
END
```

2.3.4 Calculation results

Figure 2.15b shows the elastic deformation in the ellipse contact under the given parameters, working conditions, and the pressure distribution shown in Figure 2.15a. The calculation result is consistent with the theoretical analysis. A constant distance away from coordinate surface will be incorporated into the rigid displacement H0 in EHL calculation and will be determined by the load balancing equation.

2.4 Calculation of elastic deformation with multigrid integration technique

2.4.1 Principle of multigrid integration

When the multigrid integration method is used to calculate the elastic deformation, the calculation time is nearly proportional to the number of nodes, whereas in the common numerical integration, the calculation time is proportional to the square of the number of nodes. Therefore, the more the number of nodes, the faster the multigrid integration relatively.

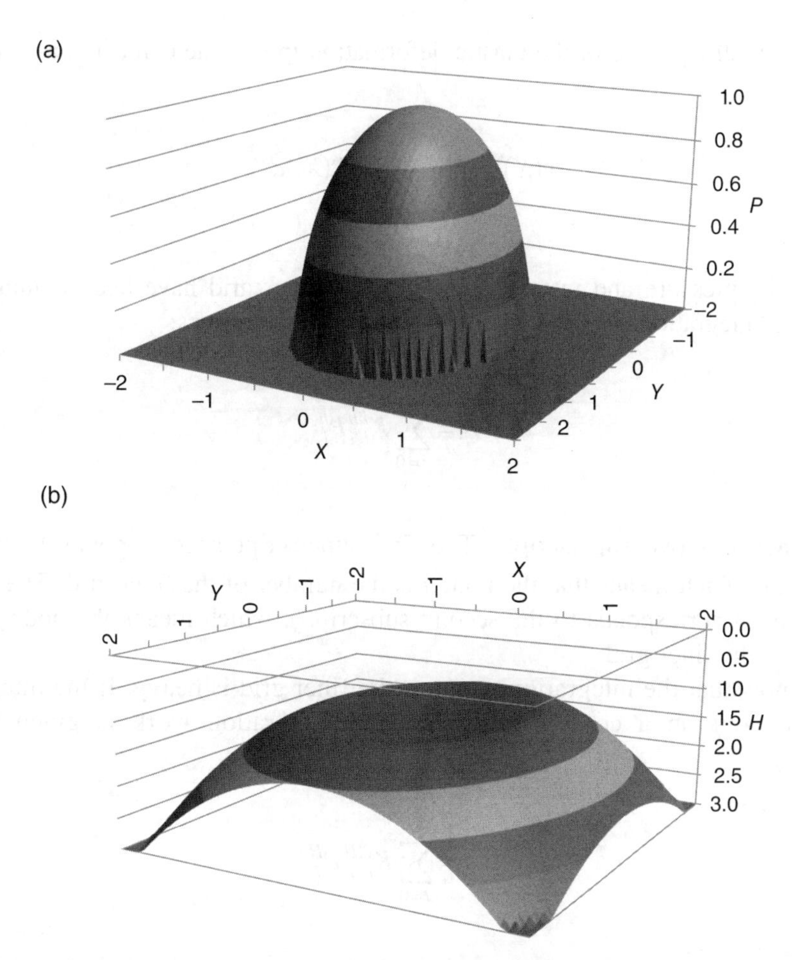

Figure 2.15 Pressure distribution and elastic deformation in ellipse contact. (a) Pressure distribution and (b) Elastic deformation

As the multigrid integration is the same for the two neighbor grid levels, we only introduce the integration method between the two grid levels in the line contact in the following.

First transfer the data from the finer grid to the coarse grid. Then, carry out integration and transfer the result back to the finer grid. Finally, modify the result on the finer grid. The loops continue until the accuracy meets the requirements on the finer grid.

Denote the variables on the finer grid with the superscript h and the number of the node number by subscripts i, j. Denote the variables on the coarser grid with the superscript H and the number of the node number by subscripts I, J. Thus, the node number i or j of the finer grid is equal to 0, 1, …, n and the node number I or J of the coarser grid is equal to 0, 1 …, N. The spacing of the coarser grid is doubled as that of the finer grid, that is, $n = 2N - 1$.

The integral equation of the elastic deformation in the line contact is as follows:

$$v(X) = \int_{X_a}^{X_b} \ln|X - X'|P(X')dX' \tag{2.29}$$

After the pressure and integral factors of the finer grid have been obtained, the numerical integration calculation formula becomes as follows:

$$v_i^h = \sum_{j=0}^{n} K_{i,j}^{hh} P_j^h \tag{2.30}$$

where, $K_{i,j}^{hh}$ has two superscripts. The first superscript h corresponds to the first subscript i, which means that the node i is the number of the finer grid. The second superscript h corresponds to the second subscript j, which means the node j is the number of the finer grid.

We know that the integration work on the finer grid is heavy. If the integration is carried out on a coarser grid, then the integration work is given by the following:

$$v_I^H = \sum_{J=0}^{N} K_{I,J}^{HH} P_J^H \tag{2.31}$$

Although the work by Equation 2.31 is less than that by Equation 2.30, the integral result on the coarser grid is clearly not as good as that on the finer grid. Therefore, the integral results must be corrected in order to obtain an accuracy as same as the integration on the finer grid. This includes several steps as follows:

2.4.1.1 Transfer pressure downwards

Although the pressure on the finer grid can be directly used as a corresponding node on the coarser grid, in order to consider the change of pressure, the pressure on the finer grid should be transmitted to the coarser grid by interpolating. The interpolation formula is:

$$P_I^H = \frac{1}{32}\left(-P_{2I-3}^h + 9P_{2I-2}^h + 16P_{2I-1}^h + 9P_{2I}^h - P_{2I+1}^h\right) \quad I = 2,3,\ldots,N-1 \tag{2.32}$$

where, the left term of the equation is the pressure on the coarser grid and the right terms are the pressures on the finer grid. For $I = 1$ or N, the above equation can be modified as follows:

$$P_I^H = \frac{1}{32}\left(16P_1^h + 18P_3^h - 2P_5^h\right) \tag{2.33}$$

$$P_N^H = \frac{1}{32}\left(-P_{n-2}^h + 18P_{n-1}^h + 16P_n^h\right) \tag{2.34}$$

2.4.1.2 Transfer integral coefficients downward

$$K_{I,J}^{HH} = K_{2I-1,2J-1}^{hh} \tag{2.35}$$

where, the left term of the equation is the integral coefficient on the coarser grid and the right term is the integral coefficient on the finer mesh.

2.4.1.3 Integrate on coarser grid

$$v_I^H = \sum_{J=0}^{N} K_{I,J}^{HH} P_J^H \tag{2.36}$$

Note that although the form of Equation 2.36 is the same as Equation 2.31, the pressure and the integral coefficients have been transferred downwards from the finer grid rather than upwards from the coarser grid.

2.4.1.4 Transfer integral value back

As fewer nodes integrals on the finer grid have not been calculated yet, we can obtain them by interpolation. First, map the node values on the coarser gird nodes to the finer grid nodes, which coincide with the coarser ones.

$$\tilde{v}_{2I-1}^h = v_I^H \tag{2.37}$$

Then, for those nodes on the finer grid that do not coincide with the coarse grid nodes, the following interpolation is used.

$$\tilde{v}_{2I}^{h} = \frac{1}{16}\left(-v_{I-1}^{H} + 8v_{I}^{H} + 8v_{I+1}^{H} - v_{I+2}^{H}\right) \tag{2.38}$$

For the node $i = 2$ or $n - 1$, the average value of two adjacent nodes may be used as its value.

2.4.1.5 Correction of finer grid

Correction of the integral values on the finer grid is done by three ways: to correct integral coefficients, to correct the integral value on the mapping node, and to correct the interpolated nodes.

a. To correct integral coefficients
   First, calculate the coefficient interpolation values. Then, obtain the correction differences by subtracting the fine grid integral coefficients with the interpolated ones. Because the integral values on the mapping nodes and on the interpolated nodes are different, we must consider these differences by the following formula:

$$\tilde{K}_{2I-1,2J-1}^{hh} = \frac{1}{16}\left(9K_{I+1,J}^{HH} + 9K_{I-1,J}^{HH} - K_{I+3,J}^{HH} - K_{I-3,J}^{HH}\right) \tag{2.39}$$

Because the neighboring integration nodes are not suitable for the high-order interpolation, the inner interpolation calculation formula can be used, that is,

$$\tilde{K}_{1,2J-1}^{hh} = \frac{1}{8}\left(9K_{2,J}^{HH} - K_{4,J}^{HH}\right) \tag{2.40}$$

$$\tilde{K}_{2,2J-1}^{hh} = \frac{1}{16}\left(9K_{1,J}^{HH} + 9K_{3,J}^{HH} - K_{3,J}^{HH} - K_{5,J}^{HH}\right) \tag{2.41}$$

$$\tilde{K}_{3,2J-1}^{hh} = \frac{1}{16}\left(9K_{2,J}^{HH} + 9K_{4,J}^{HH} - K_{2,J}^{HH} - K_{6,J}^{HH}\right) \tag{2.42}$$

For the mapping nodes, the difference of the integral coefficients is calculated by using the following formula:

$$\Delta\tilde{K}_{i,j}^{hh} = K_{i,j}^{hh} - \tilde{K}_{i,j}^{hh} \tag{2.43}$$

And for the interpolated nodes, the difference of the integral coefficients is calculated by using the following formula:

$$\Delta\tilde{K}_{i,j}^{hh} = \begin{cases} 0 & \text{mapping node} \\ K_{i,j}^{hh} - \tilde{K}_{i,j}^{hh} & \text{interpolating node} \end{cases} \tag{2.44}$$

b. To correct the integral coefficients on the mapping nodes by interpolating.

$$v_{2I-1}^{h} = \widetilde{v}_{I}^{H} + \sum_{j=1}^{M} \Delta\widetilde{K}_{i,j}^{hh} p_j \Delta x \tag{2.45}$$

c. To correct the integral coefficients on the interpolated nodes by interpolating

$$v_{2I}^{h} = \widetilde{v}_{2I}^{h} + \sum_{j=1}^{M} \Delta\hat{K}_{i,j}^{hh} p_j \Delta x \tag{2.46}$$

where, $M \geq 3 + 2\ln(n)$, M is an integer.

2.4.2 *Calculation programs and examples*

2.4.2.1 Calculation program and example in line contact

Source program

The subroutines of the multigrid integration method are briefly introduced as follows.

In the main program, set the coordinates, the surface shape, the pressure distribution, etc. It calls the subroutine SUBAK to calculate the elastic deformation coefficient first and then the subroutine DISP to calculate the elastic deformation by using multigrid integration technology.

a. Subroutine DISP: its function is to calculate the elastic deformation. Substitute the summing number M by $M \geq 3 + 2\ln(n)$ to control the calculation loops, the number of nodes N, the total multigrid node number NW, the mesh layers KMAX, the node spacing DX, and the pressure array P1. Then, bring back the elastic deformation.
b. Subroutine DOWNP: according to Equations 2.32–2.35, transfer the node parameters downward, such as pressure, integral coefficients, etc. to the coarser grid. In the subroutine, we need to calculate the start position of each unit, that is, K1 and K2.
c. Subroutine WCOS: according to Equation 2.37, map the value of pressure onto the corresponding nodes of different grids.
d. Subroutine WINT: according to Equation 2.38, interpolate the non-corresponding grid nodes of the coarser grids.
e. Subroutine WI: according to Equation 2.36, obtain the numerical integration value on the coarser grid.
f. Subroutine CORR: according to Equation 2.45, correct the integral values of the mapping nodes or according to Equation 2.46, correct the integral value on the interpolated nodes.

g. Subroutine AKIN: according to Equations 2.39–2.42, interpolate the integral coefficients on the coarser grid.
h. Subroutine AKCO: according to Equation 2.35, map the integral coefficient upwards.

The source codes are as follows:

```
      DIMENSION X(1100),P(1100),H(1100),W(2200)
      COMMON /COMAK/AK(0:1100)
      DATA NW,PAI/2200,3.14159265/
      OPEN(8,FILE='OUT.DAT',STATUS='UNKNOWN')
      N=129
      CALL SUBAK(N)
      DX=3./(N-1)
      DO I=1,N
      X(I)=-1.5+(I-1)*DX
      P(I)=0.0
      IF(ABS(X(I)).LE.1.0)P(I)=SQRT(1.-X(I)*X(I))
      ENDDO
      K=3
      CALL DISP(N,NW,K,DX,P,W)
      WX=0.5*PAI*DX*ALOG(DX)
      DO I=1,N
      W(I)=WX
      DO J=1,N
      IJ=IABS(I-J)
      W(I)=W(I)+AK(IJ)*P(J)*DX
      ENDDO
      ENDDO
      DO 30 I=1,N
      H(I)=1.24+0.5*X(I)*X(I)-W(I)/PAI
   30 CONTINUE
      DO I=1,N
      WRITE(8,40)X(I),P(I),H(I)
      ENDDO
   40 FORMAT(1X,6(E12.6,1X))
      STOP
      END
      SUBROUTINE DISP(N,NW,KMAX,DX,P1,W)
      DIMENSION P1(N),W(NW),P(2200),AK1(0:50),AK2(0:50)
      COMMON /COMAK/AK(0:1100)
      DATA NMAX,KMIN/2200,1/
      N2=N
      M=3+2*ALOG(FLOAT(N))
      K1=N+KMAX
      DO 10 I=1,N
   10 P(K1+I)=P1(I)
      DO 20 KK=KMIN,KMAX-1
      K=KMAX+KMIN-KK
      N1=(N2+1)/2
```

```
      CALL DOWNP(NMAX,N1,N2,K,P)
   20 N2=N1
      DX1=DX*2**(KMAX-KMIN)
      CALL WI(NMAX,N1,KMIN,KMAX,DX,DX1,P,W)
      DO 30 K=KMIN+1,KMAX
      N2=2*N1-1
      DX1=DX1/2.
      CALL AKCO(M+5,KMAX,K,AK1)
      CALL AKIN(M+6,AK1,AK2)
      CALL WCOS(NMAX,N1,N2,K,W)
      CALL CORR(NMAX,N2,K,M,1,DX1,P,W,AK1)
      CALL WINT(NMAX,N2,K,W)
      CALL CORR(NMAX,N2,K,M,2,DX1,P,W,AK2)
   30 N1=N2
      DO 40 I=1,N
   40 W(I)=W(K1+I)
      RETURN
      END
      SUBROUTINE DOWNP(NMAX,N1,N2,K,P)
      DIMENSION P(NMAX)
      K1=N1+K-1
      K2=N2+K-1
      DO 10 I=3,N1-2
      I2=2*I+K2
   10 P(K1+I)=(16.*P(I2)+9.*(P(I2-1)+P(I2+1))-(P(I2-3)+P(I2+3)))/32.
      P(K1+2)=0.25*(P(K2+3)+P(K2+5))+0.5*P(K2+4)
      P(K1+N1-1)=0.25*(P(K2+N2-2)+P(K2+N2))+0.5*P(K2+N2-1)
      RETURN
      END
      SUBROUTINE WCOS(NMAX,N1,N2,K,W)
      DIMENSION W(NMAX)
      K1=N1+K-1
      K2=N2+K
      DO 10 I=1,N1
      II=2*I-1
   10 W(K2+II)=W(K1+I)
      RETURN
      END
      SUBROUTINE WINT(NMAX,N,K,W)
      DIMENSION W(NMAX)
      K2=N+K
      DO 10 I=4,N-3,2
      II=K2+I
   10 W(II)=(9.*(W(II-1)+W(II+1))-(W(II-3)+W(II+3)))/16.
      I1=K2+2
      I2=K2+N-1
      W(I1)=0.5*(W(I1-1)+W(I1+1))
      W(I2)=0.5*(W(I2-1)+W(I2+1))
      RETURN
      END
      SUBROUTINE CORR(NMAX,N,K,M,I1,DX,P,W,AK)
```

```
      DIMENSION P(NMAX),W(NMAX),AK(0:M)
      K1=N+K
      IF(I1.EQ.2)GOTO 20
      DO 10 I=1,N,2
      II=K1+I
      J1=MAX0(1,I-M)
      J2=MIN0(N,I+M)
      DO 10 J=J1,J2
      IJ=IABS(I-J)
   10 W(II)=W(II)+AK(IJ)*DX*P(K1+J)
      RETURN
   20 DO 30 I=2,N,2
      II=K1+I
      J1=MAX0(1,I-M)
      J2=MIN0(N,I+M)
      DO 30 J=J1,J2
      IJ=IABS(I-J)
   30 W(II)=W(II)+AK(IJ)*DX*P(K1+J)
      RETURN
      END
      SUBROUTINE WI(NMAX,N,KMIN,KMAX,DX,DX1,P,W)
      DIMENSION P(NMAX),W(NMAX)
      COMMON /COMAK/AK(0:1100)
      K1=N+1
      K=2**(KMAX-KMIN)
      C=ALOG(DX)
      DO 10 I=1,N
      II=K1+I
      W(II)=0.0
      DO 10 J=1,N
      IJ=K*IABS(I-J)
   10 W(II)=W(II)+(AK(IJ)+C)*DX1*P(K1+J)
      RETURN
      END
      SUBROUTINE AKCO(KA,KMAX,K,AK1)
      DIMENSION AK1(0:KA)
      COMMON /COMAK/AK(0:1100)
      J=2**(KMAX-K)
      DO 10 I=0,KA
      II=J*I
   10 AK1(I)=AK(II)
      RETURN
      END
      SUBROUTINE AKIN(KA,AK1,AK2)
      DIMENSION AK1(KA),AK2(KA)
      DO 10 I=4,KA-3
   10 AK2(I)=(9.*(AK1(I-1)+AK1(I+1))-(AK1(I-3)+AK1(I+3)))/16.
      AK2(1)=(9.*AK1(2)-AK1(4))/8.
      AK2(2)=(9.*(AK1(1)+AK1(3))-(AK1(3)+AK1(5)))/16.
      AK2(3)=(9.*(AK1(2)+AK1(4))-(AK1(2)+AK1(6)))/16.
      DO 20 I=1,KA
```

```
20 AK2(I)=AK1(I)-AK2(I)
   DO 30 I=1,KA-1,2
   I1=I+1
   AK1(I)=0.0
30  AK1(I1)=AK2(I1)
   RETURN
   END
   SUBROUTINE SUBAK(MM)
   COMMON /COMAK/AK(0:1100)
      DO 10 I=0,MM
10 AK(I)=(I+0.5)*(ALOG(ABS(I+0.5))-1.)-(I-0.5)*(ALOG(ABS(I-0.5))-1.)
   RETURN
   END
```

Example

In the main program, a cylindrical surface and the Hertz pressure distribution are given. Then, the elastic deformation can be obtained by the multigrid integration method. The contact area is a horizontal line as shown in Figure 2.16. This shows that the result is consistent with the result of contact mechanics.

2.4.2.2 Calculation program and example in point contact

Source program

Similar to a line contact problem, one more dimensional calculation is added in the point contact. Its source code is directly as follows.

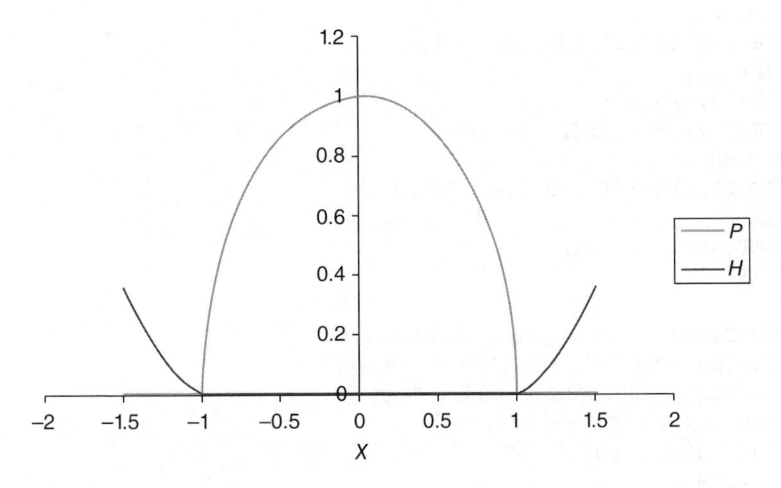

Figure 2.16 Multigrid integration method to calculate elastic deformation in line contact

```
      DIMENSION X(65),Y(65),P(65,65),H(65,65)
      DIMENSION W(150,150),P0(150,150)
      COMMON /COM1/ENDA,A1,A2,A3,Z,C1,C2,HM0/COMK/KL
     & /COMAK/AK(0:100,0:100)
      DATA N,NM,LEM,PAI1/65,150,3,0.2026423/
      OPEN(8,FILE='FILM.DAT',STATUS='UNKNOWN')
      OPEN(10,FILE='PRESS.DAT',STATUS='UNKNOWN')
      CALL SUBAK(N)
  DX=3.0/(N-1.)
      DO 10 I=1,N
      X(I)=-1.5+(I-1)*DX
 10   Y(I)=-1.5+(I-1)*DX
      DO 20 I=1,N
      D=1.-X(I)*X(I)
      DO 20 J=1,N
      C=D-Y(J)*Y(J)
      IF(C.LE.0.0)P(I,J)=0.0
 20   IF(C.GT.0.0)P(I,J)=SQRT(C)
      CALL DISP(N,NM,LEM,DX,P,P0,W)
      DO 30 I=1,N
      W1=0.5*X(I)*X(I)
      DO 30 J=1,N
      H(I,J)=W1+0.5*Y(J)*Y(J)+W(I,J)*PAI1
 30   CONTINUE
  CALL OUPT(N,X,Y,H,P)
  STOP
  END
      SUBROUTINE OUPT(N,X,Y,H,P)
      DIMENSION X(N),Y(N),H(N,N),P(N,N)
      NN=(N+1)/2
      A=0.0
      WRITE(8,*)N,NN
      WRITE(8,110)A,(Y(I),I=1,N)
      DO I=1,N
      WRITE(8,110)X(I),(H(I,J),J=1,N)
      END DO
      WRITE(10,*)N,NN
      WRITE(10,110)A,(Y(I),I=1,N)
      DO I=1,N
      WRITE(10,110)X(I),(P(I,J),J=1,N)
      END DO
 110  FORMAT(66(1X,E12.6))
      RETURN
      END
      SUBROUTINE DISP(N,NM,LEM,DX,P0,P,W)
      DIMENSION W(NM,NM),P(NM,NM),P0(N,N),
     & AK1(0:100,0:100),AK2(0:100,0:100)
      COMMON /COMAK/AK(0:100,0:100)
      DATA LEN,KA/1,100/
      LE1=N+LEM
      DO 10 I=1,N
```

```
      DO 10 J=1,N
10    P(LE1+I,LE1+J)=P0(I,J)
      N2=N
      M1=3+2.0*ALOG(FLOAT(N))
      M2=2
      DO 20 LEL=LEN,LEM-1
      LE=LEM+LEN-LEL
      N1=(N2+1)/2
      CALL DOWNP(NM,N1,N2,LE,P)
20    N2=N1
      DX1=DX*(2**(LEM-LEN))**2
      CALL WI(NM,N1,LEN,LEM,DX1,P,W)
      DO 30 LE=LEN+1,LEM
      N2=2*N1-1
      DX1=DX1/4.
      CALL AKCO(KA,M1+2,LEM,LE,AK1)
      CALL WCOS(NM,N1,N2,LE,W)
      CALL AKIN(KA,M1,AK1,AK2,1)
      CALL CORR(NM,N2,KA,LE,M1,M2,1,DX1,P,W,AK2)
      CALL WINT(NM,N2,LE,2,W)
      CALL AKIN(KA,M1,AK1,AK2,2)
      CALL CORR(NM,N2,KA,LE,M1,M2,2,DX1,P,W,AK2)
      CALL WINT(NM,N2,LE,3,W)
      CALL CORR(NM,N2,KA,LE,M1,M2,3,DX1,P,W,AK2)
30    N1=N2
      LE1=N+LEM
      DO 40 I=1,N
      DO 40 J=1,N
40    W(I,J)=W(LE1+I,LE1+J)
      RETURN
      END
C***********************************************************************
      SUBROUTINE DOWNP(NM,N1,N2,LE,P)
      DIMENSION P(NM,NM)
      LE1=N1+LE-1
      LE2=N2+LE
      DO 20 J=1,N2
      NJ=LE2+J
      DO 10 I=1,N1
      I2=LE2+2*I-1
      P1=P(I2,NJ)
      P2=0.
      P3=0.
      P4=0.
      P5=0.
      IF(I.GT.1)P2=P(I2-1,NJ)
      IF(I.LT.N1)P3=P(I2+1,NJ)
      IF(I.GT.2)P4=P(I2-3,NJ)
      IF(I.LT.N1-1)P5=P(I2+3,NJ)
10    P(LE1+I,NJ)=(16.*P1+9.*(P2+P3)-(P4+P5))/32.
20    CONTINUE
```

```
      DO 40 I=1,N1
      NI=LE1+I
      DO 30 J=1,N1
      J2=LE2+2*J-1
      P1=P(NI,J2)
      P2=0.
      P3=0.
      P4=0.
      P5=0.
      IF(J.GT.1)P2=P(NI,J2-1)
      IF(J.LT.N1)P3=P(NI,J2+1)
      IF(J.GT.2)P4=P(NI,J2-3)
      IF(J.LT.N1-1)P5=P(NI,J2+3)
30    P(NI,LE1+J)=(16.*P1+9.*(P2+P3)-(P4+P5))/32.
40    CONTINUE
      RETURN
      END
C*****************************************************************
      SUBROUTINE WCOS(NM,N1,N2,LE,W)
      DIMENSION W(NM,NM)
      LE1=N1+LE-1
      LE2=N2+LE
      DO 10 I=1,N1
      I1=LE1+I
      I2=LE2+2*I-1
      DO 10 J=1,N1
      J1=LE1+J
      J2=LE2+2*J-1
10    W(I2,J2)=W(I1,J1)
      RETURN
      END
C*****************************************************************
      SUBROUTINE WINT(NM,N,LE,KG,W)
      DIMENSION W(NM,NM)
      LE2=N+LE
      IJ1=LE2+2
      IJ2=LE2+N-1
      IF(KG.EQ.3)GOTO 22
      DO 20 J=1,N,2
      NJ=LE2+J
      DO 10 I=4,N-3,2
      II=LE2+I
      W(II,NJ)=(9.*(W(II-1,NJ)+W(II+1,NJ))
     & -(W(II-3,NJ)+W(II+3,NJ)))/16.
10    W(NJ,II)=(9.*(W(NJ,II-1)+W(NJ,II+1))
     & -(W(NJ,II-3)+W(NJ,II+3)))/16.
      W(IJ1,NJ)=0.5*(W(IJ1-1,NJ)+W(IJ1+1,NJ))
      W(IJ2,NJ)=0.5*(W(IJ2-1,NJ)+W(IJ2+1,NJ))
      W(NJ,IJ1)=0.5*(W(NJ,IJ1-1)+W(NJ,IJ1+1))
20    W(NJ,IJ2)=0.5*(W(NJ,IJ2-1)+W(NJ,IJ2+1))
      RETURN
```

```
22  DO 40 I=2,N,2
    NI=LE2+I
    DO 30 J=4,N-3,2
    JJ=LE2+J
30  W(JJ,NI)=(9.*(W(JJ-1,NI)+W(JJ+1,NI))
  & -(W(JJ-3,NI)+W(JJ+3,NI)))/16.
    W(IJ1,NI)=0.5*(W(IJ1-1,NI)+W(IJ1+1,NI))
40  W(IJ2,NI)=0.5*(W(IJ2-1,NI)+W(IJ2+1,NI))
    RETURN
    END
C*************************************************************
    SUBROUTINE CORR(NM,N,KA,LE,M1,M2,KG,DX,P,W,AK)
    DIMENSION P(NM,NM),W(NM,NM),AK(0:KA,0:KA)
    LE1=N+LE
    IF(KG.NE.1)GOTO 30
    DO 20 I=1,N,2
    II=LE1+I
    K1=MAX0(1,I-M1)
    K2=MIN0(N,I+M1)
    DO 20 J=1,N,2
    JJ=LE1+J
    L1=MAX0(1,J-M1)
    L2=MIN0(N,J+M1)
    W1=0.
    DO 10 K=K1,K2
    IK=IABS(I-K)
    DO 10 L=L1,L2
    JL=IABS(J-L)
    W1=W1+AK(IK,JL)*P(LE1+K,LE1+L)
10  CONTINUE
    W(II,JJ)=W(II,JJ)+DX*W1
20  CONTINUE
    RETURN
30  IF(KG.NE.2)GOTO 52
    DO 50 I=2,N,2
    II=LE1+I
    K1=MAX0(1,I-M1)
    K2=MIN0(N,I+M1)
    DO 50 J=1,N,2
    JJ=LE1+J
    L1=MAX0(1,J-M2)
    L2=MIN0(N,J+M2)
    W1=0.
    W2=0.
    DO 40 K=K1,K2
    IK=IABS(I-K)
    DO 40 L=L1,L2
    JL=IABS(J-L)
    W1=W1+AK(IK,JL)*P(LE1+K,LE1+L)
    W2=W2+AK(IK,JL)*P(LE1+L,LE1+K)
40  CONTINUE
```

```
      W(II,JJ)=W(II,JJ)+DX*W1
50    W(JJ,II)=W(JJ,II)+DX*W2
      RETURN
52    DO 70 I=2,N,2
      II=LE1+I
      K1=MAX0(1,I-M1)
      K2=MIN0(N,I+M1)
      DO 70 J=2,N,2
      JJ=LE1+J
      L1=MAX0(1,J-M2)
      L2=MIN0(N,J+M2)
      W1=0.
      DO 60 K=K1,K2
      IK=IABS(I-K)
      DO 60 L=L1,L2
      JL=IABS(J-L)
60    W1=W1+AK(IK,JL)*P(LE1+K,LE1+L)
70    W(II,JJ)=W(II,JJ)+DX*W1
      RETURN
      END
C*****************************************************************
      SUBROUTINE WI(NM,N,LEN,LEM,DX,P,W)
      DIMENSION P(NM,NM),W(NM,NM)
      COMMON /COMAK/AK(0:100,0:100)
      LE1=N+1
      KAK=2**(LEM-LEN)
      DO 40 I=1,N
      II=LE1+I
      DO 40 J=1,N
      JJ=LE1+J
      H0=0.0
      DO 30 K=1,N
      KK=LE1+K
      IK=KAK*IABS(I-K)
      DO 30 L=1,N
      LL=LE1+L
      JL=KAK*IABS(J-L)
30    H0=H0+AK(IK,JL)*P(KK,LL)
40    W(II,JJ)=H0*DX
      RETURN
      END
C*****************************************************************
      SUBROUTINE AKCO(KA,M,LEM,LE,AK1)
      DIMENSION AK1(0:KA,0:KA)
      COMMON /COMAK/AK(0:100,0:100)
      IJ=2**(LEM-LE)
      DO 10 I=0,M
      II=IJ*I
      DO 10 J=0,M
      JJ=IJ*J
10    AK1(I,J)=AK(II,JJ)
```

```
      RETURN
      END
C*************************************************************
      SUBROUTINE AKIN(KA,M,AK1,AK2,KG)
      DIMENSION AK1(0:KA,0:KA),AK2(0:KA,0:KA)
      IF(KG.NE.1)GOTO 30
      DO 10 I=1,M,2
      I1=I-1
      I2=I+1
      I3=IABS(I-3)
      I4=I+3
      DO 10 J=0,M,2
      AK2(I,J)=(9.*(AK1(I1,J)+AK1(I2,J))
     & -(AK1(I3,J)+AK1(I4,J)))/16.
10    AK2(J,I)=AK2(I,J)
      DO 20 I=1,M,2
      I1=I-1
      I2=I+1
      I3=IABS(I-3)
      I4=I+3
      DO 20 J=1,M,2
20    AK2(I,J)=(9.*(AK2(I1,J)+AK2(I2,J))
     & -(AK2(I3,J)+AK2(I4,J)))/16.
      DO 22 I=0,M
      DO 22 J=0,M
22    AK2(I,J)=AK1(I,J)-AK2(I,J)
      DO 24 I=0,M,2
      DO 24 J=0,M,2
24    AK2(I,J)=0.0
      RETURN
30    DO 40 I=0,M
      I1=IABS(I-1)
      I2=I+1
      I3=IABS(I-3)
      I4=I+3
      DO 40 J=0,M
40    AK2(I,J)=AK1(I,J)-(9.*(AK1(I1,J)+AK1(I2,J))
     & -(AK1(I3,J)+AK1(I4,J)))/16.
      RETURN
      END
     SUBROUTINE SUBAK(MM)
      COMMON /COMAK/AK(0:100,0:100)
      S(X,Y)=X+SQRT(X**2+Y**2)
      DO 10 I=0,MM
      XP=I+0.5
      XM=I-0.5
      DO 10 J=0,I
      YP=J+0.5
      YM=J-0.5
      A1=S(YP,XP)/S(YM,XP)
      A2=S(XM,YM)/S(XP,YM)
```

```
      A3=S(YM,XM)/S(YP,XM)
      A4=S(XP,YP)/S(XM,YP)
      AK(I,J)=XP*ALOG(A1)+YM*ALOG(A2)+
     & XM*ALOG(A3)+YP*ALOG(A4)
10    AK(J,I)=AK(I,J)
      RETURN
      END
```

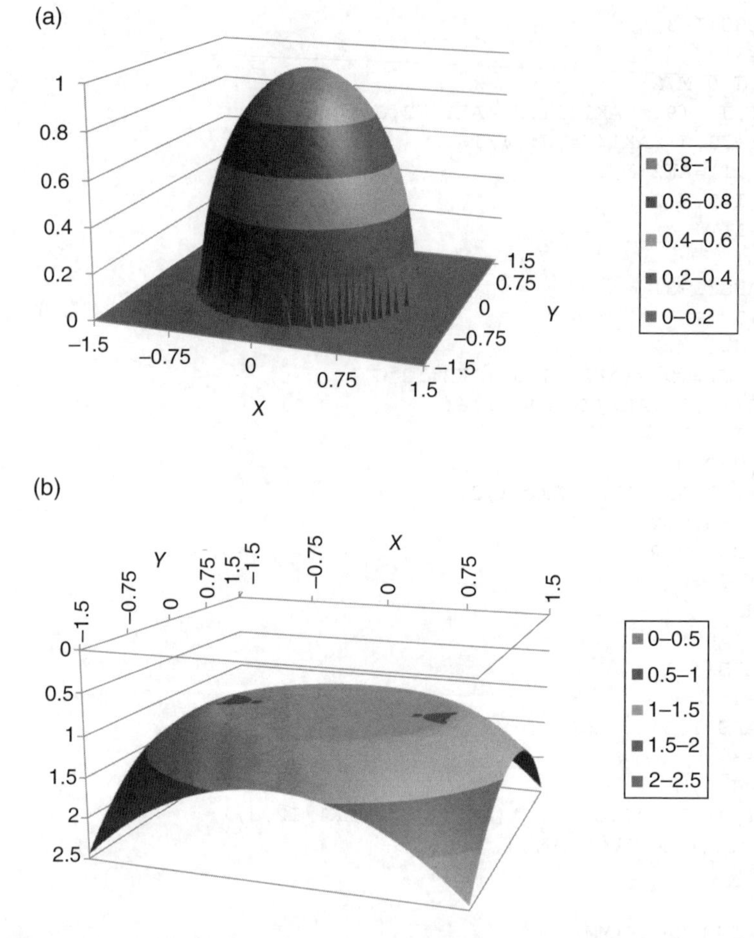

Figure 2.17 Pressure and elastic deformation calculated with multigrid integration method in point contact. (a) Hertz pressure distribution and (b) Elastic deformation

Example

Regarding the point contact, the calculation of the Hertz contact stress distribution of the elastic deformation is shown in Figure 2.17. In the contact area, the spherical surface becomes a plane. This result is consistent with the theoretical solution.

3

Numerical calculation method and program for energy equation

Temperature will significantly change the viscosity of lubricant so that it will influence the pressure and load carrying capacity. Therefore, temperature is an important factor that influences lubrication properties. In addition, the variation of the clearance shape caused by the thermal effect on lubrication surfaces can also influence lubrication properties. Furthermore, high temperature may cause lubrication failure that damages the surface materials. Therefore, the limit of the lubricant temperature is usually in the range 120–140°C.

3.1 Energy equation

In lubrication calculation, the general energy equation can be written as follows.

$$\rho c_p\left(u\frac{\partial T}{\partial x}+v\frac{\partial T}{\partial y}+w\frac{\partial T}{\partial z}\right)=k\left(\frac{\partial^2 T}{\partial x^2}+\frac{\partial^2 T}{\partial y^2}+\frac{\partial^2 T}{\partial z^2}\right)-\frac{T}{\rho}\frac{\partial \rho}{\partial T}\left(u\frac{\partial p}{\partial x}+v\frac{\partial p}{\partial y}+w\frac{\partial p}{\partial z}\right)+\Phi \tag{3.1}$$

where, Φ is the dissipation work which can be simplified as $\Phi=\eta\left[\left(\frac{\partial u}{\partial z}\right)^2+\left(\frac{\partial v}{\partial z}\right)^2\right]$.

Numerical Calculation of Elastohydrodynamic Lubrication: Methods and Programs, First Edition.
Ping Huang.

Equation 3.1 is much complicated, and in order to obtain the temperature distribution in the lubricating film, we often simplify the general energy equation.

3.1.1 *One-dimensional energy equation*

If we neglect the terms in the y direction, we can simplify Equation 3.1 to be the one-dimensional energy equation.

$$\rho c_p\left(u\frac{\partial T}{\partial x}+w\frac{\partial T}{\partial z}\right)=k\left(\frac{\partial^2 T}{\partial x^2}+\frac{\partial^2 T}{\partial z^2}\right)-\frac{T}{\rho}\frac{\partial \rho}{\partial T}\left(u\frac{\partial p}{\partial x}+w\frac{\partial p}{\partial z}\right)+\eta\left(\frac{\partial u}{\partial z}\right)^2 \tag{3.2}$$

If we assume that T, p, and η do not change across the film thickness and ρ is independent of temperature, after being integrated across the film thickness, the energy Equation 3.2 can be transformed into the commonly used one-dimensional energy equation and is as follows:

$$q_x\frac{\partial T}{\partial x}=\frac{\eta u_s^2}{J\rho c_p h}+\frac{h^3}{12\eta J\rho c_p}\left(\frac{\partial p}{\partial x}\right)^2 \tag{3.3}$$

where, q_x is the flux per width, $q_x=\frac{u_s h}{2}-\frac{h^3}{12\eta}\frac{\partial p}{\partial x}$.

Because Equation 3.3 is a first-order partial differential equation, its common temperature boundary condition is $T|_{x=x_0}=T_0$.

3.1.2 *Two-dimensional energy equation*

The two-dimensional energy equation is similar to the one-dimensional one. In lubrication calculation, if we assume that T, p, and η do not change across the film thickness and ρ is independent of temperature, after being integrated across the film thickness the energy Equation 3.1 can be transformed into the commonly used two-dimensional energy equation of lubrication as follows:

$$q_x\frac{\partial T}{\partial x}+q_y\frac{\partial T}{\partial y}=\frac{\eta U^2}{J\rho c_\rho h}+\frac{h^3}{12\eta J\rho c_\rho}\left[\left(\frac{\partial p}{\partial x}\right)^2+\left(\frac{\partial p}{\partial y}\right)^2\right] \tag{3.4}$$

where, q_x is the flux per width in the x direction, $q_x=(u_s h/2)-(h^3/12\eta)(\partial p/\partial x)$ and q_y is the flux per width in the y direction, $q_y=-(h^3/12\eta)(\partial p/\partial y)$.

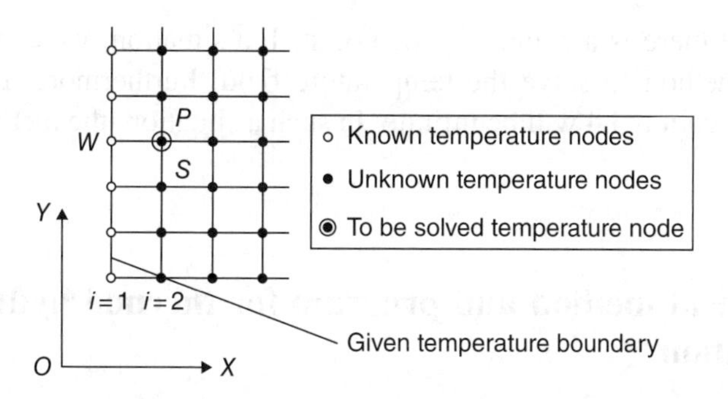

Figure 3.1 Mesh of temperature computation

Temperature analysis of lubrication is a two-dimensional problem, but Equation 3.4 is a first-order differential equation. Therefore, only one boundary condition in the x or y direction needs to be given. Generally, the temperature $T_{1,j}$ at the inlet is given as shown in Figure 3.1, but the boundary condition in the y direction depends on the circumstance. Usually, for a symmetric problem, we have $\partial T/\partial y = 0$ at the centerline. Therefore, the other boundary conditions in the y direction need not to be given again. The two-dimensional thermal lubrication is both handled in this way.

The characteristic of temperature calculation of a lubricant film is that the lubricant enters the inlet under the supplied temperature. Then, the temperature gradually rises with flow. Such a problem is called the initial value problem. Usually, we use the marching method to solve an initial value problem. The basic steps of the method are as follows.

As shown in Figure 3.1, divide the solution domain first. Then, select one row at which temperature has been known in y direction as the initial value, for example, temperature at $i = 1$ where oil is supplied. Therefore, the temperature T_{1j} on all nodes of Row 1 is known.

After T_{1j} has been known, with the discrete differential formula or the three-point parabola formula, we can calculate the value of $(\partial T/\partial y)_{1j}$. Because the pressure has been solved from the Reynolds equation, we can also calculate $(\partial p/\partial x)_{1j}$ and $(\partial p/\partial y)_{1j}$ as well as q_{1j} and q_{1j}. Substituting these values into Equation 3.4, we can obtain $(\partial T/\partial x)_{1j}$. Thus, we can calculate temperature T_{2j} of each node of Row $i = 2$.

After T_{2j} has been determined, repeat the steps as mentioned above to obtain T_{3j}. Thus, we can finally obtain the temperature of all the nodes.

The following points should be paid attention while using the marching method to solve the temperature field. First, the forward direction must be the same as the direction of lubricant flow. If the surface velocity is along the x direction, we should meet the condition $q_x > 0$. However, when the supplied oil is subjected to a high pressure or a great pressure gradient in the inlet area, $q_x < 0$ may occur in the convergence clearance.

This means that there is a counter flow. For such a situation, we cannot simply use the marching method to solve the temperature field. Furthermore, if $q_x=0$, $\partial T/\partial x$ derived from Equation 3.4 will be infinite. In such a situation, the method applied will be a failure.

3.2 Numerical method and program for thermal hydrodynamic lubrication

Although this book discusses EHL problems, it is a basic technique to calculate the temperature of a thermal hydrodynamic lubrication problem, which we will discuss in the following text

3.2.1 One-dimensional thermal hydrodynamic lubrication

3.2.1.1 Basic equations

Energy equation

First, nondimensionalize Equation 3.3. Thus, we have:

$$\frac{\partial T*}{\partial X}=\frac{1}{Q_x}\left\{\frac{2\eta*}{H}+\frac{6H}{\eta*}\left(\frac{\partial P}{\partial X}\right)^2\right\} \tag{3.5}$$

where, $Q_x=\frac{H}{2}-\frac{H^3}{2}\frac{\partial P}{\partial X}$.

Discrete Equation 3.5, we have:

$$\frac{T_i^*-T_{i-1}^*}{X_i-X_{i-1}}=\frac{1}{Q_x}\left\{\frac{2\eta_i^*}{H_i}+\frac{6H_i}{\eta_i^*}\left(\frac{P_i-P_{i-1}}{X_i-X_{i-1}}\right)^2\right\} \tag{3.6}$$

where, $Q_x=\frac{H_i}{2}-\frac{H_i^3}{2}\frac{P_i-P_{i-1}}{X_i-X_{i-1}}$.

Viscosity–temperature equation (Barus equation)

According to the viscosity–temperature equation in Chapter 1, we have its nondimensional form:

$$\eta_i^*=\exp[-\beta(T_i-T_0)] \tag{3.7}$$

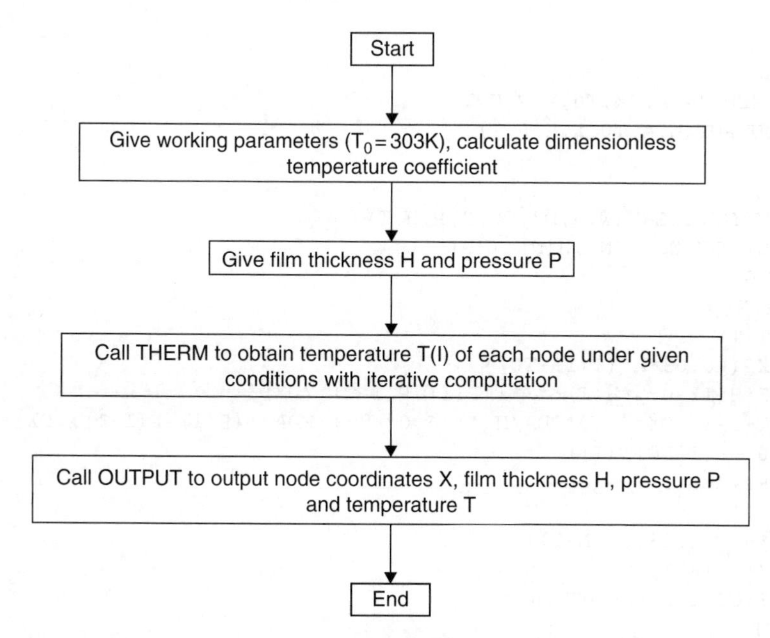

Figure 3.2 Calculation diagram of one-dimensional temperature without pressure coupling

3.2.1.2 Calculation diagram

In order to obtain the temperature, we need the value of viscosity at each node. As the viscosity is directly influenced by the temperature, we need to carry out iteration to obtain it (Figure 3.2).

3.2.1.3 Program

```
PROGRAM LINETHERM
DIMENSION X(200),P(200),H(200),T(200)
DATA U,AL,EDA0,RO,C,AJ,H1,H2/1.0,0.01,0.05,890.0,1870.0,4.184,5.5E
-6,5.E-6/
OPEN(8,FILE='OUT.DAT',STATUS='UNKNOWN')
N=129
A=U*AL*EDA0/2.0/AJ/RO/C/H2**2
T0=303.0/A
DX=1./(N-1.0)
HH=H1/H2
DH=HH-1.0
DO I=1,N
X(I)=(I-1)*DX
H(I)=HH-DH*X(I)
P(I)=-(-1.0/(H(I))+HH/(HH+1.0)/H(I)**2+1.0/(HH+1.0))/DH
T(I)=T0
ENDDO
```

```
      P(1)=0.0
      P(N)=0.0
      CALL THERM(N,A,DX,T0,X,P,H,T)
      CALL OUTPUT(N,A,T0,X,H,P,T)
      STOP
      END
      SUBROUTINE THERM(N,A,DX,T0,X,P,H,T)
      DIMENSION X(N),P(N),H(N),T(N)
   10 ERT=0.0
      DO I=2,N
      TOLD=T(I)
      EDA=EXP(-0.03*A*(T(I)-T0))
      QX=0.5*H(I)-0.5*H(I)**3*((P(I)-P(I-1))/DX)
      T(I)=T(I-1)+DX*(2.0*EDA/H(I)+6.0*H(I)/EDA*((P(I)-P(I-1))/DX)**2)/QX
      T(I)=0.5*(TOLD+T(I))
      ERT=ERT+ABS(T(I)-TOLD)
      ENDDO
      ERT=A*ERT/(303.0*(N-1))
      WRITE(*,*)ERT
      IF(ERT.GT.1.E-6)GOTO 10
      RETURN
      END
      SUBROUTINE OUTPUT(N,A,T0,X,H,P,T)
      DIMENSION X(N),H(N),P(N),T(N)
      DO I=1,N
      T(I)=A*(T(I)-T0)
      END DO
      DO I=1,N
      WRITE(8,30)X(I),H(I),P(I),T(I)
      ENDDO
   30 FORMAT(4(1X,E12.6))
      RETURN
      END
```

3.2.1.4 Example

As given in the program, the working parameters are the velocity U = 1 m·s^{-1}, the length Al = 0.01 m, the initial viscosity of lubricant EDA0 = 0.05 Pa·s, the lubricant density RO = 890 kg·m^{-3}, the specific heat of lubricant C = 1870 J·kg^{-1}·K^{-1}, the mechanical equivalent of heat AJ = 4.184 J cal^{-1}, the maximum thickness $H1 = 5.5 \times 10^{-6}$ m and the minimum film thickness $H2 = 5 \times 10^{-6}$ m. The calculation result of lubrication for the given pressure is shown in Figure 3.3. The temperature calculated is linear to the coordinates, and the maximum temperature rise is about 5°C. Theoretically, because the viscosity influences the pressure, the viscosity obtained from viscosity–temperature equation should be substituted into the Reynolds equation to calculate the pressure and then to solve the temperature again until convergency is met. However, if the temperature is not much high, we can ignore this step.

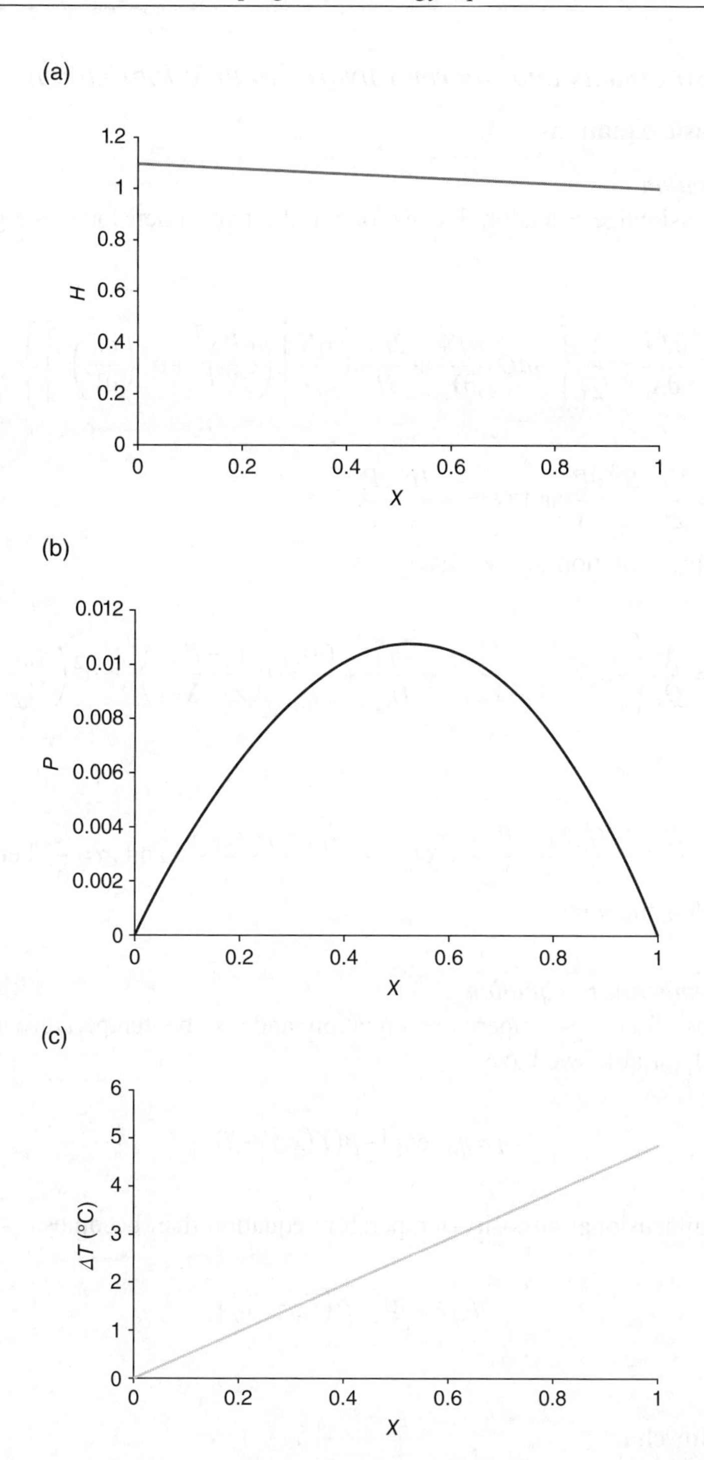

Figure 3.3 Result of temperature of hydrodynamic lubrication for given pressure. (a) Film thickness, (b) pressure distribution and (c) temperature distribution

3.2.2 *Two-dimensional thermal hydrodynamic lubrication*

3.2.2.1 Basic equations

Energy equation

To nondimensionlize Equation 3.4, we obtain the two-dimensional energy equation as follows:

$$\frac{\partial T*}{\partial X}=\frac{1}{Q_x}\left\{-\alpha Q_y\frac{\partial T*}{\partial Y}+\frac{2\eta*}{H}+\frac{6H}{\eta*}\left[\left(\frac{\partial P}{\partial X}\right)^2+\alpha^2\left(\frac{\partial P}{\partial Y}\right)^2\right]\right\} \tag{3.8}$$

where, $Q_x=\frac{H}{2}-\frac{H^3}{2}\frac{\partial P}{\partial X}$ and $Q_y=-\frac{H^3}{2}\frac{\partial P}{\partial Y}$.

Discretizing Equation 3.8, we have:

$$\frac{T^*_{i,j}-T^*_{i-1,j}}{X_i-X_{i-1}}=\frac{1}{Q_x}\left\{-\alpha Q_y\frac{T^*_{i,j}-T^*_{i,j-1}}{Y_j-Y_{j-1}}+\frac{2\eta^*_{i,j}}{H_{i,j}}+\frac{6H_{i,j}}{\eta^*_{i,j}}\left[\left(\frac{P_i-P_{i-1}}{X_i-X_{i-1}}\right)^2+\alpha^2\left(\frac{P_{i,j}-P_{i,j-1}}{Y_j-Y_{j-1}}\right)^2\right]\right\} \tag{3.9}$$

where, $Q_x=\frac{H_{i,j}}{2}-\frac{H^3_{i,j}}{2}\frac{P_{i,j}-P_{i-1,j}}{X_i-X_{i-1}}$; $Q_y=-\frac{H^3_{i,j}}{2}\frac{P_{i,j}-P_{i,j-1}}{Y_j-Y_{j-1}}$; and $\alpha=\frac{a}{b}$, here, a is the length and b is the width.

Viscosity–temperature equation

Using Barus viscosity–temperature equation and as the temperature is the two-dimensional variable, we have:

$$\eta=\eta_0\ \exp[-\beta(T(x,y)-T_0)] \tag{3.10}$$

The nondimensional viscosity–temperature equation then becomes:

$$\eta^*_{i,j}=\exp\left[-\beta\left(T_{i,j}-T_0\right)\right] \tag{3.11}$$

3.2.2.2 Flowchart

In order to obtain temperature, we need the viscosity at each node. As the viscosity is directly influenced by temperature, we need to carry out iteration. The flowchart to

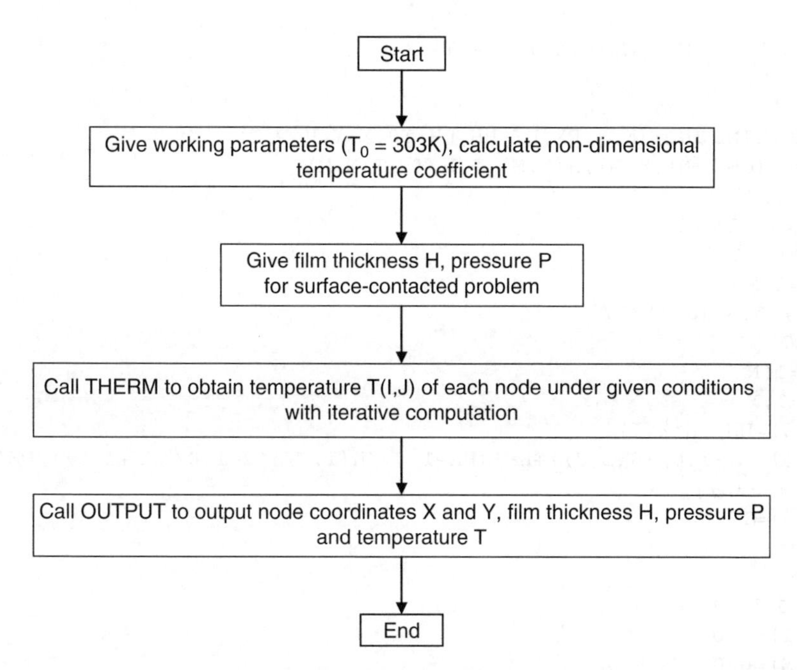

Figure 3.4 Flowchart to calculate two-dimensional temperature without pressure coupling

calculate two-dimensional temperature without pressure coupling is shown in Figure 3.4.

3.2.2.3 Program

```
PROGRAM SURFACETHERM
DIMENSION X(200),Y(200),P(20000),H(20000),T(20000)
DATA
U,ALX,ALY,EDA0,RO,C,AJ,H1,H2/
1.0,0.01,0.01,0.05,890.0,1870.0,4.184,1.1E-6,1.E-6/
OPEN(7,FILE='FILM.DAT',STATUS='UNKNOWN')
OPEN(8,FILE='PRESSURE.DAT',STATUS='UNKNOWN')
OPEN(9,FILE='TEM.DAT',STATUS='UNKNOWN')
N=129
M=65
A=U*ALX*EDA0/2.0/AJ/RO/C/H2**2
T0=303.0/A
DX=1./(N-1.0)
DY=1./(M-1.0)
HH=H1/H2
DH=HH-1.0
ALFA1=ALX/ALY
CALL INIT(N,M,DX,DY,HH,DH,T0,X,Y,H,P,T)
CALL THERM(N,M,A,ALFA1,DX,DY,T0,X,Y,P,H,T)
```

```
   CALL OUTPUT(N,M,A,T0,X,Y,H,P,T)
   STOP
   END
   SUBROUTINE INIT(N,M,DX,DY,HH,DH,T0,X,Y,H,P,T)
   DIMENSION X(N),Y(M),H(N,M),P(N,M),T(N,M)
   DO I=1,N
   X(I)=(I-1)*DX
   ENDDO
   DO J=1,M
   Y(J)=-0.5+(J-1)*DY
   ENDDO
   DO I=1,N
   DO J=1,M
   H(I,J)=HH-DH*X(I)
   P(I,J)=-(-1.0/(H(I,J))+HH/(HH+1.0)/H(I,J)**2+1.0/(HH+1.0))/DH*(1.0-
4.0*Y(J)*Y(J))
   T(I,J)=T0
   ENDDO
   ENDDO
   DO I=1,N
   P(I,1)=0.0
   P(I,M)=0.0
   ENDDO
   DO J=1,M
   P(1,J)=0.0
   P(N,J)=0.0
   ENDDO
   RETURN
   END
   SUBROUTINE THERM(N,M,A,ALFA1,DX,DY,T0,X,Y,P,H,T)
   DIMENSION X(N),Y(M),H(N,M),P(N,M),T(N,M)
10 ERT=0.0
   DO I=2,N
   DO J=M/2+1,1,-1
   TOLD=T(I,J)
   EDA=EXP(-0.03*A*(T(I,J)-T0))
   DPDX=(P(I,J)-P(I-1,J))/DX
   IF(J.EQ.M/2+1)THEN
   DPDY=0.0
   DTDY=0.0
   ELSE
   DPDY=(P(I,J+1)-P(I,J))/DY
   DTDY=(T(I,J+1)-T(I,J))/DY
   ENDIF
   QX=0.5*H(I,J)-0.5*H(I,J)**3*DPDX
   QY=-0.5*H(I,J)**3*DPDY
   AA=-0.5*ALFA1*QY*DTDY
   AB=2.0*EDA/H(I,J)
   AC=6.0*H(I,J)/EDA*(DPDX**2+ALFA1**2*DPDY**2)
```

```
   BA=QX/DX-ALFA1*QY/DY
   BB=QX/DX*T(I-1,J)-ALFA1*QY/DY*T(I,J+1)
   T(I,J)=(BB+AB+AC)/BA
   T(I,J)=0.7*TOLD+0.3*T(I,J)
   ERT=ERT+ABS(T(I,J)-TOLD)
   ENDDO
   ENDDO
   ERT=A*ERT/(303.0*(N-1)*(M-1))
   WRITE(*,*)ERT
   IF(ERT.GT.1.E-8)GOTO 10
   DO I=2,N
   DO J=1,M/2
   T(I,M-J+1)=T(I,J)
   ENDDO
   ENDDO
   RETURN
   END
   SUBROUTINE OUTPUT(N,M,A,T0,X,Y,H,P,T)
   DIMENSION X(N),Y(M),H(N,M),P(N,M),T(N,M)
   DO I=1,N
   DO J=1,M
   T(I,J)=A*(T(I,J)-T0)
   END DO
   ENDDO
   WRITE(7,30)X(1),(Y(J),J=1,M)
   WRITE(8,30)X(1),(Y(J),J=1,M)
   WRITE(9,30)X(1),(Y(J),J=1,M)
   DO I=1,N
   WRITE(7,30)X(I),(H(I,J),J=1,M)
   WRITE(8,30)X(I),(P(I,J),J=1,M)
   WRITE(9,30)X(I),(T(I,J),J=1,M)
   ENDDO
30 FORMAT(130(1X,E12.6))
   RETURN
   END
```

3.2.2.4 Example

As given in the program, the working parameters are the velocity $U = 1\ m{\cdot}s^{-1}$, the slider length $ALX = 0.01$ m, the slider width $ALY = 0.01$ m, the initial viscosity of the lubricant $EDA0 = 0.05$ Pa·s, the lubricant density $RO = 890\ kg\ m^{-3}$, the specific heat of the lubricant $C = 1870\ J{\cdot}kg^{-1}{\cdot}K^{-1}$, the mechanical equivalent of heat $AJ = 4.184\ J{\cdot}cal^{-1}$, the maximum film thickness $H1 = 1.1 \times 10^{-6}$ m, and the minimum film thickness $H2 = 1 \times 10^{-6}$ m. The calculated results for two-dimensional lubrication are shown in Figure 3.5. The results show that the temperature calculated is not linear to

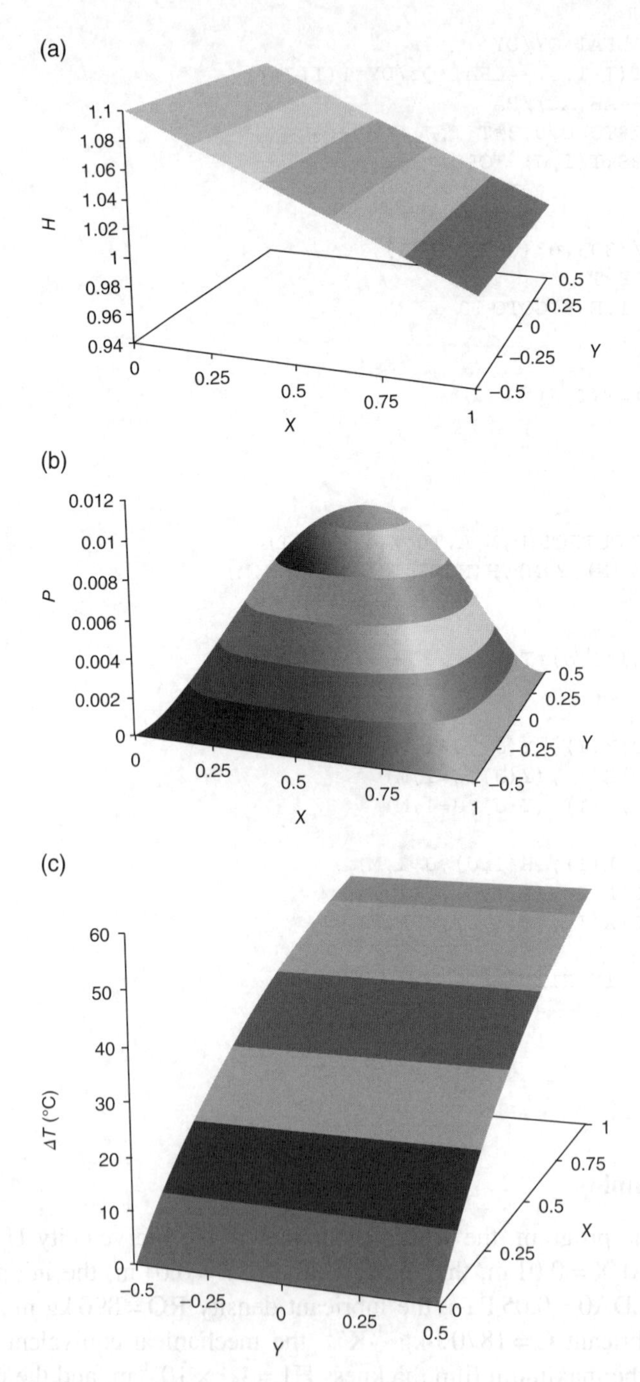

Figure 3.5 Result of temperature for two-dimensional hydrodynamic lubrication. (a) Film thickness, (b) pressure distribution, and (c) temperature distribution

the coordinate anymore and the maximum temperature rise is about 50°C. Theoretically, as the viscosity influences the pressure, the viscosity obtained from the viscosity–temperature equation should be substituted into the Reynolds equation to obtain pressure and then the temperature should be solved again until convergency is obtained. Here we have explained only the calculation of temperature under the current pressure without iteration.

4

Numerical calculation method and program for isothermal EHL in line contact

4.1 Basic equations and nondimensional equations

4.1.1 Basic equations

The basic equations of isothermal EHL in the line contact include the Reynolds equation, the film thickness equation, the elastic deformation equation, the viscosity–pressure equation, the density–pressure equation and the load equation.

Reynolds equation

$$\frac{d}{dx}\left(\frac{\rho h^3}{\eta}\frac{dp}{dx}\right) = 12u_s\frac{d(\rho h)}{dx} \tag{4.1}$$

Film thickness equation

$$h(x) = h_0 + \frac{x^2}{2R} + v(x) \tag{4.2}$$

Elastic deformation equation

$$v(x) = -\frac{2}{\pi E}\int_{x_0}^{x_e} p(s)\ln(s-x)^2 ds + c \tag{4.3}$$

Numerical Calculation of Elastohydrodynamic Lubrication: Methods and Programs, First Edition.
Ping Huang.

Viscosity–pressure equation

$$\eta = \eta_0 \exp\left\{ (\ln \eta_0 + 9.67)\left[-1 + \left(1 + \frac{p}{p_0}\right)^z \right] \right\} \tag{4.4}$$

Density–pressure equation

$$\rho = \rho_0 \left(1 + \frac{0.6p}{1 + 1.7p}\right) \tag{4.5}$$

Load equation

$$w - \int_{x_0}^{x_e} p(x)dx = 0 \tag{4.6}$$

Because the elastic deformation in the film-thickness equation and the viscosity in the viscosity–pressure equation are dependent on pressure, in order to obtain a new pressure distribution, first calculate the film thickness and the viscosity with the given initial pressure distribution (such as the Hertzian contact stress distribution) and then substitute them in Reynolds equation. According to the new pressure distribution, iteratively calculate the elastic deformation and the film thickness. Repeat these steps until the difference between the sums of the two neighbor pressures is less than the given error. Thus, the final distribution of pressure and film thickness with the elastic deformation can be obtained.

4.1.2 Nondimensional equations

The nondimensional equations of isothermal EHL in the line contact are as follows:
Reynolds equation

$$\frac{d}{dX}\left(\varepsilon \frac{dP}{dX}\right) - \frac{d(\rho^* H)}{dX} = 0 \tag{4.7}$$

The boundary conditions are as follows:

Inlet boundary condition $P(X_0) = 0$
Outlet boundary condition $P(X_e) = 0 \quad \frac{dP(X_e)}{dX} = 0$

Film thickness equation

$$H(X)=H_0+\frac{X^2}{2}-\frac{1}{\pi}\int_{X_0}^{X_e}\ln|X-X'|p(X')dX' \tag{4.8}$$

Viscosity–pressure equation

$$\rho^*=1+\frac{0.6p}{1+1.7p} \tag{4.9}$$

Density–pressure equation

$$\eta^*=\exp\left\{[\ln(\eta_0)+9.67]\left[-1+\left(1+\frac{p_H P}{p_0}\right)^z\right]\right\} \tag{4.10}$$

Load equation

$$W=\int_{X_0}^{X_e}PdX=\frac{\pi}{2} \tag{4.11}$$

It should be noted that the load w in the program of EHL in the line contact is per length.

4.1.3 *Discrete equations*

With the central and the forward discrete differential formulas, the discrete form of the nondimensional Reynolds equation, Equation 4.7, can be obtained. The nondimensional film thickness equation and load equation can also be written in the discrete forms by numerical integral. These equations are as follows.

Reynolds equation

$$\frac{\varepsilon_{i-1/2}P_{i-1}-(\varepsilon_{i-1/2}+\varepsilon_{i+1/2})P_i+\varepsilon_{i+1/2}P_{i+1}}{\Delta X^2}=\frac{\rho_i^*H_i-\rho_{i-1}^*H_{i-1}}{\Delta X} \tag{4.12}$$

where, $\varepsilon_{i\pm1/2}=(1/2)(\varepsilon_i+\varepsilon_{i\pm1})$ and $\Delta X=X_i-X_{i-1}$.

The boundary conditions of Equation 4.12 are as follows: in the inlet, set $P(X_0)=0$ and in the outlet, set the negative pressure as zero to obtain $P(X_e)=0$ and $[P(X_e)-P(X_e-1)]/\Delta X=0$.

Film thickness equation

$$H_i=H_0+\frac{X_i^2}{2}-\frac{1}{\pi}\sum_{j=1}^{n}K_{ij}P_j \tag{4.13}$$

Load equation

$$\Delta X\sum_{i=1}^{n}P_i=\frac{\pi}{2} \tag{4.14}$$

4.2 Numerical calculation method and program

4.2.1 Iteration method

4.2.1.1 Iteration method of pressure

By using the discrete Reynolds equation (4.12) directly, the iteration method of pressure is given in Subroutine ITER. According to the discrete film thickness equation (4.13), H_i and H_{i-1} in Equation 4.12 all contain P_i, that is, $-1/\pi K_{ii}P_i$ and $-1/\pi K_{i-1,i}P_i$. Based on the deformation equation, we know that $K_{ii}=\Delta X(a_0+\ln\Delta X)$ and $K_{i-1,i}=\Delta X(a_1+\ln\Delta X)$. Therefore, after adding $(\rho_i^* a_0(1/\pi)-\rho_{i-1}^* a_1(1/\pi))P_i/\Delta X$ to the both sides of Equation 4.12, the equation remains correct. Thus, the discrete Reynolds equation can be written as:

$$\begin{aligned}&\varepsilon_{i-1/2}P_{i-1}-\left[\varepsilon_{i-1/2}+\varepsilon_{i+1/2}+\left(\rho_{i-1}^* a_1\frac{1}{\pi}-\rho_i^* a_0\frac{1}{\pi}\right)\Delta X\right]P_i+\varepsilon_{i+1/2}P_{i+1}\\&=\Delta X\left[\rho_i^* H_i-\rho_{i-1}^* H_{i-1}+\left(\rho_i^* a_0\frac{1}{\pi}-\rho_{i-1}^* a_1\frac{1}{\pi}\right)P_i\right]\end{aligned} \tag{4.15}$$

Rearrange it as follows:

$$P_i=\frac{\varepsilon_{i-1/2}P_{i-1}+\varepsilon_{i+1/2}P_{i+1}-\Delta X\left[\rho_i^* H_i-\rho_{i-1}^* H_{i-1}+\left(\rho_i^* a_0\frac{1}{\pi}-\rho_{i-1}^* a_1\frac{1}{\pi}\right)P_i\right]}{\varepsilon_{i-1/2}+\varepsilon_{i+1/2}+\left(\rho_{i-1}^* a_1\frac{1}{\pi}-\rho_i^* a_0\frac{1}{\pi}\right)\Delta X} \tag{4.16}$$

4.2.1.2 Correction of H_0 and load balance condition

The dichotomy method given in Section 1.6 uses the present program. The minimum film thickness equation of EHL used here is:

$$H_{\min}=1.6\left(\frac{R}{b}\right)G^{*0.6}U^{*0.7}W^{*-0.13} \tag{4.17}$$

The initial ΔH_0 is as follows:

$$\Delta H_0=0.005H_{\min} \tag{4.18}$$

The load difference is as follows:

$$\Delta W=\frac{\pi}{2}-\sum_{i=1}^{n}P_i\Delta X_i \tag{4.19}$$

The correction of the rigid film thickness can be written as:

$$H_0 = H_0 \pm \Delta H_0 \tag{4.20}$$

where, if $\Delta W < 0$, "+" is used and if $\Delta W > 0$, "−" is used.

The correction method has been fully described in Section 1.6.

4.2.2 *Program and example*

4.2.2.1 Nomenclature of program variables

N	node number
W	load w, unit N
R	equivalent radius of the cylinder, m
US	velocity, m/s
E1	equivalent modulus of elasticity, Pa
EDA0	initial viscosity, Pa · s
ALFA	viscosity–pressure coefficient
P0	pressure coefficient p_0 in viscosity–pressure equation
B	half width of Hertzian contact, unit m
PH	maximum Hertzian contact pressure p_H, Pa
A1/A2	parameters to calculate viscosity
A3	parameter to calculate density
ENDA	λ in nondimensional Reynolds equation
U	nondimensional velocity
EPS(I)	ε_i in discrete Reynolds equation
EDA(I)	nondimensional viscosity $\eta*$ of node I
RO(I)	nondimensional velocity $\rho*$ of node I
H(I)	nondimensional film thickness H of node I
P(I)	nondimensional pressure P of node I
POLD(I)	nondimensional pressure P before iteration of node I
AK(I)	a_i in elastic deformation equation of node I.

4.2.2.2 Calculation program

The program is given in Figure 4.1.

4.2.2.3 Main subroutines

Subroutine EHL is used to realize the iteration calculation of EHL in the line contact under given conditions. The diagram of Subroutine EHL is given in Figure 4.2.

Subroutine HREE is used to realize the calculation of H(I), RO(I), EDA(I), and call VI subroutine to calculate the elastic deformation. The diagram of Subroutine HREE is given in Figure 4.3.

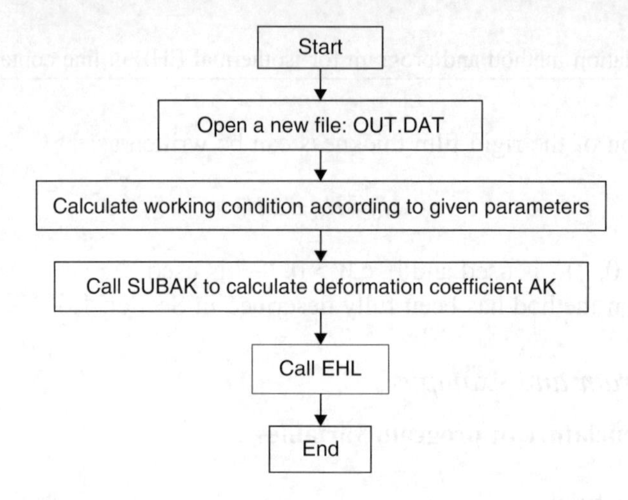

Figure 4.1 Diagram of isothermal EHL in line contact

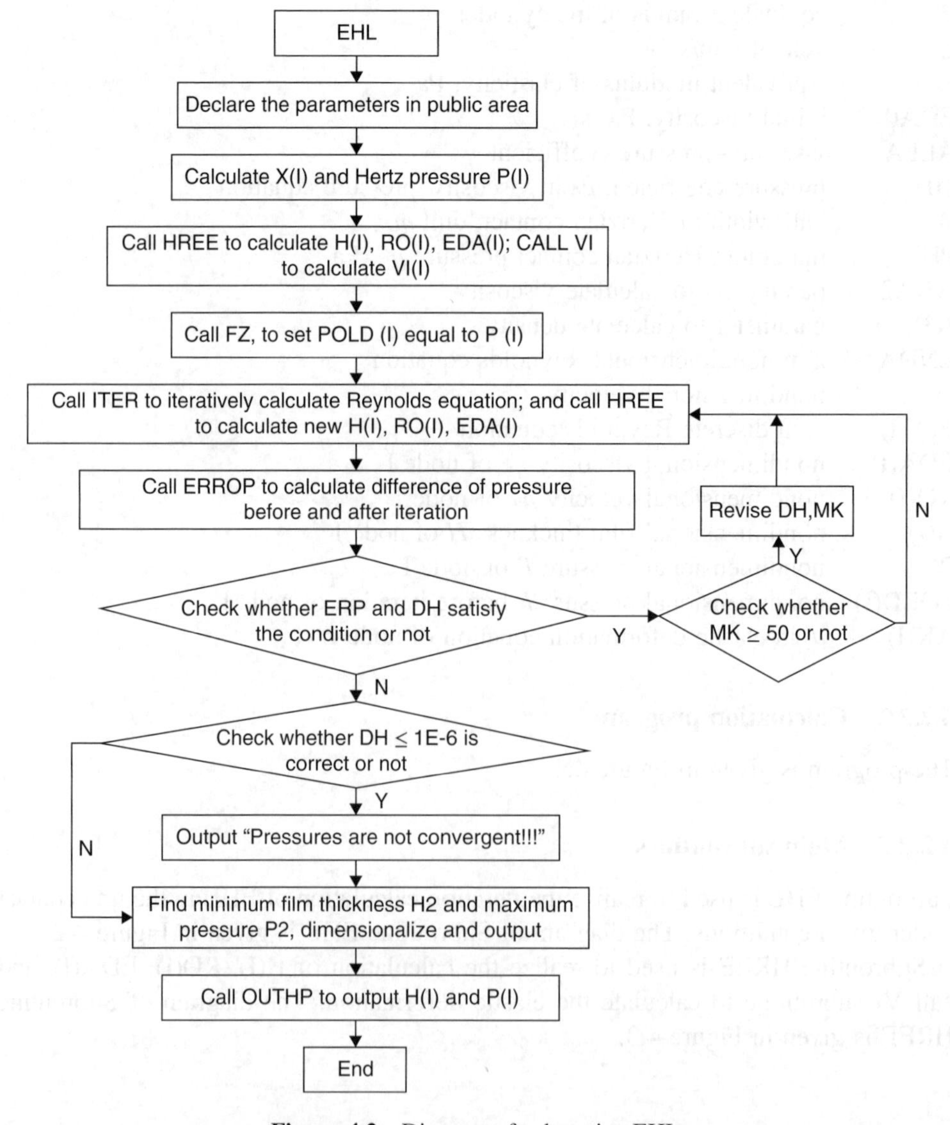

Figure 4.2 Diagram of subroutine EHL

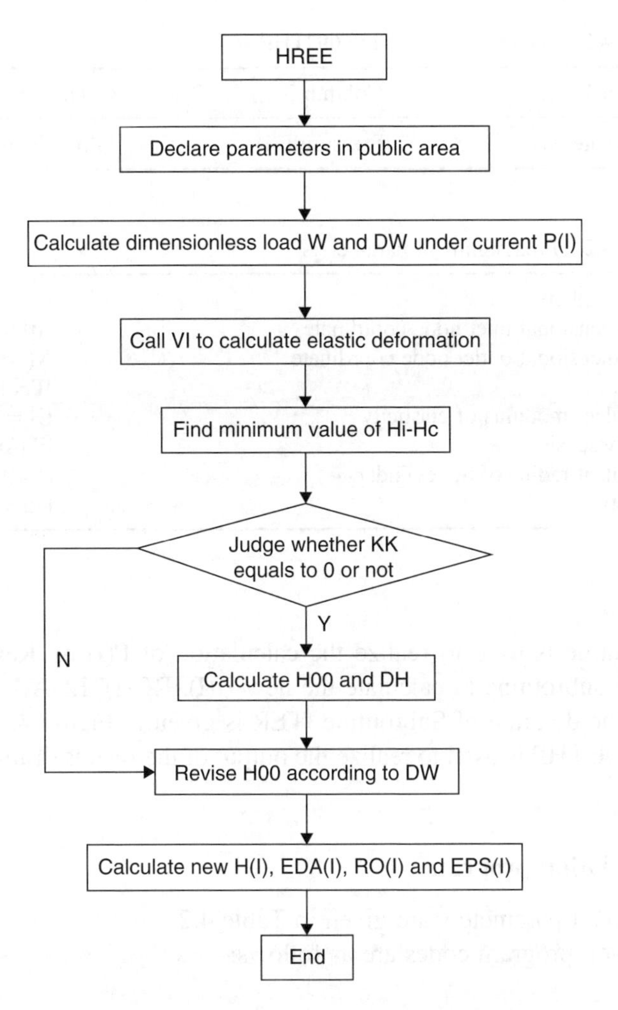

Figure 4.3 Diagram of subroutine HREE

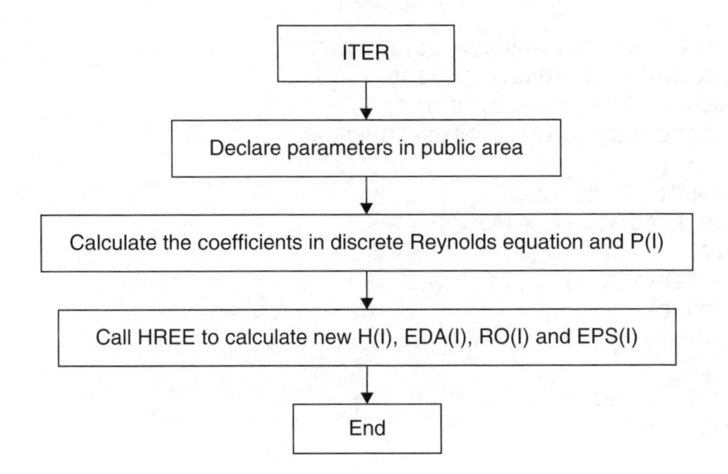

Figure 4.4 Diagram of subroutine ITER

Table 4.1 Format of subroutine OUTHP.

Column 1	Column 2	Column 3
Coordinate X(I)	Pressure P(I)	Film thickness H(I)

Table 4.2 Preassignment parameters.

Node numbers	$N = 130$
Nondimensional inlet node coordinate	X0 = −4.0
Nondimensional outlet node coordinate	XE = 1.5
Load	$W = 1.0E5$
Equivalent modulus of elasticity	E1 = 2.2E11
Initial viscosity	EDA0 = 0.05
Equivalent radius of the cylinder	$R = 0.05$
Velocity	US = 1.5

ITER subroutine is used to realize the calculation of P(I) by Reynolds equation and call HREE subroutine to calculate the new H(I), RO(I), EDA(I), and the elastic deformation. The diagram of Subroutine ITER is given in Figure 4.4.

Subroutine OUTHP is used to realize the output of the results. Table 4.1 shows the output format.

4.2.2.4 Calculation program

The preassignment parameters are given in Table 4.2.

The calculation program codes are as follows:

```
PROGRAM LINEEHL
      COMMON /COM1/ENDA,A1,A2,A3,Z,HM0,DH/COM2/EDA0/COM4/X0,XE/COM3/E1,
PH,B,R
      DATA PAI,Z,P0/3.14159265,0.68,1.96E8/
      DATA N,X0,XE,W,E1,EDA0,R,US/130,-
4.0,1.5,1.0E5,2.2E11,0.05,0.05,1.5/
      OPEN(8,FILE='OUT.DAT',STATUS='UNKNOWN')
      W1=W/(E1*R)
      PH=E1*SQRT(0.5*W1/PAI)
      A1=(ALOG(EDA0)+9.67)
      A2=PH/P0
      A3=0.59/(PH*1.E-9)
      B=4.*R*PH/E1
      ALFA=Z*A1/P0
      G=ALFA*E1
      U=EDA0*US/(2.*E1*R)
      CC1=SQRT(2.*U)
```

```
      AM=2.*PAI*(PH/E1)**2/CC1
      ENDA=3.*(PAI/AM)**2/8.
      HM0=1.6*(R/B)**2*G**0.6*U**0.7*W1**(-0.13)
      WRITE(*,*)N,X0,XE,W,E1,EDA0,R,US
      CALL SUBAK(N)
      CALL EHL(N)
      STOP
      END
      SUBROUTINE EHL(N)
       DIMENSION X(1100),P(1100),H(1100),RO(1100),POLD(1100),EPS(1100),EDA
(1100),V(1100)
      COMMON /COM1/ENDA,A1,A2,A3,Z,HM0,DH/COM4/X0,XE
      COMMON /COM3/E1,PH,B,RR
      MK=1
      DX=(XE-X0)/(N-1.0)
      DO 10 I=1,N
      X(I)=X0+(I-1)*DX
      IF(ABS(X(I)).GE.1.0)P(I)=0.0
      IF(ABS(X(I)).LT.1.0)P(I)=SQRT(1.-X(I)*X(I))
10    CONTINUE
      CALL HREE(N,DX,X,P,H,RO,EPS,EDA,V)
      CALL FZ(N,P,POLD)
14    KK=19
      CALL ITER(N,KK,DX,X,P,H,RO,EPS,EDA,V)
      MK=MK+1
      CALL ERROP(N,P,POLD,ERP)
      WRITE(*,*)'ERP=',ERP
      IF(ERP.GT.1.E-5.AND.DH.GT.1.E-6)THEN
      IF(MK.GE.50)THEN
      MK=1
      DH=0.5*DH
      ENDIF
      GOTO 14
      ENDIF
      IF(DH.LE.1.E-6)WRITE(*,*)'Pressures are not convergent!!!'
      H2=1.E3
      P2=0.0
      DO 106 I=1,N
      IF(H(I).LT.H2)H2=H(I)
      IF(P(I).GT.P2)P2=P(I)
106   CONTINUE
      H3=H2*B*B/RR
      P3=P2*PH
110   FORMAT(6(1X,E12.6))
120   CONTINUE
      WRITE(*,*)'P2,H2,P3,H3=',P2,H2,P3,H3
      CALL OUTHP(N,X,P,H)
      RETURN
      END
      SUBROUTINE OUTHP(N,X,P,H)
```

```
      DIMENSION X(N),P(N),H(N)
      DO 10 I=1,N
      WRITE(8,20)X(I),P(I),H(I)
10    CONTINUE
20    FORMAT(1X,6(E12.6,1X))
      RETURN
      END
      SUBROUTINE HREE(N,DX,X,P,H,RO,EPS,EDA,V)
      DIMENSION X(N),P(N),H(N),RO(N),EPS(N),EDA(N),V(N)
      COMMON /COM1/ENDA,A1,A2,A3,Z,HM0,DH/COM2/EDA0/COMAK/AK(0:1100)
      DATA KK,PAI1,G0/0,0.318309886,1.570796325/
      IF(KK.NE.0)GOTO 3
      H00=0.0
3     W1=0.0
      DO 4 I=1,N
4     W1=W1+P(I)
      C3=(DX*W1)/G0
      DW=1.-C3
      CALL VI(N,DX,P,V)
      HMIN=1.E3
      DO 30 I=1,N
      H0=0.5*X(I)*X(I)+V(I)
      IF(H0.LT.HMIN)HMIN=H0
      H(I)=H0
30    CONTINUE
      IF(KK.NE.0)GOTO 32
      KK=1
      DH=0.005*HM0
      H00=-HMIN+HM0
32    IF(DW.LT.0.0)H00=H00+DH
      IF(DW.GT.0.0)H00=H00-DH
      DO 60 I=1,N
      H(I)=H00+H(I)
      EDA(I)=EXP(A1*(-1.+(1.+A2*P(I))**Z))
      RO(I)=(A3+1.35*P(I))/(A3+P(I))
      EPS(I)=RO(I)*H(I)**3/(ENDA*EDA(I))
60    CONTINUE
      RETURN
      END
      SUBROUTINE ITER(N,KK,DX,X,P,H,RO,EPS,EDA,V)
      DIMENSION X(N),P(N),H(N),RO(N),EPS(N),EDA(N),V(N)
      COMMON /COMAK/AK(0:1100)
      DATA PAI1/0.318309886/
      DO 100 K=1,KK
      D2=0.5*(EPS(1)+EPS(2))
      D3=0.5*(EPS(2)+EPS(3))
      DO 70 I=2,N-1
      D1=D2
      D2=D3
      IF(I.NE.N-1)D3=0.5*(EPS(I+1)+EPS(I+2))
```

```
      D8=RO(I)*AK(0)*PAI1
      D9=RO(I-1)*AK(1)*PAI1
      D10=1.0/(D1+D2+(D9-D8)*DX)
      D11=D1*P(I-1)+D2*P(I+1)
      D12=(RO(I)*H(I)-RO(I-1)*H(I-1)+(D8-D9)*P(I))*DX
      P(I)=(D11-D12)*D10
      IF(P(I).LT.0.0)P(I)=0.0
70    CONTINUE
      CALL HREE(N,DX,X,P,H,RO,EPS,EDA,V)
100   CONTINUE
      RETURN
      END
      SUBROUTINE VI(N,DX,P,V)
      DIMENSION P(N),V(N)
      COMMON /COMAK/AK(0:1100)
      PAI1=0.318309886
      C=ALOG(DX)
      DO 10 I=1,N
      V(I)=0.0
      DO 10 J=1,N
      IJ=IABS(I-J)
10    V(I)=V(I)+(AK(IJ)+C)*DX*P(J)
      DO I=1,N
      V(I)=-PAI1*V(I)
      ENDDO
      RETURN
      END
      SUBROUTINE SUBAK(MM)
      COMMON /COMAK/AK(0:1100)
      DO 10 I=0,MM
10    AK(I)=(I+0.5)*(ALOG(ABS(I+0.5))-1.)-(I-0.5)*(ALOG(ABS(I-0.5))-1.)
      RETURN
      END
      SUBROUTINE FZ(N,P,POLD)
      DIMENSION P(N),POLD(N)
      DO 10 I=1,N
10    POLD(I)=P(I)
      RETURN
      END
      SUBROUTINE ERROP(N,P,POLD,ERP)
      DIMENSION P(N),POLD(N)
      SD=0.0
      SUM=0.0
      DO 10 I=1,N
      SD=SD+ABS(P(I)-POLD(I))
      POLD(I)=P(I)
10    SUM=SUM+P(I)
      ERP=SD/SUM
      RETURN
      END
```

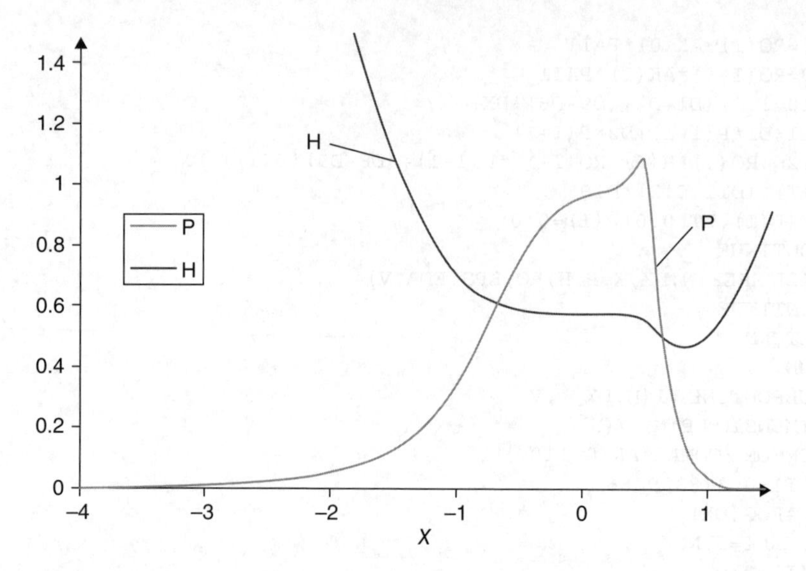

Figure 4.5 Pressure distribution and film thickness of isothermal EHL in line contact

4.2.2.5 Result

Based on the working conditions in the program, the pressure and the film thickness distribution are shown in Figure 4.5.

5

Newton–Raphson method and programs to solve EHL problems in line contact

5.1 Basic equations

To discretize nondimensional EHL equations in the line contact, they can be written as follows [3]:

Reynolds equation

$$f_i = H_i^3 \left(\frac{dP}{dX}\right)_i - A\bar{\eta}_i \left(H_i - \frac{\bar{\rho}_e H_e}{\bar{\rho}_i}\right) = 0 \tag{5.1}$$

Film thickness equation

$$H_i = H_0 + \frac{x_i^2}{2} + \sum_{j=1}^{N} K_{ij} P_j \tag{5.2}$$

Viscosity–pressure equation

$$\eta = \eta_0 \ \exp\left\{(\ln \eta_0 + 9.67)\left[\left(1 + \frac{p}{p_0}\right)^z\right]\right\} \tag{5.3}$$

Numerical Calculation of Elastohydrodynamic Lubrication: Methods and Programs, First Edition.
Ping Huang.

Density–pressure equation

$$\rho = \rho_0 \left(1 + \frac{0.6p}{1 + 1.7p} \right) \tag{5.4}$$

Load equation

$$W = \sum_{i=1}^{N} P_i \Delta X = \frac{\pi}{2} \tag{5.5}$$

where, A is the nondimensional coefficient, $A = 3\pi^2 U^* / 4W^{*2}$, $\bar{\rho}_e H_e$ the nondimensional product of the density and the outlet film thickness, N the node number of the lubrication region, and $i = 1, 2, \ldots, N$; $j = 1, 2, \ldots, N$.

5.2 Newton–Raphson iteration method

5.2.1 Coefficient matrix

Houport and Hamrock used the Newton–Raphson method to derive the aforementioned equations to calculate the increment matrix of variables [3]. As the initial equations are nearly linear, the results are approximate. Therefore, the final solution is obtained through iteration until the precision is met. The corresponding matrix to solve the incremental is as follows:

$$C = \begin{bmatrix} \frac{\partial f_1}{\partial(\bar{\rho}_e H_e)} & \frac{\partial f_1}{\partial P_2} & \cdots & \frac{\partial f_1}{\partial P_N} & \frac{\partial f_1}{\partial H_0} \\ \frac{\partial f_2}{\partial(\bar{\rho}_e H_e)} & \frac{\partial f_2}{\partial P_2} & \cdots & \frac{\partial f_2}{\partial P_N} & \frac{\partial f_2}{\partial H_0} \\ \vdots & \vdots & \cdots & \vdots & \vdots \\ \frac{\partial f_N}{\partial(\bar{\rho}_e H_e)} & \frac{\partial f_N}{\partial P_2} & \cdots & \frac{\partial f_N}{\partial P_N} & \frac{\partial f_N}{\partial H_0} \\ 0 & \frac{\partial W}{\partial P_2} & \cdots & \frac{\partial W}{\partial P_N} & 0 \end{bmatrix} \begin{bmatrix} \Delta(\bar{\rho}_e H_e) \\ \Delta P_2 \\ \vdots \\ \Delta P_N \\ \Delta H_0 \end{bmatrix} = \begin{bmatrix} -f_1 \\ -f_2 \\ \vdots \\ -f_N \\ \Delta W \end{bmatrix} \tag{5.6}$$

where, f_i is the discrete function given by Equation 5.1.

The matrix C of coefficients C_{ij} in Equation 5.6 can be solved by the following equations:

$$C_{i,1} = \frac{\partial f_i}{\partial(\rho_e H_e)} = \frac{A\bar{\eta}_i}{\bar{\rho}_i} \tag{5.7}$$

$$C_{i,j} = \frac{\partial f_i}{\partial P_j} = 3H_i^2 \left(\frac{dP}{dx}\right)_i K_{ij} + H_i^3 \frac{\partial (dP/dx)_i}{\partial P_j} - A\frac{\partial \bar{\eta}_i}{\partial P_j}\left(H_i - \frac{\bar{\rho}_e H_e}{\bar{\rho}_i}\right) - A\bar{\eta}_i\left(K_{ij} - \bar{\rho}_e H_e \frac{\partial (1/\bar{\rho}_i)}{\partial P_j}\right) \tag{5.8}$$

$$C_{i,N+1} = \frac{\partial f_i}{\partial H_0} = 3H_i^2 \left(\frac{dP}{dX}\right)_i - A\bar{\eta}_i \tag{5.9}$$

$$C_{N+1,j} = \frac{\partial W}{\partial P_j} = \Delta X \tag{5.10}$$

5.2.2 *Calculation of variables in equations*

In the coefficients given by the expressions, some variables, such as A, H_i, $\bar{\eta}_i$, $\bar{\rho}_i$, and $\bar{\rho}_e H_e$ can be calculated directly by using Equations 5.1–5.5. The derivatives in Equation 5.6 can be obtained by deriving Equations 5.1–5.5 and using the elastic deformation formula as follows:

$$\left(\frac{dP}{dx}\right)_i = \frac{P_{i+1} - P_i}{\Delta X} \tag{5.11}$$

$$\frac{\partial (dP/dx)_i}{\partial P_j} = \delta_{i+1,j} - \delta_{i,j} \tag{5.12}$$

$$\frac{\partial \bar{\eta}_i}{\partial P_j} = z p_{\mathrm{H}} P_i / p_0 (\ln \eta_0 + 9.67)(1 + p_{\mathrm{H}} P_i / p_0)^z \bar{\eta}_i \delta_{ij} \tag{5.13}$$

$$\frac{\partial (1/\bar{\rho}_i)}{\partial P_j} = \frac{0.6 \times 10^{-9} p_{\mathrm{H}}}{(1 + 2.3 \times 10^{-9} p_{\mathrm{H}} P_i)^2} \delta_{ij} \tag{5.14}$$

$$K_{ij} = \Delta X[(i-j+1/2)(\ln(i-j+1/2)-1) - (i-j-1/2)(\ln|i-j-1/2|-1) + \ln\Delta X]/\pi \tag{5.15}$$

where, δ_{ij} is the Kronecker function, that is, if $i = j$, $\delta_{ij} = 1$; otherwise $\delta_{ij} = 0$.

It should be pointed out that the Newton–Raphson method is an incremental iterative method. One of the advantages of the Newton–Raphson method is that if the initial solution is close enough to the real solution, its converging process is very fast. However, its disadvantage is that it is quite easy to diverge if the initial solution is far from the real solution. In addition, it can only be used to solve the EHL problem in the line contact.

5.3 Numerical method and program of Newton–Raphson method

Theoretically, after the variable increments have been obtained by the Newton–Raphson method, they are just added to the original variables to obtain the new ones, that is:

$$
\begin{aligned}
(\bar{\rho}_e H_e)^{k+1} &= (\bar{\rho}_e H_e)^k + \Delta\bar{\rho}_e H_e \\
P_2^{k+1} &= P_2^k + \Delta P_2 \\
&\vdots \\
P_N^{k+1} &= P_N^k + \Delta P_N \\
{H_0}^{k+1} &= {H_0}^k + \Delta H_0
\end{aligned}
\tag{5.16}
$$

However, as the outlet boundary condition has been undetermined yet, we must constantly change the outlet boundary to determine $\bar{\rho}_e H_e$ in the calculation [4].

5.3.1 *Coefficient treatment in nonlubricated region*

According to the Reynolds boundary conditions, set the pressure to be zero if the solved pressure is negative on the node x_m, because this node does not belong to the lubrication region. Therefore, the corresponding point is invalid to Reynolds equation. If so, the matrix coefficients should be treated as follows, that is, for each node $i \geq m$, set:

1. $f_i = 0; C_{ij} = 0$
2. $C_{ji} = 0$
3. $C_{ii} = 1$
   where, $j = 1, 2, \ldots, N+1$.

5.3.2 *Determination of $\bar{\rho}_e H_e$*

The first formula in Equation 5.16 is used to determine $\bar{\rho}_e H_e$. However, let us consider that the corresponding node x_m has been determined by using the outlet boundary condition; therefore, the first correction formula of Equation 5.16 cannot be used, but the following formula can be directly used to determine it.

$$\bar{\rho}_e H_e = \bar{\rho}_m H_m \tag{5.17}$$

5.3.3 Determination and correction of initial rigid film thickness

In order to calculate the film thickness, we must first give an initial value to H_0. However, the Newton–Raphson method is very sensitive to the initial value. Therefore, if the given initial H0 is far away from the real solution, it is very easy to make the iteration to be divergent. The following effective method can be used to determine the initial H0.

First, use the known minimum film thickness formula, for example, the minimum thickness formula given by Dowson *et al* [1]. For H_0':

$$H_0' \approx \frac{2.65G^{*0.54}U^{*0.7}(R/b)^2}{W^{*0.13}} \tag{5.18}$$

Then, use the following formula to calculate the film thickness with no rigid film thickness H0 at each node.

$$H_i = \frac{X_i^2}{2} + \sum_{j=1}^{n} K_{ij}P_j \tag{5.19}$$

After obtaining the minimum film thickness $H_{\min}$, we can calculate the initial rigid film thickness H_0^0 by using the following formula:

$$H_0^0 = H_0' - H_{\min} \tag{5.20}$$

It may be uncertain to use the last formula in Equation 5.20 to determine H0. Therefore, we have used the difference of the load and the sum of pressure to weightedly correct H_0. The formula is as follows:

$$H_0^{k+1} = H_0^k - \left(\frac{\pi}{2} - \sum_{i=1}^{N} P_i \Delta X\right) \times \Delta H_0 \tag{5.21}$$

5.3.4 Calculation program

```
      DIMENSION X(101),P(101),H(101),RO(101),EDA(101),POLD(101),V(101)
      DIMENSION D(0:101),C(102,102),B(102),IP(102),DP(102)
      COMMON /COMAK/AK(0:1100)
      DATA X0,XE,E,RO0,EDA0,U0,W0,Z,R/-
     2.0,1.5,2.21E11,1.0,0.02,1.5,0.5E5,0.68,0.02/
      DATA N,KG/101,0/
      DATA PAI,G,PAI1/3.14159265,5000.0,0.318309886/
      OPEN(8,FILE='OUT.DAT',STATUS='UNKNOWN')
```

```
      N1=N-1
      N2=N+1
      W=W0/E/R
      U=EDA0*U0/E/R
      AKK=0.75*PAI*PAI*U/(W*W)
      PH=E*SQRT(W/2.0/PAI)
      B0=R*SQRT(8.0*W/PAI)
      A1=0.6*PH*1.E-9
      A2=1.7*PH*1.E-9
      A3=5.1*PH*1.E-9
      A4=ALOG(EDA0)+9.67
      A5=-0.5/PAI
      A6=A3*Z*A4
      A7=2.3E-9*PH
      HMIN=2.65*G**0.54*U**0.7/W**0.13*(R/B0)**2
      DX=(XE-X0)/(N1)
      CX=ALOG(DX)
      DO I=1,N
      X(I)=X0+(I-1)*DX
      P(I)=0.0
      IF(ABS(X(I)).LE.1.0)P(I)=SQRT(1.0-X(I)*X(I))
      IF(X(I).LE.1.0)IE=I
      ENDDO
      CALL SUBAK(N2)
      DO I=0,N
      D(I)=-(AK(I)+CX)*DX*PAI1
      ENDDO
   10 C(N2,1)=0.0
      C(N2,N2)=0.0
      DO I=2,N
      C(N2,I)=DX
      ENDDO
      DO I=1,N1
      POLD(I)=P(I)
      DP(I)=(P(I+1)-P(I))/DX
      ENDDO
      POLD(N)=0.0
      DP(N)=(P(N)-P(N1))/DX
      CALL VI(N,DX,P,V)
      IF(KG.EQ.0)H0=100.0
      DO I=1,N
      H(I)=0.5*X(I)*X(I)+V(I)
      IF(KG.EQ.0.AND.H(I).LT.H0)H0=H(I)
      RO(I)=1.0+A1*P(I)/(1.0+A2*P(I))
      EDA(I)=EXP(A4*(-1.0+(1.0+A3*P(I))**Z))
      ENDDO
      IF(KG.EQ.0)THEN
      H0=-H0+HMIN
      ENDIF
      DO I=1,N
```

```
      H(I)=H0+H(I)
      ENDDO
      IF(KG.EQ.0)THEN
      ROEHE=RO(IE)*H(IE)
      ENDIF
      DO I=1,N
      C(I,1)=AKK*EDA(I)/RO(I)
      DO J=2,N
      IJ=IABS(I-J)
      D1=3.0*H(I)**2*DP(I)*D(IJ)
      D2=H(I)**3*(DELTA(I+1,J)-DELTA(I,J))/DX
      D3=-AKK*A6*(1.0+A3*P(I))**(Z-1.0)*EDA(I)*DELTA(I,J)*(H(I)-ROEHE/
      RO(I))
      D4=-AKK*EDA(I)*(D(IJ)+ROEHE*A1*DELTA(I,J)/(1.0+A7*P(I))**2)
      C(I,J)=D1+D2+D3+D4
      ENDDO
      C(I,N2)=3.0*H(I)*H(I)*DP(I)-AKK*EDA(I)
      B(I)=-H(I)**3*DP(I)+AKK*EDA(I)*(H(I)-ROEHE/RO(I))
      ENDDO
      B(N2)=0.5*PAI
      DO I=1,N
      B(N2)=B(N2)-P(I)*C(N2,I)
      ENDDO
      DO I=87,N
      IF(P(I).LE.0.0)THEN
      II=I
      GOTO 30
      ENDIF
      ENDDO
30    DO I=II,N
      P(I)=0.0
      B(I)=0.0
      DO J=1,N2
      C(I,J)=0.0
      C(J,I)=0.0
      ENDDO
      C(I,I)=1.0
      ENDDO
      CALL INV(N2,C,IP,IDET)
      IF(IDET.EQ.0)GOTO 20
      DO J=1,N2
      DP(J)=0.0
      DO I=1,N2
      DP(J)=DP(J)+C(J,I)*B(I)
      ENDDO
      ENDDO
      ROEHE=RO(II)*H(II)
      DO I=2,N1
      P(I)=P(I)+DP(I)
      IF(P(I).LT.0.0)P(I)=0.0
```

```
    ENDDO
    H0=H0-B(N2)*DP(N2)
    KG=KG+1
    ER=0.0
    SUM=0.0
    DO I=1,II
    SUM=SUM+P(I)
    ER=ER+ABS(P(I)-POLD(I))
    ENDDO
    ER=ER/SUM
    WRITE(*,*)'ER=',ER
    KG=KG+1
    IF(ER.GT.1.E-5.AND.KG.LT.100)GOTO 10
20  DO I=1,N
    WRITE(8,100)X(I),P(I),H(I)
100 FORMAT(6(E12.6,1X))
    ENDDO
    STOP
    END
    FUNCTION DELTA(I,J)
    DELTA=0.0
    IF(I.EQ.J)DELTA=1.0
    RETURN
    END
    SUBROUTINE INV(N,A,IP,IDET)
    DIMENSION A(N,N), IP(N)
    IDET=1
      EPS=1.E-6
      DO K=1,N
      P=0
      I0=K
      IP(K)=K
      DO I=K,N
      IF(ABS(A(I,K)).GT.ABS(P))THEN
    P=A(I,K)
    I0=I
    IP(K)=I
    ENDIF
    ENDDO
    IF(ABS(P).LE.EPS)THEN
      IDET=0
      GOTO 10
    ENDIF
    IF(I0.NE.K)THEN
    DO J=1,N
    S=A(K,J)
    A(K,J)=A(I0,J)
    A(I0,J)=S
    ENDDO
    ENDIF
```

```
      A(K,K)=1./P
      DO I=1,N
      IF(I.NE.K)THEN
      A(I,K)=-A(I,K)*A(K,K)
      DO J=1,N
      IF(J.NE.K)THEN
      A(I,J)=A(I,J)+A(I,K)*A(K,J)
      ENDIF
      ENDDO
      ENDIF
      ENDDO
      DO J=1,N
      IF(J.NE.K)THEN
      A(K,J)=A(K,K)*A(K,J)
      ENDIF
      ENDDO
      ENDDO
      DO K=N-1,1,-1
      IR=IP(K)
      IF(IR.NE.K)THEN
      DO I=1,N
      S=A(I,IR)
      A(I,IR)=A(I,K)
      A(I,K)=S
      ENDDO
      ENDIF
      ENDDO
10    RETURN
        END
      SUBROUTINE VI(N,DX,P,V)
      DIMENSION P(N),V(N)
      COMMON /COMAK/AK(0:1100)
      DATA PAI1/0.318309886/
      C=ALOG(DX)
      DO 10 I=1,N
      V(I)=0.0
      DO 10 J=1,N
      IJ=IABS(I-J)
10    V(I)=V(I)+(AK(IJ)+C)*DX*P(J)
      DO I=1,N
      V(I)=-PAI1*V(I)
      ENDDO
      RETURN
      END
      SUBROUTINE SUBAK(MM)
      COMMON /COMAK/AK(0:1100)
      DO 10 I=0,MM
10    AK(I)=(I+0.5)*(ALOG(ABS(I+0.5))-1.)-(I-0.5)*(ALOG(ABS(I-0.5))-1.)
      RETURN
      END
```

5.3.5 *Example*

The program to calculate the EHL problem in the line contact by using the Newton–Raphson method includes a main program, a function, and three subroutines. They are the elastic deformation calculation subroutine VI, the elastic coefficient calculation subroutine SUNAK, the matrix inverse subroutine INV, and the Kronecker function DELTA.

1. Main program
   The functions of the main program are as follows:
   Use DATA statement to set the initial values to the following variables: X0 = −2, XE = 1.5, E = 2.21E11Pa, RO0 = 1, EDA0 = 0.02 Pa·s, U0 = 1.5 m·s^{-1}, W0 = 0.5E5 N/m, Z = 0.68, and R = 0.02 m.
   Calculate the fixed parameters such as the nondimensional load W, the nondimensional velocity U, the Hertz contact stress PH, the contact zone half width B0, the nondimensional minimum film thickness of HMIN, and the nondimensional node distance ΔX.
   Calculate the nondimensional elastic deformation V, the film thickness H, the density RO, and the viscosity EDA of each node.
   Calculate the left matrix C and the right vector B of Equation 5.6. By using the inverse matrix subroutine INV, the left variable increment vector of Equation 5.6 can be obtained so as to correct the variables.
2. Subroutine VI is used to calculate the elastic deformation and Subroutine SUNAK to calculate the elastic coefficient which has been previously introduced in Chapter 2.
3. Inverse matrix calculation subroutine INV
   The substituted variables are the matrix order N2 (=N+1) and the matrix C, where, IC is a vector marking the inversion process. IDET will return a value to determine whether the determinant of the matrix is zero or not. If IDET is equal to

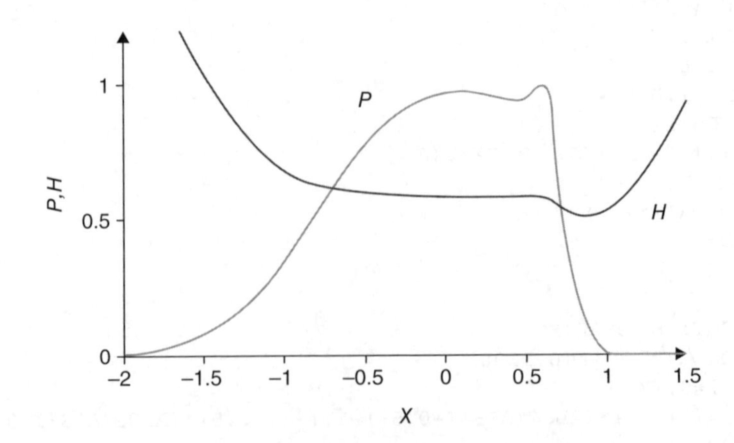

Figure 5.1 Pressure distribution and film thickness shape of EHL in line contact by using Newton–Raphson method

zero, stop the program, otherwise carry out the calculation. In addition, in order to save space, the inverse results will be put into the matrix C and return back.

4. Kronecker function DELTA

When two parameters I and J are substituted into the function, it will be 1 if I is equal to J, otherwise it will be 0.

Based on the conditions of the aforementioned program, the obtained pressure distribution and film thickness are shown in Figure 5.1.

6

Numerical calculation method and program for isothermal EHL in point contact

6.1 Basic equations for isothermal EHL in point contact

6.1.1 Basic equations

The basic equations of EHL in the point contact include the Reynolds equation, the film-thickness equation, the elastic deformation equation, the viscosity–pressure equation, and the density–pressure equation. They are as follows:

Reynolds equation

$$\frac{\partial}{\partial x}\left(\frac{\rho h^3}{\eta}\frac{\partial p}{\partial x}\right)+\frac{\partial}{\partial y}\left(\frac{\rho h^3}{\eta}\frac{\partial p}{\partial y}\right)=-12u_s\frac{\partial \rho h}{\partial x} \tag{6.1}$$

Film thickness equation

$$h(x,y)=h_0+\frac{x^2}{2R_x}+\frac{y^2}{2R_y}+v(x,y) \tag{6.2}$$

Elastic deformation equation

$$v(x,y)=\frac{2}{\pi E}\iint_{\Omega}\frac{p(s,t)}{\sqrt{(x-s)^2+(y-t)^2}}dsdt \tag{6.3}$$

Numerical Calculation of Elastohydrodynamic Lubrication: Methods and Programs, First Edition. Ping Huang.

Viscosity–pressure equation

$$\eta = \eta_0 \exp\left\{(\ln \eta_0 + 9.67)\left[\left(1 + 5.1 \times 10^{-9} p\right)^{0.68} - 1\right]\right\} \tag{6.4}$$

Density–pressure equation

$$\rho = \rho_0 \left(1 + \frac{0.6p}{1 + 1.7p}\right) \tag{6.5}$$

where, the unit of pressure p in the equation is GPa.

6.1.2 Nondimensional equations

The nondimensional Reynolds equation of isothermal EHL in the point contact is as follows:

$$\frac{\partial}{\partial X}\left[\varepsilon \frac{\partial P}{\partial X}\right] + \alpha^2 \frac{\partial}{\partial Y}\left[\varepsilon \frac{\partial P}{\partial Y}\right] - \frac{\partial(\rho^* H)}{\partial X} = 0 \tag{6.6}$$

The boundary conditions of the Reynolds equation (6.6) are as follows:

Inlet boundary condition: $P(X_0, Y) = 0$

Outlet boundary conditions: $P(X_e, Y) = 0$ and $\dfrac{\partial P(X_e, Y)}{\partial X} = 0$

Side boundary conditions: $P|_{Y=\pm 1} = 0$

Film thickness equation

$$H(X,Y) = H_0 + \frac{X^2 + Y^2}{2} + \frac{2}{\pi^2} \int_{x_0}^{x_e} \int_{y_0}^{y_e} \frac{P(S,T)\,dS\,dT}{\sqrt{(X-S)^2 + (Y-T)^2}} \tag{6.7}$$

Viscosity–pressure equation

$$\eta^* = \exp\left\{[\ln(\eta_0) + 9.67]\left[-1 + \left(1 + 5.1 \times 10^{-9} P \cdot p_{\mathrm{H}}\right)^{0.68}\right]\right\} \tag{6.8}$$

Density–pressure equation

$$\rho^* = \left(1 + \frac{0.6p}{1 + 1.7p}\right) \tag{6.9}$$

Load equation

$$\int_{x_0}^{x_e}\int_{y_0}^{y_e} P(X,Y)dXdY = \frac{2}{3}\pi \tag{6.10}$$

Compared with the line contact problem, a nondimensional coordinate in the y direction is added in the above equations. Set α to be the ratio of a to b and if the increment distance is equally divided into two directions, then $\alpha = 1$.

6.2 Numerical calculation method and program

6.2.1 *Differential equations*

The differential form of Reynolds equation (6.6) can be written as follows:

$$\frac{\varepsilon_{i-1/2,j}P_{i-1,j} + \varepsilon_{i+1/2,j}P_{i+1,j} + \varepsilon_{i,j-1/2}P_{i,j-1} + \varepsilon_{i,j+1/2}P_{i,j+1} - \varepsilon_0 P_{ij}}{\Delta X^2} = \frac{\rho^*_{ij}H_{ij} - \rho^*_{i-1,j}H_{i-1,j}}{\Delta X} \tag{6.11}$$

where, the equidistant grid is used here, that is, $\Delta Y = \Delta X$.

Film thickness equation

$$H_{ij} = H_0 + \frac{X_i^2 + Y_j^2}{2} + \frac{1}{\pi^2}\sum_{k=1}^{n}\sum_{l=1}^{n} D_{ij}^{kl}P_{kl} \tag{6.12}$$

Load equation

$$\Delta X \Delta Y \sum_{i=1}^{n}\sum_{j=1}^{n} P_{ij} = \frac{2\pi}{3} \tag{6.13}$$

Similar to the line contact problem, the same iteration method is used to solve the point contact problem. As the viscosity and the elastic deformation change with pressure, the following iteration steps are used: First, an initial pressure distribution, such as the Hertzian contact pressure, is given. Then, film thickness and viscosity are calculated and they are again substituted in the Reynolds equation again to obtain a new pressure. Iterate continuously to obtain the pressure, elastic deformation and film thickness until the pressure difference between the pressure distributions is less than the error that is set, which is about 10E−5 or less. After the iteration has been finished, calculate the final pressure distribution and film thickness with elastic deformation.

6.2.2 Iteration method

First, Equation 6.11 can be written in the residual form as follows:

$$\begin{aligned}&\varepsilon_{i-1/2,j}P_{i-1,j}+\varepsilon_{i+1/2,j}P_{i+1,j}+\varepsilon_{i,j-1/2}P_{i,j-1}+\varepsilon_{i,j+1/2}P_{i,j+1}-\varepsilon_0 P_{ij}\\&-\Delta X\left(\rho_i^* H_{i,j}-\rho_{i-1}^* H_{i-1,j}\right)=0\end{aligned} \tag{6.14}$$

Note that both the film thickness $H_{ij}=H_1+D_0P_{ij}$ and $H_{i-1,j}=H_2+D_2P_{ij}$ contain P_{ij}, P_{ij} should be moved to the left of the iteration equation to take part in the iteration. In the present program, although only parts of P_{ij} are extracted, the results show better effect. H_{ij} and $H_{i-1,j}$ can be written as follows:

$$\begin{aligned}H_{ij}&=H_1+D_1P_{ij}-a_1P_{ij}+a_1P_{ij}=H_{ij}-a_1P_{ij}+a_1P_{ij}\\H_{i-1,j}&=H_2+D_2P_{ij}-a_2P_{ij}+a_2P_{ij}=H_{i-1,j}-a_2P_{ij}+a_2P_{ij}\end{aligned} \tag{6.15}$$

Let

$$r_{ij}=\varepsilon_{i-1/2,j}P_{i-1,j}+\varepsilon_{i+1/2,j}P_{i+1,j}+\varepsilon_{i,j-1/2}P_{i,j-1}+\varepsilon_{i,j+1/2}P_{i,j+1}-\Delta X\left(\rho_i^* H_{ij}-\rho_{i-1}^* H_{i-1,j}\right) \tag{6.16}$$

Remove the terms with P_{i+1}^k from both sides of Equation 6.14 and use Equation 6.16; Equation 6.15 becomes:

$$\left(\varepsilon_0-\Delta X\rho_{i,j}^* a_1+\Delta X a_2\rho_{i-1,j}^*\right)P_{ij}=r_i-\Delta X\rho_i^* a_1 P_{ij}+\Delta X\rho_{i-1}^* a_2 P_{ij} \tag{6.17}$$

The final iteration equation is as follows:

$$P_{ij}=\frac{r_i-\Delta X\rho_i^* a_1 P_{ij}+\Delta X\rho_{i-1}^* a_2 P_{ij}}{\varepsilon_0-\Delta X\rho_{i,j}^* a_1+\Delta X a_2\rho_{i-1,j}^*} \tag{6.18}$$

Since the items containing P_i^{k+1} are subtracted from the numerator of Equation 6.18 and not from the denominator, the iteration effect in Equation 6.18 is better.

6.2.3 Calculation diagram

Following are some of the abbreviations used in the program:

N is the number n of nodes
W is the load w whose unit is N
R is the contact radius R whose unit is m

Us	is the tangential velocity on the up surface u_s whose unit is $m \cdot s^{-1}$ and the tangential velocity on the down surface is zero
E1	is the equivalent elastic modulus E whose unit is Pa
EDA0	is the initial viscosity η_0 whose unit is $Pa \cdot s$
ALFA	is the viscosity–pressure coefficient α
P0	is the pressure coefficient in the viscosity–pressure formula p_0
B	is the half width of the Hertz contact area b whose unit is m
PH	is the maximum Hertz contact pressure p_H whose unit is Pa
A1 and A2	are the parameters to calculate the viscosity; A3 is the parameter to calculate the density
ENDA	is λ in the nondimensional Reynolds equation
U	is the nondimensional average velocity
EPS(I,J)	is ε in the discrete Reynolds equation
EDA(I,J)	is the nondimensional viscosity $\eta*$
RO(I,J)	is the nondimensional density $\rho*$
H(I,J)	is the nondimensional film thickness H
P(I,J)	is the nondimensional pressure P
POLD(I,J)	is the previous iteration pressure P
AK(I,J)	is a_{ij} of the discrete elastic deformation

The main program is shown in Figure 6.1.

In the main program, Subroutine SUBAK is called to calculate the elastic deformation coefficient array AK in the point contact, which has been described in Section 2.1.3.

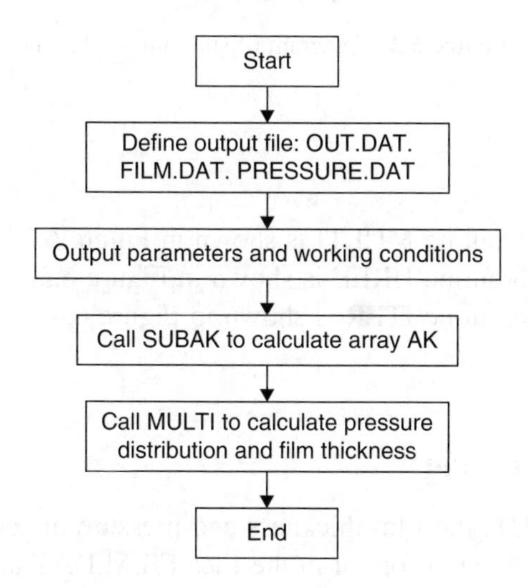

Figure 6.1 Diagram of the main program

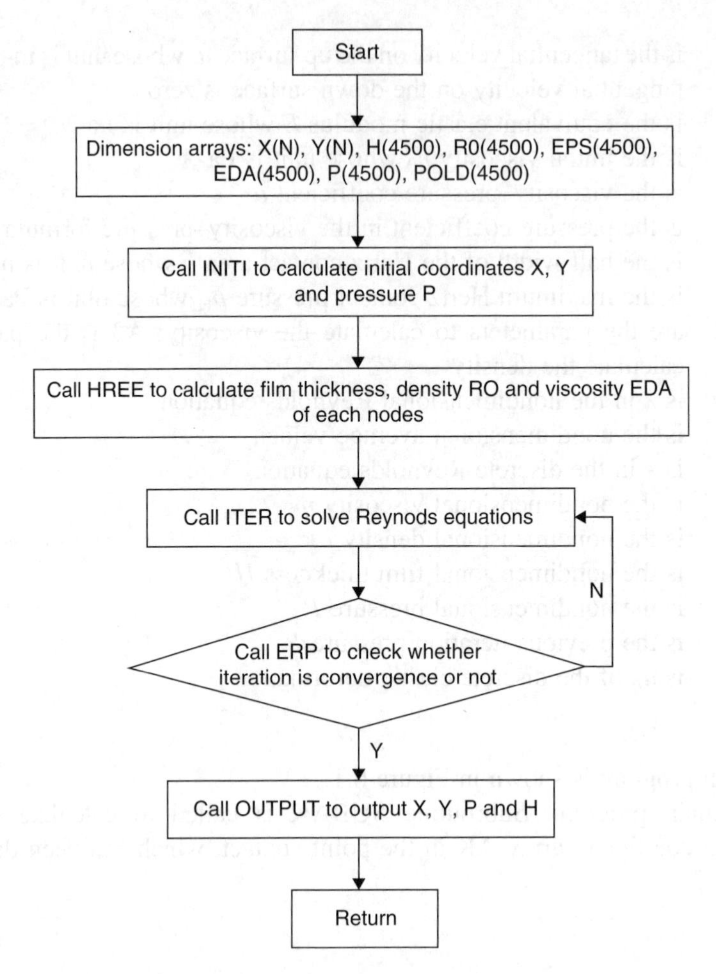

Figure 6.2 Diagram of subroutine MULTI

The diagram of Subroutine MULTI is shown in Figure 6.2.
The diagram of Subroutine HREE is shown in Figure 6.3.
The diagram of Subroutine ITER is shown in Figure 6.4.

6.2.3.1 Subroutine OUTPUT

In Subroutine OUTPUT, the film thickness and pressure of each node are output in the matrixes as the following format in the files FILM.DAT and PRESSURE.DAT, which are shown in Tables 6.1 and 6.2.

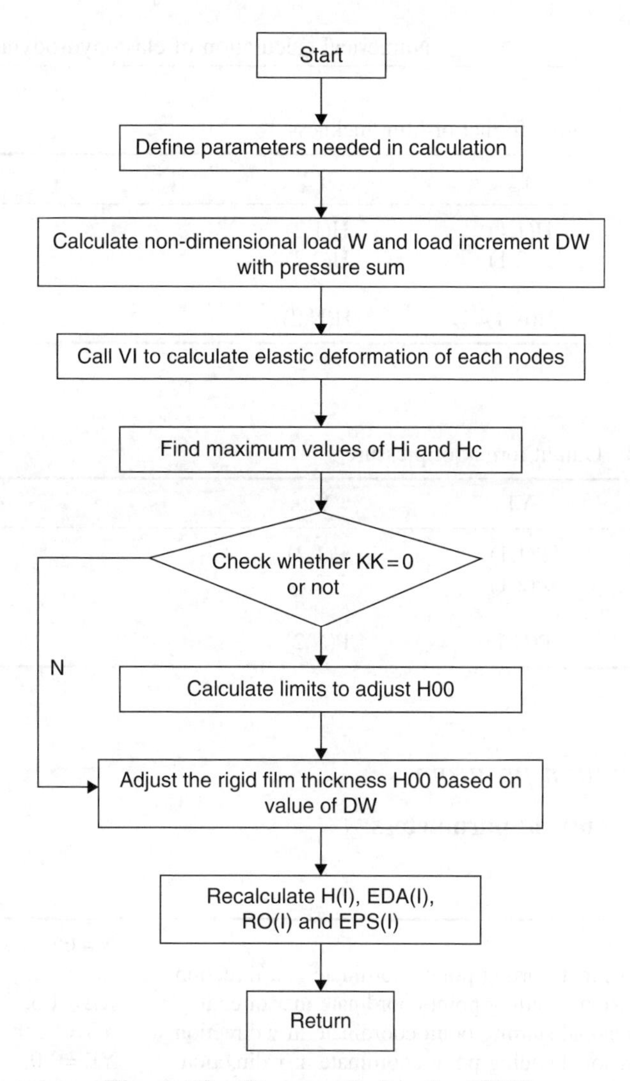

Figure 6.3 Diagram of subroutine HREE

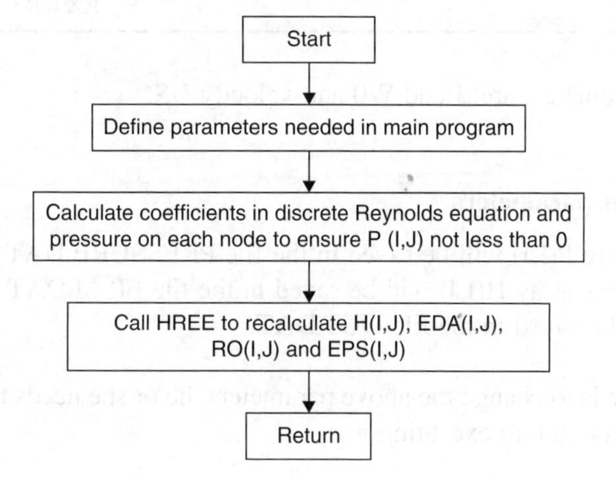

Figure 6.4 Diagram of subroutine ITER

Table 6.1 Output format of film thickness

0	Y1	Y2	…	YN
X1	H(1,1)	H(1,2)		H(1,N)
X2	H(2,1)	H(2,2)		H(2,N)
…				
XN	H(N,1)	H(N,2)		H(N,N)

Table 6.2 Output format of pressure

0	Y1	Y2	…	YN
X1	P(1,1)	P(1,2)		P(1,N)
X2	P(2,1)	P(2,2)		P(2,N)
…				
XN	P(N,1)	P(N,2)		P(N,N)

6.2.4 *Calculation program*

6.2.4.1 Preassignment parameters

Node numbers	N = 65 × 65;
Nondimensional starting point coordinate in *x* direction	X0 = −2.5;
Nondimensional ending point coordinate in *x* direction	XE = 1.5;
Nondimensional starting point coordinate in *y* direction	Y0 = −2.0;
Nondimensional ending point coordinate in *y* direction	YE = 2.0;
Equivalent modulus of elasticity	E1 = 2.21E11 Pa;
Initial viscosity	EDA0 = 0.028 Pa · s;
Radius	RX = RY = 0.05 m.

The input parameters are: Load W0 and velocity US.

6.2.4.2 Output parameters

The pressure array P(I,J) will be saved in the file PRESSURE.DAT;
The film thickness array H(I,J) will be saved in the file FILM.DAT.
Other data will be saved in the file OUT.DAT.

If the user needs to change the above parameters, he or she needs to recompile and relink the program before executing it.

```
      PROGRAM POINTEHL
      COMMON/COM1/ENDA,A1,A2,A3,Z,HM0,DH
      DATA PAI,Z/3.14159265,0.68/
      DATA N,PH,E1,EDA0,RX,US,X0,XE/33,0.8E9,2.21E11,0.05,0.02,1.0,-2.5,1.5/
      OPEN(4,FILE='OUT.DAT',STATUS='UNKNOWN')
      OPEN(8,FILE='FILM.DAT',STATUS='UNKNOWN')
      OPEN(10,FILE='PRESSURE.DAT',STATUS='UNKNOWN')
      MM=N-1
      A1=ALOG(EDA0)+9.67
      A2=5.1E-9*PH
      A3=0.59/(PH*1.E-9)
      U=EDA0*US/(2.*E1*RX)
      B=PAI*PH*RX/E1
      W0=2.*PAI*PH/(3.*E1)*(B/RX)**2
      ALFA=Z*5.1E-9*A1
      G=ALFA*E1
      HM0=3.63*(RX/B)**2*G**0.49*U**0.68*W0**(-0.073)
      ENDA=12.*U*(E1/PH)*(RX/B)**3
      WRITE(*,*)N,X0,XE,W0,PH,E1,EDA0,RX,US
      WRITE(4,*)N,X0,XE,W0,PH,E1,EDA0,RX,US
      WRITE(*,*)'          Wait please'
      CALL SUBAK(MM)
      CALL EHL(N,X0,XE)
      STOP
      END
      SUBROUTINE EHL(N,X0,XE)
     DIMENSION X(65),Y(65),H(4500),RO(4500),EPS(4500),EDA(4500),P(4500),
     POLD(4500),V(4500)
      COMMON/COM1/ENDA,A1,A2,A3,Z,HM0,DH
      DATA MK,G0/1,2.0943951/
      CALL INITI(N,DX,X0,XE,X,Y,P,POLD)
      KK=0
      CALL HREE(N,DX,KK,H00,G0,X,Y,H,RO,EPS,EDA,P,V)
14    KK=15
      CALL ITER(N,KK,DX,H00,G0,X,Y,H,RO,EPS,EDA,P,V)
      MK=MK+1
      CALL ERP(N,ER,P,POLD)
      WRITE(*,*)'ER=',ER
      IF(ER.GT.1.E-5.AND.DH.GT.1.E-6)THEN
      IF(MK.GE.20)THEN
      MK=1
      DH=0.5*DH
      ENDIF
      GOTO 14
      ENDIF
      IF(DH.LE.1.E-6)WRITE(*,*)'Pressures are not convergent!!!'
      CALL OUTPUT(N,DX,X,Y,H,P)
      RETURN
      END
      SUBROUTINE ERP(N,ER,P,POLD)
```

```
      DIMENSION P(N,N),POLD(N,N)
      ER=0.0
      SUM=0.0
      DO 10 I=1,N
      DO 10 J=1,N
      ER=ER+ABS(P(I,J)-POLD(I,J))
      POLD(I,J)=P(I,J)
      SUM=SUM+P(I,J)
10    CONTINUE
      ER=ER/SUM
      RETURN
      END
      SUBROUTINE INITI(N,DX,X0,XE,X,Y,P,POLD)
      DIMENSION X(N),Y(N),P(N,N),POLD(N,N)
      NN=(N+1)/2
      DX=(XE-X0)/(N-1.)
      Y0=-0.5*(XE-X0)
      DO 5 I=1,N
      X(I)=X0+(I-1)*DX
      Y(I)=Y0+(I-1)*DX
5     CONTINUE
      DO I=1,N
      D=1.-X(I)*X(I)
      DO J=1,NN
      C=D-Y(J)*Y(J)
      IF(C.LE.0.0)P(I,J)=0.0
      IF(C.GT.0.0)P(I,J)=SQRT(C)
      POLD(I,J)=P(I,J)
      ENDDO
      ENDDO
      RETURN
      END
      SUBROUTINE HREE(N,DX,KK,H00,G0,X,Y,H,RO,EPS,EDA,P,V)
      DIMENSION X(N),Y(N),P(N,N),H(N,N),RO(N,N),EPS(N,N),EDA(N,N),V(N,N)
      COMMON/COM1/ENDA,A1,A2,A3,Z,HM0,DH/COMAK/AK(0:65,0:65)
      DATA PAI,PAI1/3.14159265,0.2026423/
      NN=(N+1)/2
      CALL VI(N,DX,P,V)
      HMIN=1.E3
      DO 30 I=1,N
      DO 30 J=1,NN
      RAD=X(I)*X(I)+Y(J)*Y(J)
      W1=0.5*RAD
      H0=W1+V(I,J)
      IF(H0.LT.HMIN)HMIN=H0
30    H(I,J)=H0
      IF(KK.EQ.0)THEN
      KK=1
      DH=0.01*HM0
      H00=-HMIN+HM0
```

```
      ENDIF
      W1=0.0
      DO 32 I=1,N
      DO 32 J=1,N
32    W1=W1+P(I,J)
      W1=DX*DX*W1/G0
      DW=1.-W1
      IF(DW.LT.0.0)H00=H00+DH
      IF(DW.GT.0.0)H00=H00-DH
      DO 60 I=1,N
      DO 60 J=1,NN
      H(I,J)=H00+H(I,J)
      EDA1=EXP(A1*(-1.+(1.+A2*P(I,J))**Z))
      EDA(I,J)=EDA1
      RO(I,J)=(A3+1.34*P(I,J))/(A3+P(I,J))
60    EPS(I,J)=RO(I,J)*H(I,J)**3/(ENDA*EDA1)
      DO 70 J=NN+1,N
      JJ=N-J+1
      DO 70 I=1,N
      H(I,J)=H(I,JJ)
      RO(I,J)=RO(I,JJ)
      EDA(I,J)=EDA(I,JJ)
70    EPS(I,J)=EPS(I,JJ)
      RETURN
      END
      SUBROUTINE ITER(N,KK,DX,H00,G0,X,Y,H,RO,EPS,EDA,P,V)
      DIMENSION X(N),Y(N),P(N,N),H(N,N),RO(N,N),EPS(N,N),EDA(N,N),V(N,N)
      COMMON/COMAK/AK(0:65,0:65)
      DATA KG1,PAI/0,3.14159265/
      IF(KG1.NE.0)GOTO 2
      KG1=1
      AK00=AK(0,0)
      AK10=AK(1,0)
2     NN=(N+1)/2
      DO 100 K=1,KK
      DO 70 J=2,NN
      J0=J-1
      J1=J+1
      D2=0.5*(EPS(1,J)+EPS(2,J))
      DO 70 I=2,N-1
      I0=I-1
      I1=I+1
      D1=D2
      D2=0.5*(EPS(I1,J)+EPS(I,J))
      D4=0.5*(EPS(I,J0)+EPS(I,J))
      D5=0.5*(EPS(I,J1)+EPS(I,J))
      D8=2.0*RO(I,J)*AK00/PAI**2
      D9=2.0*RO(I0,J)*AK10/PAI**2
      D10=D1+D2+D4+D5+D8*DX-D9*DX
      D11=D1*P(I0,J)+D2*P(I1,J)+D4*P(I,J0)+D5*P(I,J1)
```

```
      D12=(RO(I,J)*H(I,J)-D8*P(I,J)-RO(I0,J)*H(I0,J)+D9*P(I,J))*DX
      P(I,J)=(D11-D12)/D10
      IF(P(I,J).LT.0.0)P(I,J)=0.0
70    CONTINUE
      DO 80 J=1,NN
      JJ=N+1-J
      DO 80 I=1,N
80    P(I,JJ)=P(I,J)
      CALL HREE(N,DX,KK,H00,G0,X,Y,H,RO,EPS,EDA,P,V)
100   CONTINUE
      RETURN
      END
      SUBROUTINE VI(N,DX,P,V)
      DIMENSION P(N,N),V(N,N)
      COMMON/COMAK/AK(0:65,0:65)
      PAI1=0.2026423
      DO 40 I=1,N
      DO 40 J=1,N
      H0=0.0
      DO 30 K=1,N
      IK=IABS(I-K)
      DO 30 L=1,N
      JL=IABS(J-L)
30    H0=H0+AK(IK,JL)*P(K,L)
40    V(I,J)=H0*DX*PAI1
      RETURN
      END
      SUBROUTINE SUBAK(MM)
      COMMON/COMAK/AK(0:65,0:65)
      S(X,Y)=X+SQRT(X**2+Y**2)
      DO 10 I=0,MM
      XP=I+0.5
      XM=I-0.5
      DO 10 J=0,I
      YP=J+0.5
      YM=J-0.5
      A1=S(YP,XP)/S(YM,XP)
      A2=S(XM,YM)/S(XP,YM)
      A3=S(YM,XM)/S(YP,XM)
      A4=S(XP,YP)/S(XM,YP)
      AK(I,J)=XP*ALOG(A1)+YM*ALOG(A2)+XM*ALOG(A3)+YP*ALOG(A4)
10    AK(J,I)=AK(I,J)
      RETURN
      END
      SUBROUTINE OUTPUT(N,DX,X,Y,H,P)
      DIMENSION X(N),Y(N),H(N,N),P(N,N)
      A=0.0
      WRITE(8,110)A,(Y(I),I=1,N)
      DO I=1,N
      WRITE(8,110)X(I),(H(I,J),J=1,N)
```

```
    ENDDO
    WRITE(10,110)A,(Y(I),I=1,N)
    DO I=1,N
    WRITE(10,110)X(I),(P(I,J),J=1,N)
    ENDDO
110 FORMAT(66(E12.6,1X))
    RETURN
    END
```

6.2.5 *Example*

Under the given working conditions in the program ($p_H = 0.7 \times 10^9$, $u_s = 1.0\,\text{m} \cdot \text{s}^{-1}$), the pressure distribution and the film thickness are shown in Figure 6.5.

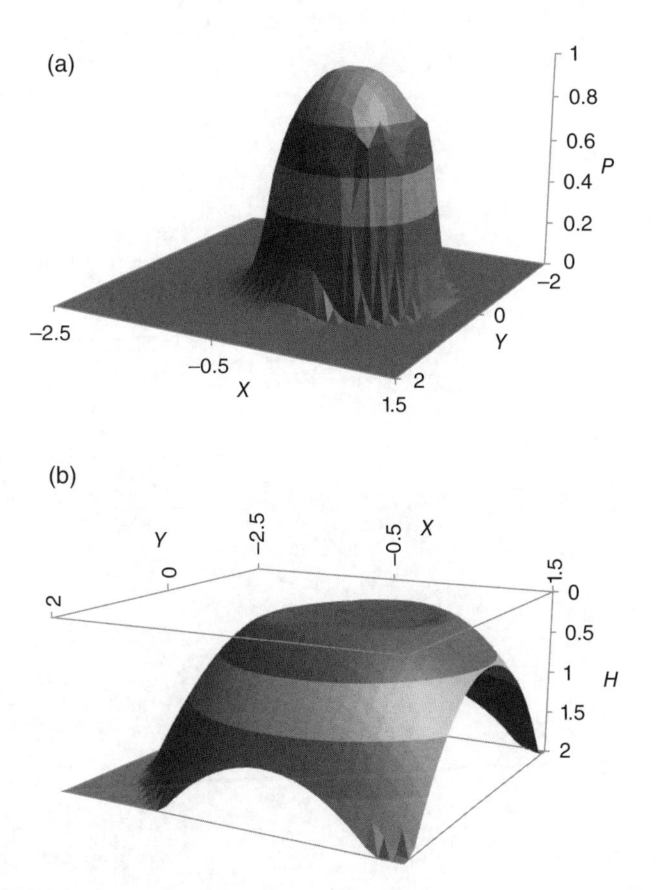

Figure 6.5 Pressure distribution and film thickness of EHL in point contact. (a) Pressure distribution and (b) film thickness

7

Numerical calculation method and programs of multigrid method for isothermal EHL

The multigrid method is a very powerful tool to solve EHL problems [5]. Furthermore, with the multigrid integration method, it is of the rapid speed characteristic. In this chapter, we will briefly introduce its calculation steps.

7.1 Basic principles of multigrid method

The multigrid method has been put forward for solving large algebraic equations for a long time. With the iteration method to solve algebraic equations, the deviations of an approximate solution to the exact real solution can be decomposed into a variety of frequency deviation components. The basic idea of the multigrid method is to carry out iteration among the fine and coarse grids so as to eliminate all the deviative components [5, 6]. The higher frequency components can be quickly eliminated in the fine grids, while the lower frequency components can only be eliminated in the coarse grids.

7.1.1 Grid structure

Take the one-dimensional problem with three uniform grids as an example. The finer grid has 17 nodes, the middle 9, and the coarser 5, as shown in Figure 7.1.

Usually, the node number of the coarsest grid is set to be equal to 3, named as the first grid. Then, if the grid is the *K*th grid, its node number is equal to $n = 2^K + 1$.

Numerical Calculation of Elastohydrodynamic Lubrication: Methods and Programs, First Edition.
Ping Huang.

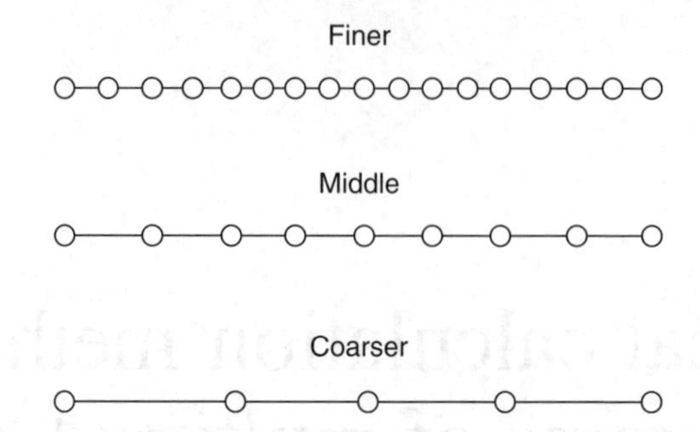

Figure 7.1 One-dimensional uniform multigrid structure

For example, in Figure 7.1, K of the finer grid is equal to 4 so that its node number n is equal to $2^4 + 1 = 17$. In addition, the uniform grid is appropriate, although it should not be. For a two-dimensional problem, the increments of the coordinates in the two directions may not be equal, that is, $\Delta x \neq \Delta y$. However, the uniform grid will be much convenient.

7.1.2 Equation discrete

If the solution region is Ω, the solving equation is generally written as follows:

$$L\mathbf{p} = f \tag{7.1}$$

where, L is an operator, which can be differential, integral, or other operator; $\mathbf{p}$ is the variable vector to be solved; and f is the known vector on the right-hand side.

When using a numerical method to solve Equation 7.1, first divide Ω into a mesh. Then, discrete Equation 7.1 into the algebraic equations. For the multigrid method, the grids must be given in each grid. For the kth grid, the discrete formula is recorded as follows:

$$L^k \mathbf{p}^k = f^k \tag{7.2}$$

where, $\mathbf{p}^k = \{p^k\} = \left[p_1^k, p_2^k, \ldots, p_{nk-1}^k\right]$; and $f^k = \{f^k\} = \left[f_1^k, f_2^k, \ldots, f_{nk-1}^k\right]$.

7.1.3 Restriction and extension

To apply the multigrid method to the EHL problem, one should choose one iterative method, such as Gauss–Seidel iterative method, to obtain the approximate solution from the algebraic equation (7.2). The iterative process generally iterates several times on one grid and then transfers the result to another grid. In the coarsest grid, a large number of iterations are usually carried out. Because the number of nodes of the coarsest grid is very few, the wasting iteration time is little. Between the two adjacent grids, the process that the result of a finer grid is transferred to a coarser grid is called restriction. This is achieved by a restriction operator. The contrary process is called the extension. This is achieved by an extension operator. Some simple restriction and extension processes are as follows:

7.1.3.1 Mapping operator

A mapping operator is a special operator, which can be used as a restriction or extension operator. It transfers the result of one grid directly to the corresponding nodes of the adjacent grid, as shown in Figure 7.2.

7.1.3.2 Weighted restriction operator

The weighted restriction operator will transfer the present result to the adjacent coarser grid after weighting the result of the corresponding and neighbour nodes, as shown in Figure 7.3. The simple weighted restriction operator is suitable for the linear problem. For a strong nonlinear problem, the high-order weighted operator can be used as described in the following:

7.1.3.3 Weighted interpolation operator

As shown in Figure 7.4, the extension operator will transfer the result to the finer grid by mapping and weighting both.

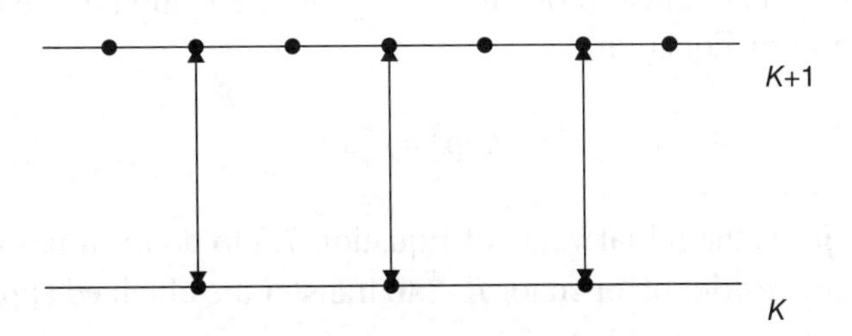

Figure 7.2 Mapping operator

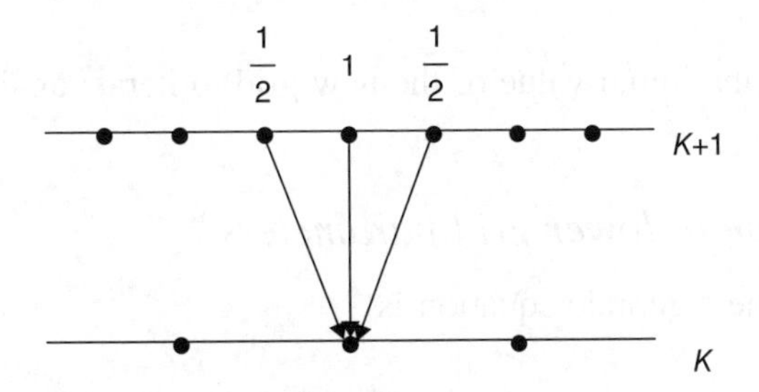

Figure 7.3 Weighted restriction operator

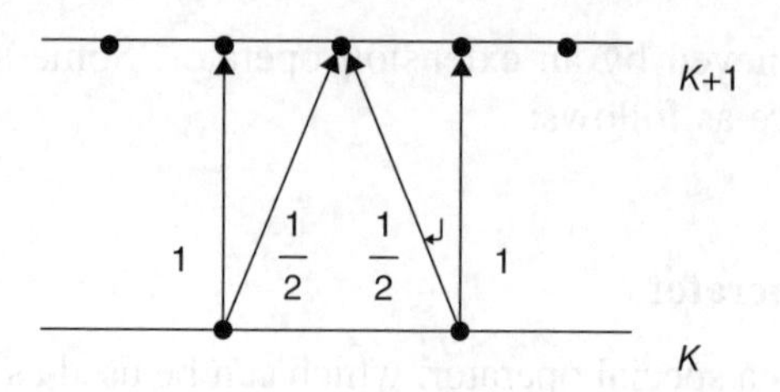

Figure 7.4 Weighted interpolation operator

The simple weighted extension operator given above is used for the linear or the approximately linear problem. For the strongly nonlinear problem, one can choose the high-order weighted extension operator as described in the following.

7.2 Nonlinear full approximation scheme of multigrid method

In the multigrid method, to solve a linear problem, a coarse grid modification is usually selected, but the full approximation scheme (FAS) is used for a nonlinear problem. Because the EHL problem is nonlinear, we will discuss FAS in detail although FAS is also suitable to a linear problem.

7.2.1 Parameter transformation downwards

If we apply FAS to a nonlinear problem, the algebraic equation of the kth grid can be written in the form of Equation 7.3.

$$L^k \mathbf{p}^k = f^k \tag{7.3}$$

If $k \neq 1$, take $\bar{\mathbf{p}}^k$ as the initial value of Equation 7.3 to do m_1 times relaxation iteration. Then use the restriction operator I_k^{k-1} to transfer the obtained approximate solution $\tilde{\mathbf{p}}^k$ to the next coarser grid. That is

$$\bar{\mathbf{p}}^{k-1} = I_k^{k-1} \tilde{\mathbf{p}}^k \tag{7.4}$$

Then, take $\tilde{\mathbf{p}}^k$ as the initial value of the new grid to iterate again.

7.2.2 Correction of lower grid parameters

In the k–1th grid, the algebraic equation is

$$L^{k-1} \mathbf{p}^{k-1} = f^{k-1} \tag{7.5}$$

The key factor of FAS is to determine $\boldsymbol{f}^{k-1}$ in Equation 7.5, because the goal to solve Equation 7.5 is to modify the approximate solution $\tilde{\mathbf{p}}^k$ of Equation 7.3. In order to obtain $\boldsymbol{f}^{k-1}$, we must analyze Equation 7.4 and $\tilde{\mathbf{p}}^k$ first.

From Equation 7.3, subtracting $L^k\tilde{\mathbf{p}}^k$ from both right-hand and left-hand sides, we have

$$L^k\mathbf{p}^k - L^k\tilde{\mathbf{p}}^k = \boldsymbol{f}^k - L^k\tilde{\mathbf{p}}^k \tag{7.6}$$

where, the right hand side of the above equation is denoted as $\boldsymbol{r}^k$, which is the error of the equation. It is equal to

$$\boldsymbol{r}^k = \boldsymbol{f}^k - L^k\tilde{\mathbf{p}}^k \tag{7.7}$$

Because the error on the k–1th grid is quite different from the error on the kth grid, with transformation of $\tilde{\mathbf{p}}^k$, the error $\boldsymbol{r}^k$ has also been transferred to the $k-1$th grid. The error can be calculated by the following formula.

$$L^{k-1}\mathbf{p}^{k-1} - L^{k-1}\left(I_k^{k-1}\tilde{\mathbf{p}}^k\right) = I_k^{k-1}\boldsymbol{r}^k \tag{7.8}$$

The above transformation is accurate for a linear problem, but is approximate for a nonlinear problem. Substituting Equation 7.5 into the above, we have

$$L^{k-1}\mathbf{p}^{k-1} = L^{k-1}\left(I_k^{k-1}\tilde{\mathbf{p}}^k\right) + I_k^{k-1}\left(\boldsymbol{f}^k - L^k\tilde{\mathbf{p}}^k\right) \tag{7.9}$$

Comparing Equation 7.3 with Equation 7.4, we can see that the right-hand side of Equation 7.5 should be

$$\boldsymbol{f}^{k-1} = L^{k-1}\left(I_k^{k-1}\tilde{\mathbf{p}}^k\right) + I_k^{k-1}\left(\boldsymbol{f}^k - L^k\tilde{\mathbf{p}}^k\right) \tag{7.10}$$

From Equation 7.10 it is known that only on the finest grid, that is, $k=m$, the numerical calculation of the right-hand side items of the equation can be directly obtained from the original equation. However, on the coarser grid, all the right-hand side items of the equation contain the errors of the approximate solution.

After having obtained $\boldsymbol{f}^{k-1}$ of Equation 7.4 from Equation 7.10, the algebraic equations of the $k-1$th grid can be determined. Then, set $k=k-1$, the calculation can be carried out continuously on the finer grid. If $k \neq 1$, carry out iteration m_1 times and if $k=1$, m_0 times.

7.2.3 Parameter transformation upwards

If the kth grid iteration process has been finished, the obtained $\tilde{\mathbf{p}}^k$ will be sent to the upper grid to obtain an approximate solution. Usually, we do not transfer $\tilde{\mathbf{p}}^k$ to the grid to be iterated directly, but modify $\tilde{\mathbf{p}}^k$ on the original grid first, and then transfer to the finer grid with interpolation. Therefore, the result combining $\tilde{\mathbf{p}}^k$ with $\tilde{\mathbf{p}}^{k+1}$ is taken as the initiaty of the $k+1$th grid to carry out iteration m_2 times until to the finest grid. The process can be expressed as the following formula.

$$\bar{\mathbf{p}}^{k+1} = \tilde{\mathbf{p}}^{k+1} + I_k^{k+1}\left(\tilde{\mathbf{p}}^k - I_{k+1}^k \tilde{\mathbf{p}}^{k+1}\right) \tag{7.11}$$

7.2.4 V and W loops

The multigrid method is an iterative process, by using restriction and extension operators alternatively on different grids to iterate Equation 7.3. The V loop and W loop are the two typical processes. Figure 7.5 shows a V loop with $k_{\max} = 4$, while Figure 7.6 shows a W loop with $k_{\max} = 4$. Here, $k_{\max}$ is the maximum number of the grids; m_0, m_1, and m_2 are the numbers of iterations at the bottom, the top, or the middle grids respectively.

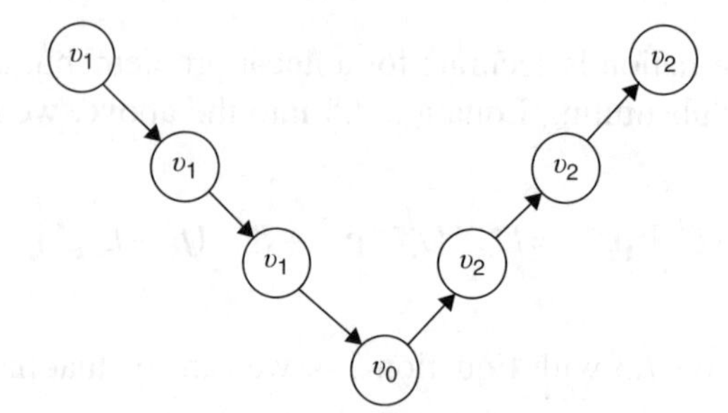

Figure 7.5 V loop ($k_{\max} = 4$)

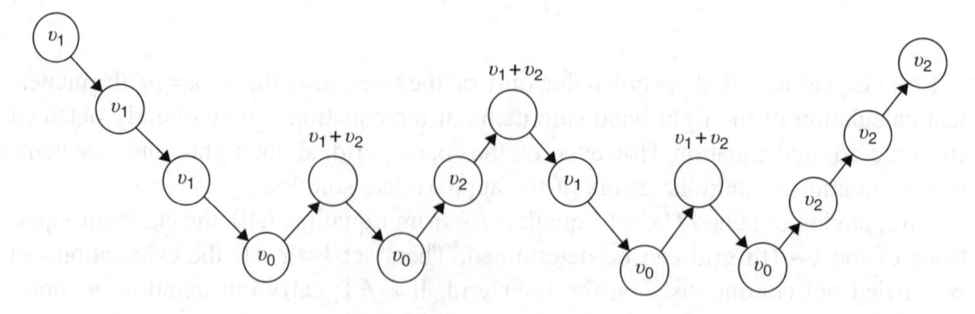

Figure 7.6 W loop ($k_{\max} = 4$)

In Figures 7.5 and 7.6, the sideway and downward arrows indicate the restriction operators, the sideway and upward arrows represent the extension operators, and the circles represent iteration. In addition, the number of v_1, v_2, and v_0 are numbers of restriction, extension and the iteration at the corresponding grid.

7.3 Key factors to solve EHL problem with multigrid method

7.3.1 *Iteration methods*

The iterative process of the multi grid method includes the pressure correction and load balancing by adjusting the rigid film thickness. All these calculations are carried out on the same grid. For the pressure correction, the Gauss-Seidel iteration method is commonly used in the low pressure area. The Jacobi bipolar iteration will be chosen in the high pressure area. The both iterative methods can be written as

$$\bar{P}_i = \tilde{P}_i + c_1\delta_i \tag{7.12}$$

where, c_1 is the relaxation factor; δ_i is the pressure correction quantity; $\bar{P}_i$ and $\tilde{P}_i$ are the pressures before and after iteration respectively.

For the kth grid, the solving equation can be simply written as

$$L_k(P_i) = 0 \tag{7.13}$$

For the Gauss–Seidel iteration

$$\delta_i = \left(\frac{\partial L_k}{\partial P_i}\right)^{-1} \bar{\gamma}_i \tag{7.14}$$

For the Jacobi bipolar iteration

$$\delta_i = \left(\frac{\partial L_k}{\partial P_i} - \frac{\partial L_k}{\partial P_{i-1}}\right)^{-1} \tilde{\gamma}_i \tag{7.15}$$

where,

$$\bar{\gamma}_i = -\frac{\left[\varepsilon_{i-1/2}\bar{P}_{i-1} - \left(\varepsilon_{i-1/2} + \varepsilon_{i+1/2}\right)\bar{P}_i + \varepsilon_{i+1/2}\bar{P}_{i+1}\right]}{\delta^2} + \frac{\bar{\rho}_i\tilde{H}_i - \bar{\rho}_{i-1}\tilde{H}_{i-1}}{\delta} \tag{7.16}$$

To calculate $\tilde{\gamma}_i$, we can substitute $\tilde{P}_{i-1}$ by $\bar{P}_{i-1}$ in Equation 7.16. Because the film thickness H_i is the function of pressure, the derivative $\partial L_i/\partial P_i$ should be considered but the derivative of ε to P_i can be omited. Therefore,

$$\frac{\partial L_k}{\partial P_i} = \frac{\varepsilon_{i-1/2}+\varepsilon_{i+1/2}}{\delta^2} + \frac{1}{\pi}\frac{\left(\bar{\rho}_i K_{ii}^{\delta\delta} - \bar{\rho}_i K_{i-1,i}^{\delta\delta}\right)}{\delta} \tag{7.17}$$

where, $\partial L_i/\partial P_i$ is the derivative of L_k to P_i and $\partial L_k/\partial P_{i-1}$ is to P_{i-1}.

By using the Jacobi bipolar iteration, one must remember not only to add δ_i to pressure of the present node, but to subtract δ_i from the pressure of the front node as well. That is,

$$\begin{aligned} \bar{P}_i &= \tilde{P}_i + c_2\delta_i \\ \bar{P}_{i-1} &= \tilde{P}_{i-1} - c_2\delta_i \end{aligned} \tag{7.18}$$

where, c_2 is the relaxation factor.

The load balance condition can be achieved by modifying the rigid film thickness H_0 as follows.

$$\bar{H}_0 = \tilde{H}_0 + c_3\left[G^{\Delta} - \frac{\Delta}{\pi}\sum_{j=1}^{N-1}\left(P_j + P_{j+1}\right)\right] \tag{7.19}$$

where, c_3 is another relaxation factors; and G^{Δ} is the nondimensional load on the coarsest grid.

In order to stabilize the iteration process, the modification of the rigid film thickness is only carried out on the coarsest grid. This is also benefit to reduce the computational work.

While iterating pressure, the two different methods can be used in different zones for one problem. Although we can separate the whole region into the high and low pressure zones, we should give a criterion to divide the whole region for the two iterative methods. There are two parts in Equation 7.17 to influence pressure. They are

Pressure influence region

$$A_1 = \frac{\varepsilon_{i-1/2}+\varepsilon_{i+1/2}}{\delta^2} \tag{7.20}$$

Thickness influence region

$$A_2 = \frac{1}{\pi}\frac{\left(\bar{P}_i K_{ii}^{\delta\delta} - \bar{P}_i K_{i-1,i}^{\delta\delta}\right)}{\delta} \tag{7.21}$$

When A_1 is larger, the Gauss–Seidel method is more effective. When A_2 is larger, because the film thickness is not modified in γ_i, the Gauss–Seidel method is not effective so that it is easy to diverge. Therefore, the Jacobi bipolar iteration should be used. Our calculation experiences show that when $A_1 \geq 0.1A_2$, the Gauss–Seidel method will be adopted. While $A_1 < A_2$, the Jacobi bipolar method is more effective. Between $0.1 \times A_2 < A_1 < A_2$, either methods can be used.

7.3.2 Relaxation factors selection

In the multigrid method iterative process, there are three relaxation factors to be selected the Gauss–Seidel iterative relaxation factor c_1, the Jacobi bipolar iterative relaxation factor c_2, and the rigid film thickness iterative relaxation factor c_3. The choices of these factors usually depend on the experience. According to author's experiences, the range of the first two factors are $c_1 = 0.3 \sim 1.0$ and $c_2 = 0.1 \sim 0.6$. Actually, c_2 has a greater influence on the convergence, especially in the heavy load condition. For such a situation, c_2 shall be smaller. There is no rule to determine the region of c_3. Following, we give a method to determine c_3 by the existing empirical formulas. The film thickness usually has a relationship with the load as follows.

$$H = G^{*\alpha} U^{*\beta} W^{*\gamma} \tag{7.22}$$

where, α, β, and γ are the indexes of the empirical formula. If the load is unbalanced, the corresponding load increment and the film thickness increment have the following relationship.

$$dH = \gamma G^{*\alpha} U^{*\beta} W^{*\gamma - 1} dW \tag{7.23}$$

Because $G^{\Delta} - \Delta/\pi \sum_{j=1}^{N-1} \left(P_j + P_{j+1}\right)$ is equal to dW, the relationship of dW and dH is

$$dH = \bar{H}_0 - \tilde{H}_0 = -c_3 dW \tag{7.24}$$

From Equations 7.23 and 7.24, the relaxation factor c_3 can be determined.

Usually, iteration will be carried out for several times on each grid. On the coarsest grid, the iteration number is equal to m_0, while for the downward or upward process, the iteration number is equal to m_1 or m_2. According to the author experience, $m_1 = 2$, $m_0 = 5 \sim 20$, $m_2 = 1$ [6].

For the numerical algorithms, the consuming time usually increases exponentially with the node number. However, with the multigrid method, for the same number of iterations of a V loop, the consuming time is approximately linear to the number of nodes.

7.4 Program for EHL in line contact with multigrid method

7.4.1 Specification of program

The program solving EHL problems in the line contact with multigrid method includes one main program and eleven subroutines. Subroutine MULTI includes subroutines of HREE, ITER and SUBAK which have been introduced previously. The elastic deformation is calculated by Subroutine DISP by using the multi grid integration technique, in which subroutines DOWNP, WCOS, WINT, WI, and CORR have been included and introduced in Chapter 2. Here, only the main program and subroutine MULTI will be introduced in the following:

7.4.1.1 Main program

a. In the main program, the parameters, such as the load W, the velocity Us, the initial viscosity EDA0, the equivalent radius R, dimensional coordinates X0 and XE of the inlet and outlet, the number of nodes N, are preassigned. If these parameters are different, readers can change or input them.
b. Calculate the order LMIN from the top to bottom grids, calculate the Hertzian pressure PH, call subroutine SUBAK to calculate the elastic deformation coefficient, and finally call subroutine MULTI to calculate the pressure P(I) and the film thickness H(I) on each node.

7.4.1.2 Subroutine MULTI

a. Dimension arrays, call the KNDX to calculate the finest coordinates, the starting nodes of each grid, the initial pressure, film thickness, viscosity, density and etc.
b. The statements from Label 14–100 are the V loop (see Figure 7.5). In this paragraph, the pressure is iteratively calculated. On the top and the middle grids only carry out one iteration, but at the bottom 19 iterations.
c. Calculate the difference of the theoretical and numerical loads of each grid, which is used to modify the rigid film thickness H0.
d. After Label 100, the program will calculate the difference of the two last calculated pressures to determine whether the convergence is achieved or not. If no, return Label 14 and iterate pressure again.
e. If the convergence has been achieved or the iterative loops are more than 12 times, the program stops the iteration and outputs the present results.

7.4.1.3 Storage of node parameters

In the multigrid method program, the node number of each grid is different so the simplest way is to dimension enough different arrays of the pressure, film thickness, viscosity, and density. However, this will make the program very complicated. In the present program, we use a large array to store the same variables of different grids. For an example, the pressures of different grids are stored in the array P (1100). In order to be convenient and correct, we use a subroutine KNDX to calculate and denote the starting node of each grid. The storage manner of other parameters is the same as pressure.

7.4.2 *Calculation program*

```
      CHARACTER*1 S,S1,S2
      CHARACTER*16 FILEO
      COMMON /COM1/ENDA,A1,A2,A3,Z,C1,C2,C3,CW
      & /COM3/E1,PH,B,U1,U2,R/COM4/X0,XE
      & /COM5/H2,P2,T2,ROM,HM,FM
      DATA S1,S2/1HY,1Hy/
      PAI=3.14159265
      Z=0.68
      P0=1.96E8
      N=129
      X0=-4.
      XE=1.4
      EDA0=0.08
      W=1.5E5
      E1=2.21E11
      R=0.02
      Us=1.5
      C1=0.37
      C2=0.37
      CU=0.25
      OPEN(8,FILE='OUT.DAT',STATUS='UNKNOWN')
      WRITE(*,*)'Show the example or not (Y or N)?'
      READ(*,'(A)')S
      IF(S.EQ.S1.OR.S.EQ.S2)GOTO 10
      WRITE(*,*)'N,X0,XE,W,E,EDA0,R,US='
      READ(*,*)N,X0,XE,W,E1,EDA0,R,US
      WRITE(*,*)' Change iteration factors or not (Y or N) ?'
      READ(*,'(A)')S
      IF(S.EQ.S1.OR.S.EQ.S2)THEN
      WRITE(*,*)'C1,C2='
      READ(*,*)C1,C2
      ENDIF
10    CW=N+0.1
      LMAX=ALOG(CW)/ALOG(2.)
      N=2**LMAX+1
      LMIN=(ALOG(CW)-ALOG(SQRT(CW)))/ALOG(2.)
      LMAX=LMIN
```

```
      H00=0.0
      W1=W/(E1*R)
      PH=E1*SQRT(0.5*W1/PAI)
      A1=(ALOG(EDA0)+9.67)
      A2=PH/P0
      A3=0.59/(PH*1.E-9)
      T2=0.0
      B=4.*R*PH/E1
      ALFA=Z*A1/P0
      G=ALFA*E1
      U=EDA0*US/(2.*E1*R)
      CC1=SQRT(2.*U)
      AM=2.*PAI*(PH/E1)**2/CC1
      AL=G*SQRT(CC1)
      CW=(PH/E1)*(B/R)
      C3=1.6*(R/B)**2*G**0.6*U**0.7*W1**(-0.13)
      ENDA=3.*(PAI/AM)**2/8.
      U1=0.5*(2.+CU)*U
      U2=0.5*(2.-CU)*U
      CW=-1.13*C3
      WRITE(*,20)N,X0,XE,W,E1,EDA0,R,US,C1,C2
20    FORMAT(/////,
     & 15X,'+--------+--------+--------+-------+--------+',/,
     & 15X,'|   N    |  X0    |  XE    | W(N/M)| E(Pa) |',/,
     & 15X,'+--------+--------+--------+-------+--------+',/,
     & 15X,'| ',I4,' | ',F6.3,' | ',F6.3,' |',E8.3,'|',E8.3,'|',/,
     & 15X,'+--------+--------+--------+-------+--------+',/,
     & 15X,'|Eda(PaS)| R (M) | Us(M/S)|  C1  |  C2  |',/,
     & 15X,'+--------+--------+--------+-------+--------+',/,
     & 15X,'| ',F6.3,' | ',F6.3,' | ',F6.3,' | ',F6.3,' | ',F6.3,' |',/,
     & 15X,'+--------+--------+--------+-------+--------+',/)
      WRITE(*,40)
40    FORMAT(2X,'          Wait   Please',//)
      CALL SUBAK(N)
      CALL MULTI(N,LMIN,LMAX,H00)
      Q=2.*ROM*HM*US
      FM=FM/W
      WRITE(*,50)H2,P2,T2,Q,FM
50    FORMAT(15X,//////
     & 13X,'Minmum Film Thichness, Maximum Pressure, Maximum',/,
     & 15X,' Temperature, Flow and Friction Coefficient',//,
     & 15X,'+--------+--------+-------+--------+--------+',/,
     & 15X,'| Hmin(M) |Pmax(Pa)| Tmax(C)|Q(M**3/S|  f   |',/,
     & 15X,'+--------+--------+-------+--------+--------|',/,
     & 15X,'|',E8.3,'|',E8.3,'|',E8.3,'|',E8.3,'|',E8.3,'|',/,
     & 15X,'+--------+--------+-------+--------+--------+',///,
     & 20X,'The Fortran Program Terminated',/)
      STOP
      END
      SUBROUTINE MULTI(N,LMIN,LMAX,H00)
      DIMENSION X(1100),P(1100),H(1100),RO(1100),POLD(1100),
```

```
      & EPS(1100),EDA(1100),P0(2200),F(1100),F0(2200),R(1100),
      & R0(2200),G(10)
      COMMON /COM1/ENDA,A1,A2,A3,Z,C1,C2,C3,CW
      COMMON /COMK/K/COM3/E1,PH,B,U1,U2,RR
      & /COM5/H2,P2,T2,RM,HM,FM
      DATA MK,IT,KH,NMAX,PAI,G0/0,0,0,1100,3.14159265,1.570796325/
      LT=LMAX
      NX=N
      K=LMIN
      N0=(N-1)/2**(LMIN-1)
      CALL KNDX(K,N,N0,N1,NMAX,DX,X)
      DO 10 I=1,N
      IF(ABS(X(I)).GE.1.0)P(I)=0.0
10    IF(ABS(X(I)).LT.1.0)P(I)=SQRT(1.-X(I)*X(I))
12    CALL HREE(N,DX,H00,G0,X,P,H,RO,EPS,EDA,F,0)
      IF(KH.NE.0)GOTO 14
      KH=1
      GOTO 12
14    CALL FZ(N,P,POLD)
      DO 100 L=LMIN,LMAX
      K=L
      G(K)=PAI/2.
      DO 18 I=1,N
      R(I)=0.0
      F(I)=0.0
      R0(N1+I)=0.0
18    F0(N1+I)=0.0
20    KK=2
      CALL ITER(N,KK,DX,H00,G0,X,P,H,RO,EPS,EDA,F,R,0)
      KK=1
      CALL ITER(N,KK,DX,H00,G0,X,P,H,RO,EPS,EDA,F,R,1)
      G(K-1)=G(K)
      DO 24 I=1,N
      IF(I.LT.N)G(K-1)=G(K-1)-0.5*DX*(P(I)+P(I+1))
24    P0(N1+I)=P(I)
      N2=N
      K=K-1
      CALL KNDX(K,N,N0,N1,NMAX,DX,X)
      CALL TRANS(N,N2,P,H,RO,EPS,EDA,R)
      CALL ITER(N,KK,DX,H00,G0,X,P,H,RO,EPS,EDA,F,R,2)
      DO 26 I=1,N
      IF(I.LT.N)G(K)=G(K)+0.5*DX*(P(I)+P(I+1))
26    F(I)=H(I)
      G0=G(K)
      CALL HREE(N,DX,H00,G0,X,P,H,RO,EPS,EDA,F,1)
      DO 28 I=1,N
      R0(N1+I)=R(I)
28    F0(N1+I)=F(I)
      IF(K.NE.1)GOTO 20
      KK=19
      CALL ITER(N,KK,DX,H00,G0,X,P,H,RO,EPS,EDA,F,R,0)
```

```
40  DO 42 I=1,N
42  P0(N1+I)=P(I)
    N2=N1
    K=K+1
    CALL KNDX(K,N,N0,N1,NMAX,DX,X)
    G0=G(K)
    DO 50 I=2,N,2
    I1=N1+I
    I2=N2+I/2
    P(I-1)=P0(I2)
    P(I)=P0(I1)+0.5*(P0(I2)+P0(I2+1)-
    & P0(I1-1)-P0(I1+1))
50  IF(P(I).LT.0.0)P(I)=0.
    DO 52 I=1,N
    R(I)=R0(N1+I)
52  F(I)=F0(N1+I)
    CALL HREE(N,DX,H00,G0,X,P,H,RO,EPS,EDA,F,0)
    KK=1
    CALL ITER(N,KK,DX,H00,G0,X,P,H,RO,EPS,EDA,F,R,0)
    IF(K.LT.L)GOTO 40
100 CONTINUE
    MK=MK+1
    CALL ERROP(N,P,POLD,ERP)
    IF(ERP.GT.0.01*C2.AND.MK.LE.12)GOTO 14
    IF(MK.GE.10)THEN
    WRITE(*,*)'Pressures are not convergent !!!'
    READ(*,*)
    ENDIF
    FM=FRICT(N,DX,H,P,EDA)
    DO I=1,N
    H(I)=H(I)*B*B/RR
    P(I)=P(I)*PH
    ENDDO
110 FORMAT(6(1X,E12.6))
    DO I=1,N
    WRITE(8,110)X(I),P(I),H(I)
    ENDDO
    HM=P(1)
    RM=RO(1)
    DO I=2,N-1
    IF(P(I).GE.P(I-1).AND.P(I).GE.P(I+1))THEN
    HM=H(I)
    RM=RO(I)
    GOTO 120
    ENDIF
    ENDDO
120 H2=1.0
    P2=0.0
    DO I=1,N
    IF(H(I).LT.H2)H2=H(I)
    IF(P(I).GT.P2)P2=P(I)
```

```
      H(I)=H(I)*1.E6
      P(I)=P(I)*1.E-9
      ENDDO
      RETURN
      END
      SUBROUTINE HREE(N,DX,H00,G0,X,P,H,RO,EPS,EDA,F0,KG)
      DIMENSION X(N),P(N),H(N),RO(N),EPS(N),EDA(N),F0(N)
      DIMENSION W(2200)
      COMMON /COM1/ENDA,A1,A2,A3,Z,C1,C2,C3,CW/COMK/K
     & /COMAK/AK(0:1100)
      DATA KK,MK1,MK2,NW,PAI1/0,3,0,2200,0.318309886/
      IF(KK.NE.0)GOTO 3
      HM0=C3
3     W1=0.0
      DO 4 I=1,N
4     W1=W1+P(I)
      C3=(DX*W1)/G0
      DW=1.-C3
      IF(K.EQ.1)GOTO 6
      CALL DISP(N,NW,K,DX,P,W)
      GOTO 10
6     WX=W1*DX*ALOG(DX)
      DO 8 I=1,N
      W(I)=WX
      DO 8 J=1,N
      IJ=IABS(I-J)
8     W(I)=W(I)+AK(IJ)*P(J)*DX
10    HMIN=1.E3
      DO 30 I=1,N
      H0=0.5*X(I)*X(I)-PAI1*W(I)
      IF(KG.EQ.1)GOTO 20
      IF(H0+F0(I).LT.HMIN)HMIN=H0+F0(I)
      H(I)=H0
      GOTO 30
20    F0(I)=F0(I)-H00-H0
30    CONTINUE
      IF(KG.EQ.1)RETURN
      H0=H00+HMIN
      IF(KK.NE.0)GOTO 32
      H01=-HMIN+HM0
      DH=0.005*HM0
      H02=-HMIN
      H00=0.5*(H01+H02)
      KK=1
32    IF(DW.LT.0.0)THEN
      H00=AMIN1(H01,H00+DH)
      ENDIF
      IF(DW.GT.0.0)THEN
      H00=AMAX1(H02,H00-DH)
      ENDIF
50    DO 60 I=1,N
```

```
60  H(I)=H00+H(I)+F0(I)
    DO 100 I=1,N
    EDA(I)=EXP(A1*(-1.+(1.+A2*P(I))**Z))
    RO(I)=(A3+1.34*P(I))/(A3+P(I))
    EPS(I)=RO(I)*H(I)**3/(ENDA*EDA(I))
100 CONTINUE
    RETURN
    END
    SUBROUTINE ITER(N,KK,DX,H00,G0,X,P,H,RO,EPS,EDA,F0,R0,KG)
    DIMENSION X(N),P(N),H(N),RO(N),EPS(N),EDA(N),F0(N),R0(N)
    COMMON /COM1/ENDA,A1,A2,A3,Z,C1,C2,C3
    &  /COMAK/AK(0:1100)
    DATA PAI/3.14159265/
    DX1=1./DX
    DX2=DX*DX
    DX3=1./DX2
    DX4=DX1/PAI
    DXL=DX*ALOG(DX)
    AK0=DX*AK(0)+DXL
    AK1=DX*AK(1)+DXL
    DO 100 K=1,KK
    RMAX=0.0
    D2=0.5*(EPS(1)+EPS(2))
    D3=0.5*(EPS(2)+EPS(3))
    D5=DX1*(RO(2)*H(2)-RO(1)*H(1))
    D7=DX4*(RO(2)*AK0-RO(1)*AK1)
    PP=0.
    DO 70 I=2,N-1
    D1=D2
    D2=D3
    D4=D5
    D6=D7
    IF(I+2.LE.N)D3=0.5*(EPS(I+1)+EPS(I+2))
    D5=DX1*(RO(I+1)*H(I+1)-RO(I)*H(I))
    D7=DX4*(RO(I+1)*AK0-RO(I)*AK1)
    IF(KG.NE.0)GOTO 30
    DD=(D1+D2)*DX3
    IF(DD.LT.0.1*ABS(D6))GOTO 10
    RI=-DX3*(D1*P(I-1)-(D1+D2)*P(I)+D2*P(I+1))+D4+R0(I)
    DLDP=-DX3*(D1+D2)+D6
    RI=C1*RI/DLDP
    GOTO 20
10  RI=-DX3*(D1*PP-(D1+D2)*P(I)+D2*P(I+1))+D4+R0(I)
    DLDP=-DX3*(2.*D1+D2)+2.*D6
    RI=C2*RI/DLDP
    IF(I.GT.2.AND.P(I-1)-RI.GT.0.0)P(I-1)=P(I-1)-RI
20  PP=P(I)
    P(I)=P(I)+RI
    IF(P(I).LT.0.0)P(I)=0.0
    IF(K.NE.KK)GOTO 70
    IF(RMAX.LT.ABS(RI).AND.P(I).GT.0.0)RMAX=ABS(RI)
```

```
      GOTO 70
30    IF(KG.EQ.2)GOTO 40
      RO(I)=-DX3*(D1*P(I-1)-(D1+D2)*P(I)+D2*P(I+1))+D4+RO(I)
      GOTO 70
40    RO(I)=DX3*(D1*P(I-1)-(D1+D2)*P(I)+D2*P(I+1))-D4+RO(I)
70    CONTINUE
      IF(KG.NE.0)GOTO 100
      CALL HREE(N,DX,H00,G0,X,P,H,RO,EPS,EDA,F0,0)
100   CONTINUE
      RETURN
      END
      SUBROUTINE DISP(N,NW,KMAX,DX,P1,W)
      DIMENSION P1(N),W(NW),P(2200),AK1(0:50),AK2(0:50)
      COMMON /COMAK/AK(0:1100)
      DATA NMAX,KMIN/2200,1/
      N2=N
      M=3+2*ALOG(FLOAT(N))
      K1=N+KMAX
      DO 10 I=1,N
10    P(K1+I)=P1(I)
      DO 20 KK=KMIN,KMAX-1
      K=KMAX+KMIN-KK
      N1=(N2+1)/2
      CALL DOWNP(NMAX,N1,N2,K,P)
20    N2=N1
      DX1=DX*2**(KMAX-KMIN)
      CALL WI(NMAX,N1,KMIN,KMAX,DX,DX1,P,W)
      DO 30 K=KMIN+1,KMAX
      N2=2*N1-1
      DX1=DX1/2.
      CALL AKCO(M+5,KMAX,K,AK1)
      CALL AKIN(M+6,AK1,AK2)
      CALL WCOS(NMAX,N1,N2,K,W)
      CALL CORR(NMAX,N2,K,M,1,DX1,P,W,AK1)
      CALL WINT(NMAX,N2,K,W)
      CALL CORR(NMAX,N2,K,M,2,DX1,P,W,AK2)
30    N1=N2
      DO 40 I=1,N
40    W(I)=W(K1+I)
      RETURN
      END
      SUBROUTINE DOWNP(NMAX,N1,N2,K,P)
      DIMENSION P(NMAX)
      K1=N1+K-1
      K2=N2+K-1
      DO 10 I=3,N1-2
      I2=2*I+K2
10    P(K1+I)=(16.*P(I2)+9.*(P(I2-1)+P(I2+1))-
      & (P(I2-3)+P(I2+3)))/32.
      P(K1+2)=0.25*(P(K2+3)+P(K2+5))+0.5*P(K2+4)
      P(K1+N1-1)=0.25*(P(K2+N2-2)+P(K2+N2))+
```

```
     & 0.5*P(K2+N2-1)
     RETURN
     END
     SUBROUTINE WCOS(NMAX,N1,N2,K,W)
     DIMENSION W(NMAX)
     K1=N1+K-1
     K2=N2+K
     DO 10 I=1,N1
     II=2*I-1
10   W(K2+II)=W(K1+I)
     RETURN
     END
     SUBROUTINE WINT(NMAX,N,K,W)
     DIMENSION W(NMAX)
     K2=N+K
     DO 10 I=4,N-3,2
     II=K2+I
10   W(II)=(9.*(W(II-1)+W(II+1))
     & -(W(II-3)+W(II+3)))/16.
     I1=K2+2
     I2=K2+N-1
     W(I1)=0.5*(W(I1-1)+W(I1+1))
     W(I2)=0.5*(W(I2-1)+W(I2+1))
     RETURN
     END
     SUBROUTINE CORR(NMAX,N,K,M,I1,DX,P,W,AK)
     DIMENSION P(NMAX),W(NMAX),AK(0:M)
     K1=N+K
     IF(I1.EQ.2)GOTO 20
     DO 10 I=1,N,2
     II=K1+I
     J1=MAX0(1,I-M)
     J2=MIN0(N,I+M)
     DO 10 J=J1,J2
     IJ=IABS(I-J)
10   W(II)=W(II)+AK(IJ)*DX*P(K1+J)
     RETURN
20   DO 30 I=2,N,2
     II=K1+I
     J1=MAX0(1,I-M)
     J2=MIN0(N,I+M)
     DO 30 J=J1,J2
     IJ=IABS(I-J)
30   W(II)=W(II)+AK(IJ)*DX*P(K1+J)
     RETURN
     END
     SUBROUTINE WI(NMAX,N,KMIN,KMAX,DX,DX1,P,W)
     DIMENSION P(NMAX),W(NMAX)
     COMMON /COMAK/AK(0:1100)
     K1=N+1
     K=2**(KMAX-KMIN)
```

```
      C=ALOG(DX)
      DO 10 I=1,N
      II=K1+I
      W(II)=0.0
      DO 10 J=1,N
      IJ=K*IABS(I-J)
10    W(II)=W(II)+(AK(IJ)+C)*DX1*P(K1+J)
      RETURN
      END
      SUBROUTINE AKCO(KA,KMAX,K,AK1)
      DIMENSION AK1(0:KA)
      COMMON /COMAK/AK(0:1100)
      J=2**(KMAX-K)
      DO 10 I=0,KA
      II=J*I
10    AK1(I)=AK(II)
      RETURN
      END
      SUBROUTINE AKIN(KA,AK1,AK2)
      DIMENSION AK1(KA),AK2(KA)
      DO 10 I=4,KA-3
10    AK2(I)=(9.*(AK1(I-1)+AK1(I+1))
      & -(AK1(I-3)+AK1(I+3)))/16.
      AK2(1)=(9.*AK1(2)-AK1(4))/8.
      AK2(2)=(9.*(AK1(1)+AK1(3))
      & -(AK1(3)+AK1(5)))/16.
      AK2(3)=(9.*(AK1(2)+AK1(4))
      & -(AK1(2)+AK1(6)))/16.
      DO 20 I=1,KA
20    AK2(I)=AK1(I)-AK2(I)
      DO 30 I=1,KA-1,2
      I1=I+1
      AK1(I)=0.0
30    AK1(I1)=AK2(I1)
      RETURN
      END
      SUBROUTINE SUBAK(MM)
      COMMON /COMAK/AK(0:1100)
      DO 10 I=0,MM
10    AK(I)=(I+0.5)*(ALOG(ABS(I+0.5))-1.)-
      & (I-0.5)*(ALOG(ABS(I-0.5))-1.)
      RETURN
      END
      FUNCTION FRICT(N,DX,H,P,EDA)
      DIMENSION H(N),P(N),EDA(N)
      COMMON /COM3/E1,PH,B,U1,U2,R
      DATA TAU0/4.E7/
      TP=TAU0/PH
      TE=TAU0/E1
      BR=B/R
      FRICT=0.0
```

```
      DO I=1,N
      DP=0.0
      IF(I.NE.N)DP=(P(I+1)-P(I))/DX
      TAU=0.5*H(I)*ABS(DP)*(BR/TP)+
      & 2.*ABS(U1-U2)*EDA(I)/(H(I)*BR**2*TE)
      FRICT=FRICT+TAU
      ENDDO
      FRICT=FRICT*DX*B*TAU0
      RETURN
      END
      SUBROUTINE FZ(N,P,POLD)
      DIMENSION P(N),POLD(N)
      DO 10 I=1,N
   10 POLD(I)=P(I)
      RETURN
      END
      SUBROUTINE ERROP(N,P,POLD,ERP)
      DIMENSION P(N),POLD(N)
      SD=0.0
      SUM=0.0
      DO 10 I=1,N
      SD=SD+ABS(P(I)-POLD(I))
   10 SUM=SUM+P(I)
      ERP=SD/SUM
      RETURN
      END
      SUBROUTINE KNDX(K,N,N0,N1,NMAX,DX,X)
      DIMENSION X(NMAX)
      COMMON /COM4/X0,XE
      N=2**(K-1)*N0
      DX=(XE-X0)/N
      N=N+1
      N1=N+K
      DO 10 I=1,N
   10 X(I)=X0+(I-1)*DX
      RETURN
      END
      SUBROUTINE TRANS(N1,N2,P,H,RO,EPS,EDA,R)
      DIMENSION P(N2),H(N2),RO(N2),EPS(N2),EDA(N2),R(N2)
      DO 10 I=1,N1
      II=2*I-1
      P(I)=P(II)
      H(I)=H(II)
      R(I)=R(II)
      RO(I)=RO(II)
      EPS(I)=EPS(II)
   10 EDA(I)=EDA(II)
      RETURN
      END
```

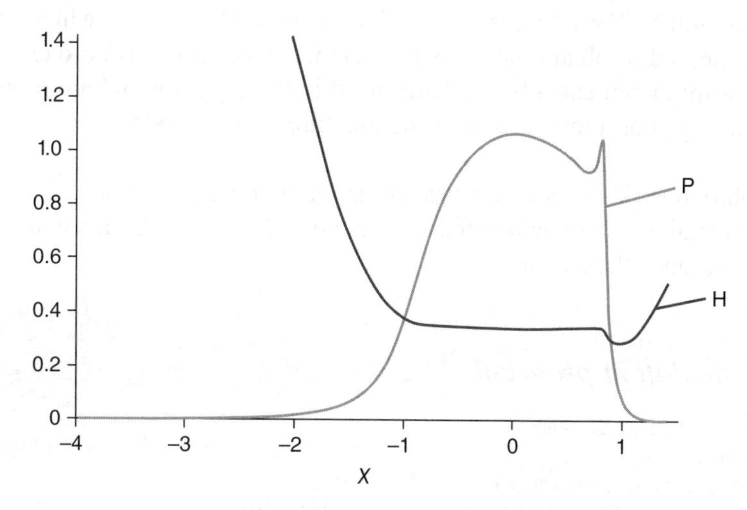

Figure 7.7 Pressure distribution and film thickness for EHL in line contact

7.4.3 Example

According to the program given, the calculation results of the film thickness and pressure distribution are shown in Figure 7.7.

7.5 Program for EHL in point contact with multigrid method

7.5.1 Specification of program

The program solving EHL problems in the point contact with the multigrid method includes 1 main program and 13 subroutines. Among them, because only the Jacobi iteration method in Subroutine ITER is different from the line contact program, we make a detailed introduction here.

1. Iteration is along the X axis (velocity) direction first. After carrying out all the nodes of one row, do the next column along the Y axis direction.
2. Because in the Jacobi iteration, the pressure increment will be added to and subtracted from two different nodes at the same time, see Equation 7.18 and the pressure corrections on the same line are related, the pressure corrections of all nodes should be finished together at the same time. In the program, we choose the way to solve the pressure increments of the entire row simultaneously. In Subroutine ITER, the statements from Label 20 to 50 are used to establish the simultaneous equations. Because the matrix of the simultaneous equations is three dimensional, in order to save the space, the coefficients of the three columns in the matrix and the right terms are placed in the matrix A with four columns.

3. Call Subroutine TRA4, that is, CALL TRA4 (MM, D, A, B), in which we use the chasing method to obtain the pressure increments of all nodes on one row.
4. The pressure increments obtained are saved in the D vector and are brought back to the calling subroutine to correct the pressure of each node.

Other subroutines in the point contact are similar to those of the line contact situation, only to add iteration, interpolation, transformation and etc in the Y direction. Therefore, we do not introduce them again.

7.5.2 Calculation program

```
      CHARACTER*16 FILENAME
      CHARACTER*1 S,S1,S2
      CHARACTER*16 CDATE,CTIME
      COMMON /COM1/ENDA,A1,A2,A3,Z,C1,C2,HM0
      DATA MK,PAI,Z/6,3.14159265,0.68/,S1,S2/1HY,1Hy/,
      & N,PH,E1,EDA0,RX,US,X0,XE,C1,C2/33,0.5E9,2.21E11,0.05,
      & 0.02,1.0,-2.5,1.5,0.31,0.31/,
      &CDATE,CTIME/'The date is','The time is'/
      OPEN(4,FILE='OUT.DAT',STATUS='UNKNOWN')
      OPEN(8,FILE='FILM.DAT',STATUS='UNKNOWN')
      OPEN(10,FILE='PRESS.DAT',STATUS='UNKNOWN')
1     FORMAT(20X,A12,I2.2,':',I2.2,':',I4.4)
2     FORMAT(20X,A12,I2.2,':',I2.2,':',I2.2,'.',I2.2)
      WRITE(*,*)'Show the example or not (Y or N)?'
      READ(*,'(A)')S
      IF(S.EQ.S1.OR.S.EQ.S2)GOTO 5
      WRITE(*,*)'N,PH,E1,EDA0,RX,US,X0,XE='
      READ(*,*)N,PH,E1,EDA0,RX,US,X0,XE
      WRITE(*,*)' Change iteration factors or not (Y or N) ?'
      READ(*,'(A)')S
      IF(S.EQ.S1.OR.S.EQ.S2)THEN
      WRITE(*,*)'C1,C2='
      READ(*,*)C1,C2
      ENDIF
5     CALL GETDAT(IYR,IMON,IDAY)
      WRITE(4,1)CDATE,IMON,IDAY,IYR
      WRITE(*,1)CDATE,IMON,IDAY,IYR
      WRITE(*,20)N,X0,XE,PH,E1,EDA0,RX,US,C1,C2
      WRITE(4,20)N,X0,XE,PH,E1,EDA0,RX,US,C1,C2
20    FORMAT(/////,
      & 15X,'+--------+--------+--------+--------+--------+',/,
      & 15X,'|   N   |  X0   |  XE   | W(N/M) | E(Pa) |',/,
      & 15X,'+--------+--------+--------+--------+--------+',/,
      & 15X,'|  ',I4,'  |',F6.3,'|',F6.3,'|',E8.3,'|',E8.3,'|',/,
      & 15X,'+--------+--------+--------+--------+--------+',/,
      & 15X,'|Eda(PaS)| R (M) | Us(M/S)|  C1  |  C2  |',/,
      & 15X,'+--------+--------+--------+--------+--------+',/,
```

```
      & 15X,'| ',F6.3,' | ',F6.3,' | ',F6.3,' | ',F6.3,' | ',F6.3,'|',/,
      & 15X,'+--------+--------+--------+--------+--------+',/)
      H00=0.0
      MM=N-1
      LMIN=ALOG(N-1.)/ALOG(2.)-1.99
      LMAX=LMIN
      U=EDA0*US/(2.*E1*RX)
      A1=ALOG(EDA0)+9.67
      A2=5.1E-9*PH
      A3=0.59/(PH*1.E-9)
      B=PAI*PH*RX/E1
      W=2.*PAI*PH/(3.*E1)*(B/RX)**2
      ALFA=Z*5.1E-9*A1
      G=ALFA*E1
      AM=W*(2.*U)**(-0.75)
      AL=G*(2.*U)**(0.25)
      HM0=3.63*(RX/B)**2*G**0.49*U**0.68*W**(-0.073)
      ENDA=12.*U*(E1/PH)*(RX/B)**3
      WRITE(*,*)'      Wait please '
      CALL SUBAK(MM)
      CALL MULTI(N,LMIN,LMAX,MK,X0,XE,H00)
      CALL GETDAT(IYR,IMON,IDAY)
      WRITE(4,1)CDATE,IMON,IDAY,IYR
      WRITE(*,1)CDATE,IMON,IDAY,IYR
      STOP
      END
C*************************************************************
      SUBROUTINE MULTI(N,LMIN,LMAX,MK,X0,XE,H00)
      DIMENSION X(100),Y(100),H(5000),RO(5000),R0(5000),
      & EPS(5000),P(5000),F(5000),R(5000),G(20)
      COMMON /COMK/KL
      DATA NMAX,G0/100,2.0943951/
      NN=(N+1)/2
      KL=LMIN
      CALL KNDX(KL,N,NS,NMAX,DX,X0,XE,X,Y)
      CALL INITI(N,X,Y,P(NS))
12    CALL HREE(N,DX,H00,G0,X,Y,H,RO,EPS,P(NS),F(NS),0)
14    DO 100 L=LMIN,LMAX
      KL=L
      G(KL)=G0
20    KK=2
      CALL ITER(N,KK,DX,H00,G0,X,Y,H,RO,EPS,P(NS),F(NS),R(NS),0)
      KK=1
      CALL CHAN(N,R0,R(NS))
      CALL ITER(N,KK,DX,H00,G0,X,Y,H,RO,EPS,P(NS),F(NS),R0,1)
      CALL GCAL(N,G1,P(NS))
      G(KL-1)=G(KL)-DX*DX*G1
      N2=N
      NS0=NS
      KL=KL-1
      CALL KNDX(KL,N,NS,NMAX,DX,X0,XE,X,Y)
```

```
      CALL TRANS(N,N2,N2*N2,H,RO,EPS,P(NS),P(NS0),R(NS),R0)
      CALL ITER(N,KK,DX,H00,G0,X,Y,H,RO,EPS,P(NS),F(NS),R(NS),2)
      CALL GCAL(N,G1,P(NS))
      G(KL)=G(KL)+DX*DX*G1
      G0=G(KL)
      CALL HREE(N,DX,H00,G0,X,Y,H,RO,EPS,P(NS),F(NS),1)
      IF(KL.NE.1)GOTO 20
      KK=15
      CALL ITER(N,KK,DX,H00,G0,X,Y,H,RO,EPS,P(NS),F(NS),R(NS),0)
 40   NS0=NS
      KL=KL+1
      N1=N
      CALL KNDX(KL,N,NS,NMAX,DX,X0,XE,X,Y)
      G0=G(KL)
      CALL TRAP(N1,N,P(NS0),P(NS))
      CALL HREE(N,DX,H00,G0,X,Y,H,RO,EPS,P(NS),F(NS),0)
      KK=1
      CALL ITER(N,KK,DX,H00,G0,X,Y,H,RO,EPS,P(NS),F(NS),R(NS),0)
      IF(KL.LT.L)GOTO 40
      CALL CHAN(N,R0,R(NS))
      CALL ITER(N,KK,DX,H00,G0,X,Y,H,RO,EPS,P(NS),F(NS),R0,1)
 100  CONTINUE
      M=M+1
      IF(M.LT.MK)GOTO 14
 120  CALL OUPT(N,X,Y,H,P(NS))
      RETURN
      END
      C***************************************************************
      SUBROUTINE OUPT(N,X,Y,H,P)
      DIMENSION X(N),Y(N),H(N,N),P(N,N)
      NN=(N+1)/2
      A=0.0
      WRITE(8,*)N,NN
      WRITE(8,110)A,(Y(I),I=1,N)
      DO I=1,N
      WRITE(8,110)X(I),(H(I,J),J=1,N)
      END DO
      WRITE(10,*)N,NN
      WRITE(10,110)A,(Y(I),I=1,N)
      DO I=1,N
      WRITE(10,110)X(I),(P(I,J),J=1,N)
      END DO
 110  FORMAT(34(1X,E12.6))
      RETURN
      END
      C***************************************************************
      SUBROUTINE TRAP(N1,N2,P1,P2)
      DIMENSION P1(N1,N1),P2(N2,N2)
      DO 10 I=1,N1-1
      II=2*I
      DO 10 J=1,N1-1
```

```
      JJ=2*J
 10   P2(II,JJ)=P2(II,JJ)+0.25*(P1(I,J)+P1(I+1,J)+
      & P1(I+1,J+1)+P1(I,J+1)-P2(II-1,JJ-1)-P2(II-1,JJ+1)-
      & P2(II+1,JJ+1)-P2(II+1,JJ-1))
      DO 20 I=1,N1-1
      II=2*I
      DO 20 J=1,N1
      JJ=2*J-1
      P2(II,JJ)=P2(II,JJ)+0.5*(P1(I,J)+P1(I+1,J)-
      & P2(II-1,JJ)-P2(II+1,JJ))
 20   P2(JJ,II)=P2(JJ,II)+0.5*(P1(J,I)+P1(J,I+1)-
      & P2(JJ,II-1)-P2(JJ,II+1))
      DO 30 I=1,N1
      II=2*I-1
      DO 30 J=1,N1
      JJ=2*J-1
 30   P2(II,JJ)=P1(I,J)
      RETURN
      END
      C*************************************************************
      SUBROUTINE INITI(N,X,Y,P)
      DIMENSION X(N),Y(N),P(N,N)
      NN=(N+1)/2
      DO 10 I=1,N
      D=1.-X(I)*X(I)
      DO 10 J=1,NN
      C=D-Y(J)*Y(J)
      IF(C.LE.0.0)P(I,J)=0.0
 10   IF(C.GT.0.0)P(I,J)=SQRT(C)
      DO 20 I=1,N
      DO 20 J=NN+1,N
      JJ=N-J+1
 20   P(I,J)=P(I,JJ)
      RETURN
      END
      C*************************************************************
      SUBROUTINE CHAN(N,R1,R2)
      DIMENSION R1(N,N),R2(N,N)
      DO 10 I=1,N
      DO 10 J=1,N
 10   R1(I,J)=R2(I,J)
      RETURN
      END
      C*************************************************************
      SUBROUTINE GCAL(N,G,P)
      DIMENSION P(N,N)
      G=0.
      DO 10 I=1,N
      DO 10 J=1,N
 10   G=G+P(I,J)
      RETURN
```

```
      END
      C***********************************************************
      SUBROUTINE KNDX(K,N,NS,NMAX,DX,X0,XE,X,Y)
      DIMENSION INDEX(6),NNDEX(6),X(NMAX),Y(NMAX)
      DATA INDEX,NNDEX/1,82,371,1460,5685,16616,
     & 9,17,33,65,129,257/
      N=NNDEX(K)
      NS=INDEX(K)
      DX=(XE-X0)/(N-1.)
      Y0=-0.5*(XE-X0)
      DO 10 I=1,N
      X(I)=X0+(I-1)*DX
 10   Y(I)=Y0+(I-1)*DX
      RETURN
      END
      C***********************************************************
      SUBROUTINE HREE(N,DX,H00,G0,X,Y,H,RO,EPS,P,F,KG)
      DIMENSION X(N),Y(N),P(N,N),H(N,N),RO(N,N),EPS(N,N),F(N,N)
      DIMENSION W(150,150),P0(150,150)
      COMMON /COM1/ENDA,A1,A2,A3,Z,C1,C2,HM0/COMK/KL
     & /COMAK/AK(0:100,0:100)
      DATA KK,NW,PAI1/0,150,0.2026423/
      NN=(N+1)/2
       CALL VI(N,DX,P,W,NW)
      HMIN=1.E3
      DO 30 I=1,N
      W1=0.5*X(I)*X(I)
      DO 30 J=1,NN
      H0=W1+0.5*Y(J)*Y(J)+W(I,J)
      IF(KG.EQ.1)GOTO 20
      IF(H0+F(I,J).LT.HMIN)HMIN=H0+F(I,J)
      H(I,J)=H0
      GOTO 30
 20   F(I,J)=H(I,J)-H00-H0
      F(I,N-J+1)=F(I,J)
 30   CONTINUE
      IF(KK.EQ.0)H00=HM0-HMIN
      KK=1
      IF(KG.EQ.1)RETURN
      H0=H00+HMIN
      IF(H0.LE.0.0)GOTO 40
      IF(KL.NE.1)GOTO 50
      W1=0.0
      DO 32 I=1,N
      DO 32 J=1,N
 32   W1=W1+P(I,J)
      W1=DX*DX*W1/G0
      DW=1.-W1
      HM0=H0*W1
 40   H00=-HMIN+HM0
```

```
50  DO 60 I=1,N
    DO 60 J=1,NN
    IF(P(I,J).LT.0.0)P(I,J)=0.0
    EDA=EXP(A1*(-1.+(1.+A2*P(I,J))**Z))
    H(I,J)=H00+H(I,J)+F(I,J)
    RO(I,J)=(A3+1.34*P(I,J))/(A3+P(I,J))
60  EPS(I,J)=RO(I,J)*H(I,J)**3/(ENDA*EDA)
    DO 70 J=NN+1,N
    JJ=N-J+1
    DO 70 I=1,N
    H(I,J)=H(I,JJ)
    RO(I,J)=RO(I,JJ)
70  EPS(I,J)=EPS(I,JJ)
    RETURN
    END
C****************************************************************
    SUBROUTINE ITER(N,KK,DX,H00,G0,X,Y,H,RO,EPS,P,F,R,KG)
    DIMENSION X(N),Y(N),P(N,N),H(N,N),RO(N,N),
    & EPS(N,N),F(N,N),R(N,N)
    DIMENSION D(200),A(1000),B(600),ID(200)
    COMMON /COM1/ENDA,A1,A2,A3,Z,C1,C2,C3
    & /COMAK/AK(0:100,0:100)
    DATA KG1,PAI1/0,0.2026423/
    IF(KG1.NE.0)GOTO 2
    KG1=1
    AK00=AK(0,0)
    AK10=AK(1,0)
    AK20=AK(2,0)
    BK00=AK00-AK10
    BK10=AK10-0.25*(AK00+2.*AK(1,1)+AK(2,0))
    BK20=AK20-0.25*(AK10+2.*AK(2,1)+AK(3,0))
2   NN=(N+1)/2
    MM=N-1
    DX1=1./DX
    DX2=DX*DX
    DX3=1./DX2
    DX4=0.3*DX2
    DO 100 K=1,KK
    DO 70 J=2,NN
    J0=J-1
    J1=J+1
    JJ=N-J+1
    IF(KG.NE.0)GOTO 20
    IA=1
8   MM=N-IA
    IF(P(MM,J0).GT.1.E-6)GOTO 20
    IF(P(MM,J).GT.1.E-6)GOTO 20
    IF(P(MM,J1).GT.1.E-6)GOTO 20
    IA=IA+1
    IF(IA.LT.N)GOTO 8
```

```
      GOTO 70
20    IF(MM.LT.N-1)MM=MM+1
      D2=0.5*(EPS(1,J)+EPS(2,J))
      DO 50 I=2,MM
      I0=I-1
      I1=I+1
      II=5*I0
      D1=D2
      D2=0.5*(EPS(I1,J)+EPS(I,J))
      D4=0.5*(EPS(I,J0)+EPS(I,J))
      D5=0.5*(EPS(I,J1)+EPS(I,J))
      P1=P(I0,JJ)
      P2=P(I1,JJ)
      P3=P(I,JJ)
      P4=P(I,JJ+1)
      P5=P(I,JJ-1)
      D3=D1+D2+D4+D5
      IF(KG.NE.0)GOTO 32
      IF(J.EQ.NN.AND.ID(I).EQ.1)P(I,J)=P(I,J)-0.5*C2*D(I)
      IF(D1.GE.DX4)GOTO 30
      IF(D2.GE.DX4)GOTO 30
      IF(D4.GE.DX4)GOTO 30
      IF(D5.GE.DX4)GOTO 30
      ID(I)=1
      IF(J.EQ.NN)P5=P4
      A(II+1)=PAI1*(RO(I0,J)*BK10-RO(I,J)*BK20)
      A(II+2)=DX3*(D1+0.25*D3)+PAI1*(RO(I0,J)*BK00-RO(I,J)*BK10)
      A(II+3)=-1.25*DX3*D3+PAI1*(RO(I0,J)*BK10-RO(I,J)*BK00)
      A(II+4)=DX3*(D2+0.25*D3)+PAI1*(RO(I0,J)*BK20-RO(I,J)*BK10)
      A(II+5)=-DX3*(D1*P1+D2*P2+D4*P4+D5*P5-D3*P3)+
      & DX1*(RO(I,J)*H(I,J)-RO(I0,J)*H(I0,J))+R(I,J)
      GOTO 50
30    ID(I)=0
      P4=P(I,J0)
      IF(J.EQ.NN)P5=P4
      A(II+1)=PAI1*(RO(I0,J)*AK10-RO(I,J)*AK20)
      A(II+2)=DX3*D1+PAI1*(RO(I0,J)*AK00-RO(I,J)*AK10)
      A(II+3)=-DX3*D3+PAI1*(RO(I0,J)*AK10-RO(I,J)*AK00)
      A(II+4)=DX3*D2+PAI1*(RO(I0,J)*AK20-RO(I,J)*AK10)
      A(II+5)=-DX3*(D1*P1+D2*P2+D4*P4+D5*P5-D3*P3)+
      & DX1*(RO(I,J)*H(I,J)-RO(I0,J)*H(I0,J))+R(I,J)
      GOTO 50
32    IF(KG.EQ.2)GOTO 40
      R(I,J)=-DX3*(D1*P1+D2*P2+D4*P4+D5*P5-D3*P3)+
      & DX1*(RO(I,J)*H(I,J)-RO(I0,J)*H(I0,J))+R(I,J)
      GOTO 50
40    R(I,J)=DX3*(D1*P1+D2*P2+D4*P4+D5*P5-D3*P3)-
      & DX1*(RO(I,J)*H(I,J)-RO(I0,J)*H(I0,J))+R(I,J)
50    CONTINUE
      IF(KG.NE.0)GOTO 70
```

```
      CALL TRA4(MM,D,A,B)
      DO 54 I=2,MM
      IF(ID(I).EQ.0)GOTO 52
      DD=D(I+1)
      IF(I.EQ.MM)DD=0
      P(I,J)=P(I,J)+C2*(D(I)-0.25*(D(I-1)+DD))
      IF(J0.NE.1)P(I,J0)=P(I,J0)-0.25*C2*D(I)
      IF(P(I,J0).LT.0.)P(I,J0)=0.0
      IF(J1.GE.NN)GOTO 54
      P(I,J1)=P(I,J1)-0.25*C2*D(I)
      GOTO 54
52    P(I,J)=P(I,J)+C1*D(I)
54    IF(P(I,J).LT.0.0)P(I,J)=0.0
70    CONTINUE
      IF(KG.NE.0)GOTO 100
      DO 80 J=1,NN
      JJ=N+1-J
      DO 80 I=1,N
80    P(I,JJ)=P(I,J)
      CALL HREE(N,DX,H00,G0,X,Y,H,RO,EPS,P,F,0)
100   CONTINUE
      RETURN
      END
      C***************************************************************
      SUBROUTINE TRANS(N1,N2,NMAX,H,RO,EPS,P1,P2,R1,R2)
      DIMENSION H(NMAX),RO(NMAX),EPS(NMAX),
      & P1(N1,N1),P2(N2,N2),R1(N1,N1),R2(N2,N2)
      DO 10 I=1,N1
      II=2*I-1
      NI1=(I-1)*N1
      NI2=2*(I-1)*N2
      DO 10 J=1,N1
      JJ=2*J-1
      NJ1=NI1+J
      NJ2=NI2+JJ
      H(NJ1)=H(NJ2)
      RO(NJ1)=RO(NJ2)
      EPS(NJ1)=EPS(NJ2)
      P1(I,J)=P2(II,JJ)
10    R1(I,J)=R2(II,JJ)
      RETURN
      END
      C***************************************************************
      SUBROUTINE TRA4(N,D,A,B)
      DIMENSION D(N),A(5,N),B(3,N)
      C=1./A(3,N)
      B(1,N)=-A(1,N)*C
      B(2,N)=-A(2,N)*C
      B(3,N)=A(5,N)*C
      DO 10 I=1,N-2
```

```
   IN=N-I
   IN1=IN+1
   C=1./(A(3,IN)+A(4,IN)*B(2,IN1))
   B(1,IN)=-A(1,IN)*C
   B(2,IN)=-(A(2,IN)+A(4,IN)*B(1,IN1))*C
10 B(3,IN)=(A(5,IN)-A(4,IN)*B(3,IN1))*C
   D(1)=0.0
   D(2)=B(3,2)
   DO 20 I=3,N
20 D(I)=B(1,I)*D(I-2)+B(2,I)*D(I-1)+B(3,I)
   RETURN
   END
C***********************************************************
   SUBROUTINE VI(N,DX,P,V,NW)
   DIMENSION P(N,N),V(NW,NW)
   COMMON /COMAK/AK(0:100,0:100)
   PAI1=0.2026423
   DO 40 I=1,N
   DO 40 J=1,N
   H0=0.0
   DO 30 K=1,N
   IK=IABS(I-K)
   DO 30 L=1,N
   JL=IABS(J-L)
30 H0=H0+AK(IK,JL)*P(K,L)
40 V(I,J)=H0*DX*PAI1
   RETURN
   END
C***********************************************************
   SUBROUTINE SUBAK(MM)
   COMMON /COMAK/AK(0:100,0:100)
   S(X,Y)=X+SQRT(X**2+Y**2)
   DO 10 I=0,MM
   XP=I+0.5
   XM=I-0.5
   DO 10 J=0,I
   YP=J+0.5
   YM=J-0.5
   A1=S(YP,XP)/S(YM,XP)
   A2=S(XM,YM)/S(XP,YM)
   A3=S(YM,XM)/S(YP,XM)
   A4=S(XP,YP)/S(XM,YP)
   AK(I,J)=XP*ALOG(A1)+YM*ALOG(A2)+
  & XM*ALOG(A3)+YP*ALOG(A4)
10 AK(J,I)=AK(I,J)
   RETURN
   END
C*******************************************************
```

7.5.3 Example

The pressure distribution and film thickness according to the above calculation program are given in Figure 7.8.

(a)

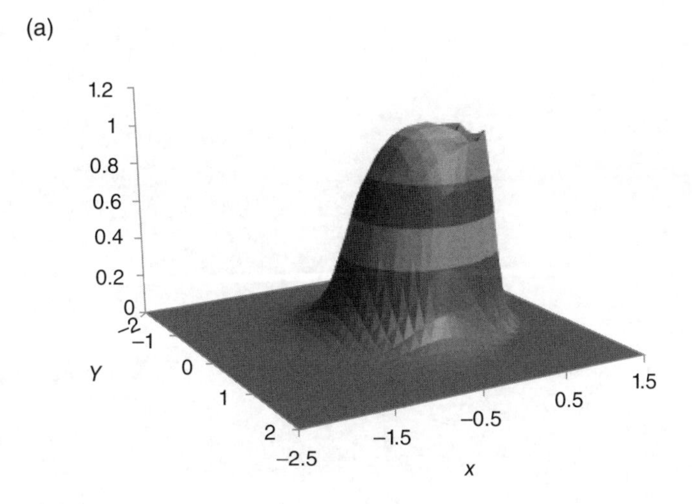

(b)

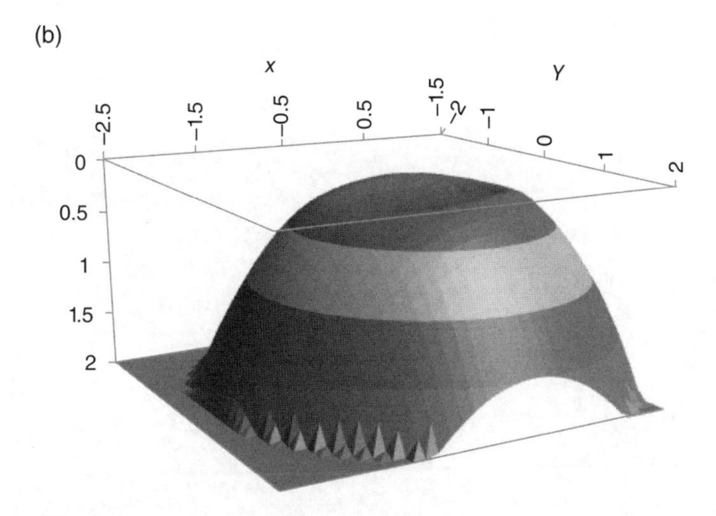

Figure 7.8 (a) Pressure distribution and (b) Film thickness for EHL with multigrid method in point contact

8

Numerical calculation method and program for isothermal EHL in ellipse contact

8.1 Basic equation

8.1.1 Reynolds equation

For isothermal EHL in the ellipse contact in steady state, Reynolds equation can be expressed in the following generalized form

$$\frac{\partial}{\partial x}\left(\frac{\rho h^3}{\eta}\frac{\partial p}{\partial x}\right)+\frac{\partial}{\partial y}\left(\frac{\rho h^3}{\eta}\frac{\partial p}{\partial y}\right)=12u_s\frac{\partial(\rho h)}{\partial x} \tag{8.1}$$

By using some nondimensional parameters, Reynolds equation can be written as

$$\frac{\partial}{\partial X}\left(\varepsilon\frac{\partial P}{\partial X}\right)+\frac{\partial}{\partial Y}\left(\varepsilon\frac{\partial P}{\partial Y}\right)-\frac{\partial(\rho^* H)}{\partial X}=0 \tag{8.2}$$

The boundary conditions are:

At the inlet $P(X_0, Y)=0$

At the outlet $P(X_e, Y)=0$ and $\dfrac{\partial P(X_e, Y)}{\partial X}=0$

At the two sides $P|_{Y=\pm 1}=0$

Numerical Calculation of Elastohydrodynamic Lubrication: Methods and Programs, First Edition.
Ping Huang.

The nondimensional discrete Reynolds equation is

$$\frac{\varepsilon_{i-1/2,j}P_{i-1,j}+\varepsilon_{i+1/2,j}P_{i+1,j}+\varepsilon_{i,j-1/2}P_{i,j-1}+\varepsilon_{i,j+1/2}P_{i,j+1}-\varepsilon_0 P_{ij}}{\Delta X^2}=\frac{\rho_{ij}^*H_{ij}-\rho_{i-1,j}^*H_{i-1,j}}{\Delta X} \tag{8.3}$$

8.1.2 Film thickness equation

The film thickness is expressed as follows.

$$h(x,y)=h_0+\frac{x^2}{2R_x}+\frac{y^2}{2R_y}+\frac{2}{\pi E}\iint_{\Omega}\frac{p(s,t)}{\sqrt{(x-s)^2+(y-t)^2}}dsdt \tag{8.4}$$

By using some nondimensional parameters, the film thickness can be written as

$$H(X,Y)=H_0+\frac{X^2}{2}+e_k\frac{Y^2}{2}+P_{\text{tr}}\int_{X_0}^{X_e}\int_{Y_0}^{Y_e}\frac{P(S,T)dSdT}{\sqrt{(X-S)^2+(Y-T)^2}} \tag{8.5}$$

The nondimensional discrete film thickness is

$$H_{ij}=H_0+\frac{X_i^2+e_kY_j^2}{2}+P_{\text{tr}}\sum_{k=1}^{n}\sum_{l=1}^{n}D_{ij}^{kl}P_{kl} \tag{8.6}$$

8.1.3 Viscosity–pressure equation

We use the viscosity–pressure equation recommended by Roelands [1]

$$\eta=\eta_0\exp\left\{(\ln\eta_0+9.67)\left[\left(1+\frac{p}{p_0}\right)^z-1\right]\right\} \tag{8.7}$$

By using some nondimensional parameters, we get

$$\eta^*=\exp\left\{(\ln\eta_0+9.67)\left[\left(1+\frac{p_{\text{H}}P}{p_0}\right)^z-1\right]\right\} \tag{8.8}$$

8.1.4 Density–pressure equation

In EHL calculations, the most commonly used density–pressure equation is

$$\rho = \rho_0 \left(1 + \frac{0.6p}{1 + 1.7p} \right) \tag{8.9}$$

By using the parameters, the nondimensional one is

$$\rho^* = 1 + \frac{0.6P \cdot p_H}{1 + 1.7P \cdot p_H} \tag{8.10}$$

8.1.5 Load balancing equation

The applied normal load must be supported by the generated hydrodynamic pressure in the elliptic contact region, so the force balance equation can be expressed by

$$\int_{x_0}^{x_e} \int_{y_0}^{y_e} p(x,y)dxdy = w \tag{8.11}$$

By substituting the nondimensional parameters into the above equation, the nondimensional form is

$$\int_{X_0}^{X_e} \int_{Y_0}^{Y_e} P(X,Y)dXdY = \frac{2\pi b}{3a} \tag{8.12}$$

After discretization, we obtain

$$W = \Delta X \Delta Y \sum_{i=1}^{n} \sum_{j=1}^{n} P_{ij} = \frac{2\pi b}{3a} \tag{8.13}$$

8.2 Calculation program

8.2.1 Calculation diagram

The calculation diagram of the isothermal EHL in the ellipse contact is shown in Figure 8.1.

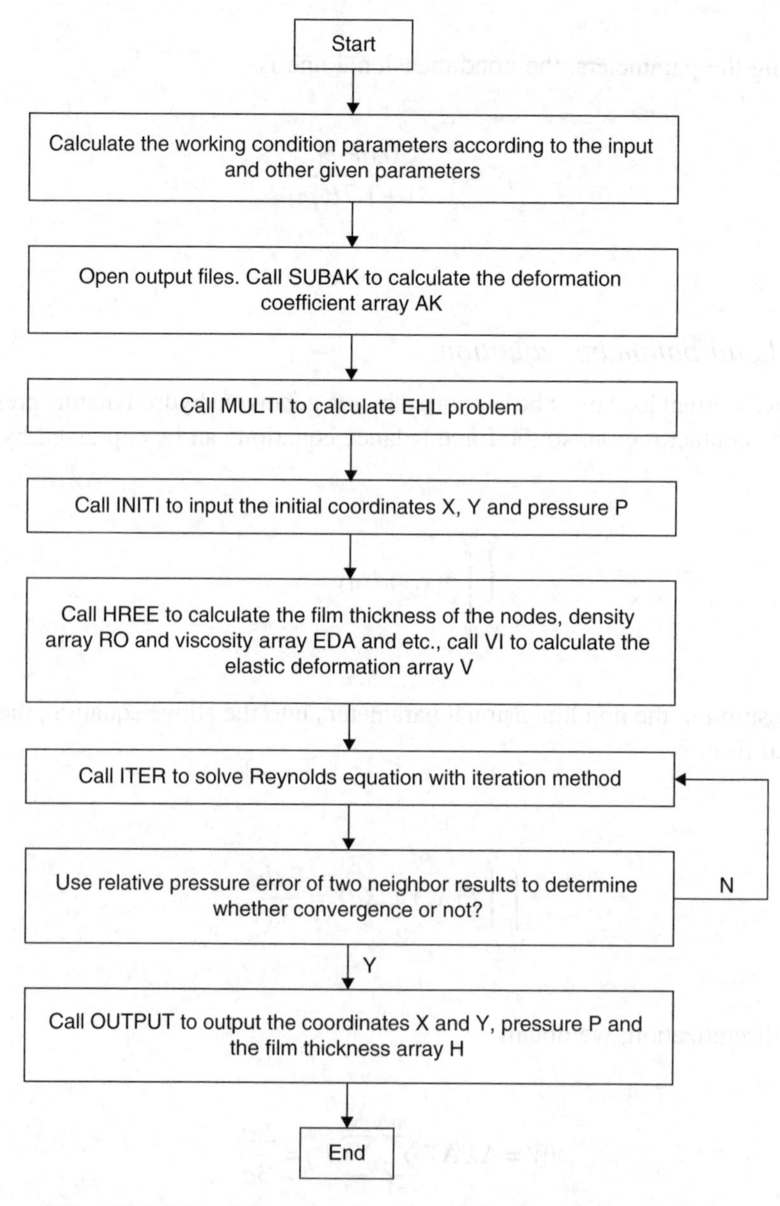

Figure 8.1 Calculation diagram of isothermal EHL in ellipse contact

8.2.2 *Calculation program*

In this program, the given parameters are listed as follows:

Number of nodes	N × N = 65 × 65
Nondimensional starting coordinate in *x* direction	X0 = −2.5
Nondimensional ending coordinate in *x* direction	XE = 1.5
Equivalent modulus of elasticity	E1 = 2.21E11 Pa
Initial plastic viscosity	EDA0 = 0.02 Pa·s
Equivalent curvature radius	RX = 0.03 m, RY = 0.06 m
Normal load	W0 = 1000 N
Average velocity	US = 3 ms^{-1}

Output data:

The pressure array P(I, J) will be saved in the file PRESSURE.DAT.
The film thickness array H(I, J) will be saved in the file FILM.DAT.

It should be noted that the interval of the nondimensional coordinate Y in the y direction will be determined in the subroutine INITI. It is symmetrically distributed along the X axis, and its length is the same as X = XE–X0. In the program, the nondimensionalization parameters of *x* and *y* are the semi-axes of the contacting ellipse in X direction. When modifying R_x and R_y, the users should be careful to select the interval of X. If $R_x < R_y$ (i.e., $a < b$), then the following approximate equation holds in a relatively wide range

$$\frac{b}{a} = \left(\frac{R_y}{R_x}\right)^{2/3} \tag{8.14}$$

For example, if $(R_y/R_x) = 5$, then $(b/a) = 2.924 \approx 3$. The length of X should not be less than 6 (the interval can be [−3, 3] or [−4, 3] or some others). Otherwise, the calculated pressure distribution will not be displayed in the result, or we may get a wrong result.

```
PROGRAM ELLIPEHL
COMMON /COM1/ENDA,A1,A2,A3,Z,HM0,DH
COMMON /COMEK/EK,BX,BY,PTR
REAL*8 RX,RY,KA,KB
DATA N,PAI,Z,E1,EDA0,RX,RY,X0,XE,W0,US/65,3.14159265,0.68,2.21E11,
0.02,0.03,0.06,-2.5,1.5,1000.,3./
OPEN(8,FILE='FILM.DAT',STATUS='UNKNOWN')
OPEN(10,FILE='PRESSURE.DAT',STATUS='UNKNOWN')
EK=RX/RY
AA=0.5*(1./RX+1./RY)
BB=0.5*ABS(1./RX-1./RY)
CALL HERTZELLIPTIC(RX,RY,KA,KB)
EA=KA*(1.5*W0/AA/E1)**(1./3.0)
```

```
    EB=KB*(1.5*W0/AA/E1)**(1./3.0)
    PH=1.5*W0/(EA*EB*PAI)
    WRITE(*,*)"N,X0,XE,W0,PH,E1,EDA0,RX,US=",N,X0,XE,W0,PH,E1,EDA0,
    RX,US
    MM=N-1
    U=EDA0*US/(E1*RX)
    A1=ALOG(EDA0)+9.67
    A2=5.1E-9*PH
    A3=0.59/(PH*1.E-9)
    BX=EB
    BY=EA
    IF(RX.GT.RY) THEN
    BX=EA
    BY=EB
    ENDIF
    W=W0/(E1*RX**2)
    PTR=3*W*(RX/BY)*(RX/BX)**2/(PAI**2)
    ALFA=Z*5.1E-9*A1
    G=ALFA*E1
    AHM=1.0-EXP(-0.68*1.03)
    HM0=3.63*(RX/BX)**2*G**0.49*U**0.68*W**(-0.073)*AHM
    ENDA=12.*U*(E1/PH)*(RX/BX)**3
    WRITE(*,*)'          Wait please'
    CALL SUBAK(MM)
    CALL MULTI(N,X0,XE)
    STOP
    END
    SUBROUTINE MULTI(N,X0,XE)
    DIMENSION X(65),Y(65),H(4500),RO(4500),EPS(4500),EDA(4500),P(4500),
    POLD(4500),V(4500)
    COMMON /COM1/ENDA,A1,A2,A3,Z,HM0,DH
    COMMON /COMEK/EK,BX,BY,PTR
    DATA MK,G00/1,2.0943951/
    G0=G00*BY/BX
    NX=N
    NY=N
    NN=(N+1)/2
    CALL INITI(N,DX,X0,XE,X,Y,P,POLD)
    CALL HREE(N,DX,G0,X,Y,H,RO,EPS,EDA,P,V)
14  KK=15
    CALL ITER(N,KK,DX,G0,X,Y,H,RO,EPS,EDA,P,V)
    MK=MK+1
    CALL ERP(N,ER,P,POLD)
    WRITE(*,*)'ER=',ER
    IF(ER.GT.1.E-5.AND.DH.GT.1.E-7)THEN
    IF(MK.GE.10)THEN
    WRITE(*,*)'ER,DH=',ER,DH
    MK=1
    DH=0.5*DH
    ENDIF
    GOTO 14
    ENDIF
```

```
      IF(DH.LE.1.E-7)WRITE(*,*)'Pressures are not convergent!!!'
      CALL OUTPUT(N,DX,X,Y,H,P)
      RETURN
      END
      SUBROUTINE ERP(N,ER,P,POLD)
      DIMENSION P(N,N),POLD(N,N)
      ER=0.0
      SUM=0.0
      NN=(N+1)/2
      DO 10 I=1,N
      DO 10 J=1,NN
      ER=ER+ABS(P(I,J)-POLD(I,J))
      SUM=SUM+P(I,J)
10    CONTINUE
      ER=ER/SUM
      DO I=1,N
      DO J=1,N
      POLD(I,J)=P(I,J)
      ENDDO
      ENDDO
      RETURN
      END
      SUBROUTINE INITI(N,DX,X0,XE,X,Y,P,POLD)
      COMMON /COMEK/EK,BX,BY,PTR
      DIMENSION X(N),Y(N),P(N,N),POLD(N,N)
      NN=(N+1)/2
      DX=(XE-X0)/(N-1.)
      Y0=-0.5*(XE-X0)
      DO 5 I=1,N
      X(I)=X0+(I-1)*DX
      Y(I)=Y0+(I-1)*DX
5     CONTINUE
      DO 10 I=1,N
      D=1.-X(I)*X(I)
      DO 10 J=1,NN
      C=D-(BX/BY)**2*Y(J)*Y(J)
      IF(C.LE.0.0)P(I,J)=0.0
10    IF(C.GT.0.0)P(I,J)=SQRT(C)
      DO 20 I=1,N
      DO 20 J=NN+1,N
      JJ=N-J+1
20    P(I,J)=P(I,JJ)
      DO I=1,N
      DO J=1,N
      POLD(I,J)=P(I,J)
      ENDDO
      ENDDO
      RETURN
      END
      SUBROUTINE HREE(N,DX,G0,X,Y,H,RO,EPS,EDA,P,V)
      DIMENSION X(N),Y(N),P(N,N),H(N,N),RO(N,N),EPS(N,N),EDA(N,N),V(N,N)
      COMMON /COM1/ENDA,A1,A2,A3,Z,HM0,DH/COMAK/AK(0:65,0:65)
```

```
      COMMON /COMEK/EK,BX,BY,PTR
      DATA KK,PAI,PAI1/0,3.14159265,0.2026423/
      NN=(N+1)/2
      CALL VI(N,DX,P,V)
      HMIN=1.E3
      DO 30 I=1,N
      DO 30 J=1,NN
      RAD=X(I)*X(I)+EK*Y(J)*Y(J)
      W1=0.5*RAD
      H0=W1+V(I,J)
      IF(H0.LT.HMIN)HMIN=H0
 30   H(I,J)=H0
      W1=0.0
      DO 40 I=1,N
      DO 40 J=1,N
 40   W1=W1+P(I,J)
      W1=DX*DX*W1/G0
      DW=1.-W1
      IF(KK.NE.0)GOTO 50
      KK=1
      DH=0.1*HM0
      H00=-HMIN+HM0
 50   IF(DW.LT.0.0)H00=H00+DH
      IF(DW.GT.0.0)H00=H00-DH
      DO 60 I=1,N
      DO 60 J=1,NN
      H(I,J)=H00+H(I,J)
      IF(P(I,J).LT.0.0)P(I,J)=0.0
      EDA1=EXP(A1*(-1.+(1.+A2*P(I,J))**Z))
      EDA(I,J)=EDA1
      RO(I,J)=(A3+1.34*P(I,J))/(A3+P(I,J))
 60   EPS(I,J)=RO(I,J)*H(I,J)**3/(ENDA*EDA1)
      DO 70 J=NN+1,N
      JJ=N-J+1
      DO 70 I=1,N
      H(I,J)=H(I,JJ)
      RO(I,J)=RO(I,JJ)
      EDA(I,J)=EDA(I,JJ)
 70   EPS(I,J)=EPS(I,JJ)
      RETURN
      END
      SUBROUTINE ITER(N,KK,DX,G0,X,Y,H,RO,EPS,EDA,P,V)
      DIMENSION X(N),Y(N),P(N,N),H(N,N),RO(N,N),EPS(N,N),EDA(N,N),V(N,N)
      COMMON /COMAK/AK(0:65,0:65)
      COMMON /COMEK/EK,BX,BY,PTR
      DATA KG1,PAI/0,3.14159265/
      IF(KG1.NE.0)GOTO 2
      KG1=1
      AK00=AK(0,0)
      AK10=AK(1,0)
 2    NN=(N+1)/2
      DO 100 K=1,KK
```

```
      DO 70 J=2,NN
      J0=J-1
      J1=J+1
      D2=0.5*(EPS(1,J)+EPS(2,J))
      DO 70 I=2,N-1
      I0=I-1
      I1=I+1
      D1=D2
      D2=0.5*(EPS(I1,J)+EPS(I,J))
      D4=0.5*(EPS(I,J0)+EPS(I,J))
      D5=0.5*(EPS(I,J1)+EPS(I,J))
      D8=PTR*RO(I,J)*AK00
      D9=PTR*RO(I0,J)*AK10
      D10=D1+D2+D4+D5+D8*DX-D9*DX
      D11=D1*P(I0,J)+D2*P(I1,J)+D4*P(I,J0)+D5*P(I,J1)
      D12=(RO(I,J)*H(I,J)-D8*P(I,J)-RO(I0,J)*H(I0,J)+D9*P(I,J))*DX
      P(I,J)=(D11-D12)/D10
      IF(P(I,J).LT.0.0)P(I,J)=0.0
70    CONTINUE
      DO 80 J=1,NN
      JJ=N+1-J
      DO 80 I=1,N
80    P(I,JJ)=P(I,J)
      CALL HREE(N,DX,G0,X,Y,H,RO,EPS,EDA,P,V)
100   CONTINUE
      RETURN
      END
      SUBROUTINE VI(N,DX,P,V)
      DIMENSION P(N,N),V(N,N)
      COMMON /COMEK/EK,BX,BY,PTR
      COMMON /COMAK/AK(0:65,0:65)
      DO 40 I=1,N
      DO 40 J=1,N
      H0=0.0
      DO 30 K=1,N
      IK=IABS(I-K)
      DO 30 L=1,N
      JL=IABS(J-L)
30    H0=H0+AK(IK,JL)*P(K,L)
40    V(I,J)=H0*DX*PTR
      RETURN
      END
      SUBROUTINE SUBAK(MM)
      COMMON /COMAK/AK(0:65,0:65)
      S(X,Y)=X+SQRT(X**2+Y**2)
      DO 10 I=0,MM
      XP=I+0.5
      XM=I-0.5
      DO 10 J=0,I
      YP=J+0.5
      YM=J-0.5
      A1=S(YP,XP)/S(YM,XP)
```

```
      A2=S(XM,YM)/S(XP,YM)
      A3=S(YM,XM)/S(YP,XM)
      A4=S(XP,YP)/S(XM,YP)
      AK(I,J)=XP*ALOG(A1)+YM*ALOG(A2)+XM*ALOG(A3)+YP*ALOG(A4)
10    AK(J,I)=AK(I,J)
      RETURN
      END
      SUBROUTINE OUTPUT(N,DX,X,Y,H,P)
      DIMENSION X(N),Y(N),H(N,N),P(N,N)
      NN=(N+1)/2
      A=0.0
      WRITE(8,110)A,(Y(I),I=1,N)
      DO I=1,N
      WRITE(8,110)X(I),(H(I,J),J=1,N)
      ENDDO
      WRITE(10,110)A,(Y(I),I=1,N)
      DO I=1,N
      WRITE(10,110)X(I),(P(I,J),J=1,N)
      ENDDO
110   FORMAT(66(E12.6,1X))
      RETURN
      END
      SUBROUTINE HERTZELLIPTIC(RX,RY,KA,KB)
      IMPLICIT NONE
      REAL*8, EXTERNAL :: EE,KE
      REAL*8 RX,RY,BPA,BMA,CTH,THT,PAI,E1,KA,KB
      DATA PAI/3.1415926/
      BPA=0.5*(1./RX+1./RY)
      BMA=0.5*ABS(1./RX-1./RY)
      CTH=BMA/BPA
      THT=ACOS(CTH)*180.0/PAI
      CALL CACUE(CTH,E1)
      KA=(2.*EE(E1)/(PAI*(1-E1**2)))**(1/3.)
      KB=KA*(1.-E1**2)**(1/2.)
      END SUBROUTINE
      SUBROUTINE CACUE(CTH,E1)
      IMPLICIT NONE
      REAL*8, EXTERNAL :: EE,KE
      REAL*8, EXTERNAL :: FAB
      INTEGER FLG,I
      REAL*8 PAI,CTH,E1,E11,E12,DX,A,B,A1,A2,A3,A4,A5,T1,T2,T3,T4,T5,ER0
      DATA PAI,DX,FLG,I,T1,T5,ER0/3.1415926,0.0001,1,1,1.E-30,1.,1.E-12/
      IF(CTH.LT.1.E-6)THEN
      E1=0.
      RETURN
      ENDIF
      IF(CTH.GT.0.9999999999)THEN
      E1=1.
      RETURN
      ENDIF
      A1=FAB(T1,CTH)
      A5=FAB(T5,CTH)
```

```
DO WHILE(FLG.EQ.1)
T3=T1+I*DX
A3=FAB(T3,CTH)
I=I+1
IF((A1*A3.LT.0.).AND.(A3*A5.LT.0.)) THEN
FLG=0
END IF
END DO
DO WHILE((T3-T1).GT.ER0)
T2=(T1+T3)/2.
A2=FAB(T2,CTH)
IF(A2.GT.0.) T1=T2
IF(A2.LT.0.) T3=T2
IF(A2.EQ.0.)THEN
E11=T2
EXIT
END IF
END DO
E11=T2
DO WHILE((T5-T3).GT.ER0)
T4=(T3+T5)/2.
A4=FAB(T4,CTH)
IF(A4.GT.0.) T5=T4
IF(A4.LT.0.) T3=T4
IF(A4.EQ.0.)THEN
E12=T2
EXIT
END IF
END DO
E12=T4
E1=E11
IF(E11.LT.E12) E1=E12
RETURN
END SUBROUTINE
REAL*8 FUNCTION FAB(E1,CTH)
IMPLICIT NONE
REAL*8 E1,CTH,T1,T2
REAL*8, EXTERNAL :: EE,KE
T1=EE(E1)
T2=KE(E1)
FAB=2*(1-E1**2)*(T1-T2)+(1.-CTH)*E1**2*T1
END FUNCTION
REAL*8 FUNCTION KE(E1)
IMPLICIT NONE
INTEGER N,I,FLG
REAL*8 E1,PAI,H,T,T1,T2,S1,S2,P,Q
PAI=3.1415926
IF(E1.EQ.1) THEN
KE=1.E10
RETURN
ENDIF
IF(E1.LT.1.E-20) THEN
```

```
KE=PAI/2.
RETURN
ENDIF
N=1
H=PAI/2.
Q=SQRT(1.-E1*E1*SIN(H)*SIN(H))
IF(Q.LT.1.E-35) Q=1.E35
Q=1./Q
T1=.5*H*(1+Q)
S1=T1
FLG=1
DO WHILE(FLG.EQ.1)
P=0.
DO I=0,N-1
T=(I+0.5)*H
Q=SQRT(1.-E1*E1*SIN(T)*SIN(T))
IF(Q.LT.1.E-35) Q=1.E35
Q=1./Q
P=P+Q
END DO
T2=(T1+H*P)/2.
S2=(4.*T2-T1)/3.
IF(ABS(S2-S1).LT.ABS(S2)*1.E-7) FLG=0
T1=T2
S1=S2
N=N+N
H=.5*H
END DO
KE=S2
RETURN
END FUNCTION
REAL*8 FUNCTION EE(E1)
IMPLICIT NONE
INTEGER N,I,FLG
REAL*8 E1,PAI,H,T,T1,T2,S1,S2,P,Q
PAI=3.1415926
N=1
H=PAI/2.
IF(E1.EQ.1) THEN
EE=1.
RETURN
ENDIF
IF(E1.LT.1.E-20) THEN
EE=PAI/2.
RETURN
ENDIF
Q=SQRT(1.-E1*E1*SIN(H)*SIN(H))
T1=.5*H*(1+Q)
S1=T1
FLG=1
DO WHILE(FLG==1)
P=0.
DO I=0,N-1
```

```
T=(I+0.5)*H
Q=SQRT(1.-E1*E1*SIN(T)*SIN(T))
P=P+Q
END DO
T2=(T1+H*P)/2.
S2=(4.*T2-T1)/3.
IF(ABS(S2-S1).LT.ABS(S2)*1.E-7) FLG=0
T1=T2
S1=S2
N=N+N
H=.5*H
END DO
EE=S2
RETURN
END FUNCTION
```

8.2.3 *Example*

Under the given working condition in the above program, the results of the film thickness and pressure distributions are shown in Figure 8.2.

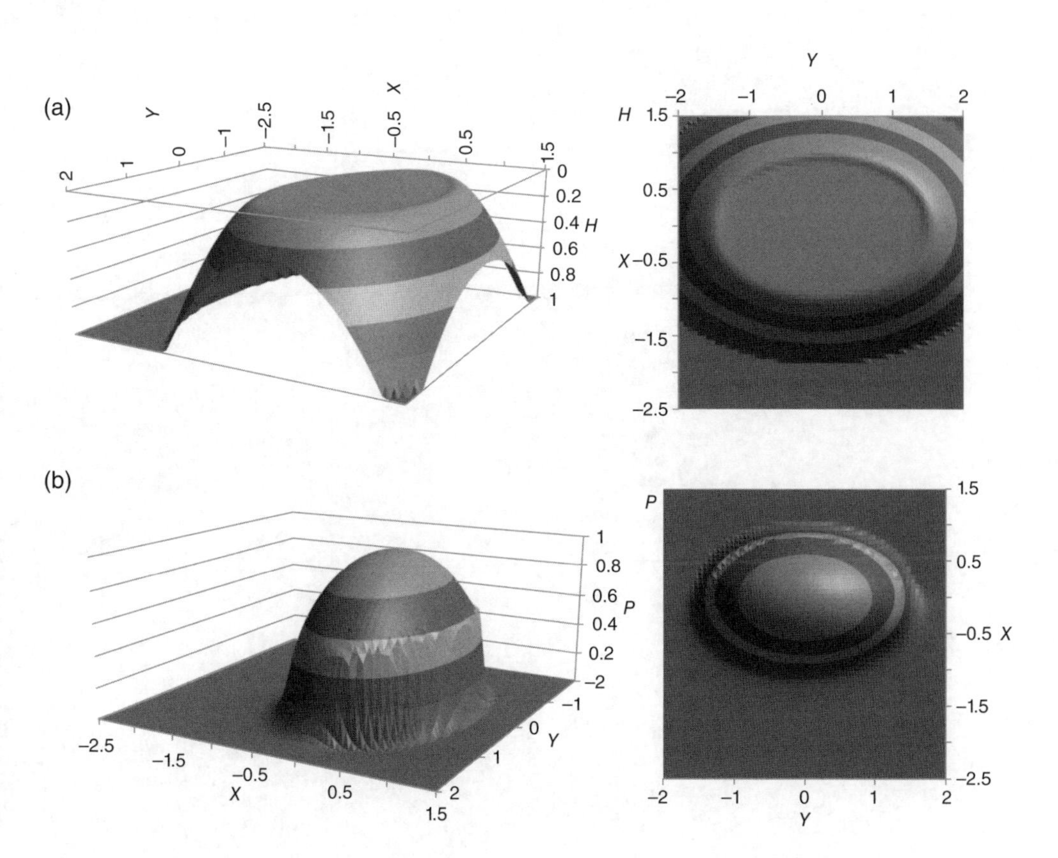

Figure 8.2 Film thickness and pressure distribution of isothermal EHL in ellipse contact. (a) Film thickness and (b) pressure distribution

9

Numerical calculation method and program for isothermal EHL in elliptical contact with two-dimensional velocities

9.1 Basic equations

The numerical calculation methods and programs for isothermal EHL in the ellipse contact have been discussed in Chapter 8, where the relative velocity of the two contact surfaces is along the x direction and not along the y direction. In practice, the main velocities of two surfaces may not coincide with the x direction. For example, in the helical gear or worm drive, the moving velocities of the two surfaces are angled to the x direction. That is, there are two-dimension velocities in the x and y directions, which make the relationship between the two sliding surfaces of the rollers more complex. Therefore, the previous methods and programs cannot be used to analyze their lubrication performances. The numerical methods and programs in such cases are discussed further.

The Reynolds equation of the isothermal EHL in the ellipse contact with two-dimension velocities is as follows:

$$\frac{\partial}{\partial x}\left(\frac{\rho h^3}{\eta}\frac{\partial p}{\partial x}\right)+\frac{\partial}{\partial y}\left(\frac{\rho h^3}{\eta}\frac{\partial p}{\partial y}\right)=12u_s\frac{\partial(\rho h)}{\partial x}+12v_s\frac{\partial(\rho h)}{\partial y} \tag{9.1}$$

The nondimensional Reynolds equation is as follows:

$$\frac{\partial}{\partial X}\left[\varepsilon\frac{\partial P}{\partial X}\right]+\frac{\partial}{\partial Y}\left[\varepsilon\frac{\partial P}{\partial Y}\right]-U_X\frac{\partial(\rho^* H)}{\partial X}-U_Y\frac{\partial(\rho^* H)}{\partial Y}=0 \tag{9.2}$$

Numerical Calculation of Elastohydrodynamic Lubrication: Methods and Programs, First Edition. Ping Huang.

where, U_X and U_Y are the nondimensional velocities in the x and y directions.

The boundary conditions of the Reynolds equation (9.2) are as follows:

Inlet boundary condition: $P(X_0, Y) = 0$
Outlet boundary conditions: $P(X_e, Y) = 0$ and $\partial P(X_e, Y)/\partial X = 0$
Side boundary condition: $P|_{Y=\pm 1} = 0$

The discrete form of the Reynolds equation (9.2) can be written as follows:

$$\frac{\varepsilon_{i-1/2,j}P_{i-1,j} + \varepsilon_{i+1/2,j}P_{i+1,j} + \varepsilon_{i,j-1/2}P_{i,j-1} + \varepsilon_{i,j+1/2}P_{i,j+1} - \varepsilon_0 P_{ij}}{\Delta X^2}$$
$$= U_X \frac{\rho^*_{ij}H_{ij} - \rho^*_{i-1,j}H_{i-1,j}}{\Delta X} + U_Y \frac{\rho^*_{ij}H_{ij} - \rho^*_{i,j-1}H_{i,j-1}}{\Delta X} \tag{9.3}$$

The film thickness equation, the deformation equation, the viscosity–pressure equation, and the density–pressure equation are similar to those in Chapter 8.

9.2 Velocity treatment

In calculation of the general isothermal EHL in the elliptical contact, the direction of the velocity coincides with the x direction. Actually, if there are two-dimension velocities, the Reynolds equation needs to be resolved in the x and y directions as shown in Equation 9.1. The term $12v_s(\partial(\rho h)/\partial y)$ in the right side of the equation is the dynamic pressure effect in the y direction.

Suppose $Q_1 = \varepsilon_{i-1/2,j}P_{i-1,j} + \varepsilon_{i+1/2,j}P_{i+1,j} + \varepsilon_{i,j-1/2}P_{i,j-1} + \varepsilon_{i,j+1/2}P_{i,j+1}$, Equation 9.3 can be written as follows:

$$Q_1 - \varepsilon_0 P_{i,j} - U_X \Delta X * \left(\rho^*_{i,j}H_{i,j} - \rho^*_{i-1,j}H_{i-1,j}\right) - U_Y \Delta X \left(\rho^*_{i,j}H_{i,j} - \rho^*_{i,j-1}H_{i,j-1}\right) = 0 \tag{9.4}$$

Then

$$H_{ij} = P_1 + P_{\mathrm{tr}} D^{ij}_{ij} P_{ij} \tag{9.5}$$

$$H_{i-1,j} = P_2 + P_{\mathrm{tr}} D^{ij}_{i-1,j} P_{ij} \tag{9.6}$$

$$H_{i,j-1} = P_3 + P_{\mathrm{tr}} D^{ij}_{i,j-1} P_{ij} \tag{9.7}$$

Or

$$P_1 = H_{ij} - P_{\mathrm{tr}} D^{ij}_{ij} P_{ij} \tag{9.8}$$

$$P_2 = H_{i-1,j} - P_{\text{tr}} D_{i-1,j}^{ij} P_{ij} \tag{9.9}$$

$$P_3 = H_{i,j-1} - P_{\text{tr}} D_{i,j-1}^{ij} P_{ij} \tag{9.10}$$

The iteration formula derived from Equation 9.4 is as follows:

$$P_{i,j} = \frac{Q_1 - U_X \Delta X\left(\rho_{i,j}^* P_1 - \rho_{i-1,j}^* P_2\right) - U_Y \Delta X\left(\rho_{i,j}^* P_1 - \rho_{i,j-1}^* P_3\right)}{\varepsilon_0 + U_X \Delta X\left(\rho_{i,j}^* P_{\text{tr}} D_{ij}^{ij} - \rho_{i-1,j}^* P_{\text{tr}} D_{i-1,j}^{ij}\right) + U_Y \Delta X\left(\rho_{i,j}^* P_{\text{tr}} D_{ij}^{ij} - \rho_{i,j-1}^* P_{\text{tr}} D_{i,j-1}^{ij}\right)} \tag{9.11}$$

Since the velocity U_y in the y direction is added to the Reynolds equation, the iteration calculation requires making some changes according to Equation 9.11, but the iteration methods are the same as in Chapter 6.

9.3 Numerical calculation method and program

9.3.1 Flowchart

The flowchart to calculate isothermal EHL in the elliptical contact with two-dimension velocities is shown in Figure 9.1 and the pressure distribution and film thickness are shown in Figure 9.2.

9.3.2 Program

9.3.2.1 Program description

The main modification of the program compared with that of Chapter 8 is in the iteration subroutine ITER. The other subroutines are almost the same as those in Chapter 8. In the following program, the statements modified are made bold. The followings are some important points that should be noted:

1. The ratio of the two-dimension velocities UX and UY is added in the DATA statement. For convenience, UX is used in the nondimensional equations. Since UX is set to be 1, we only need to modify UY. If readers need to modify UX, the difference in the nondimensional equations must be paid attention to.
2. Due to the addition of the velocity in the y direction, the iteration equation should be modified accordingly. In Subroutine ITER, we have added four variables D6–D9 and made some modifications in the other three variables D10–D12 to meet the need, which are shown as Equations 9.4–9.8. Then, the pressure

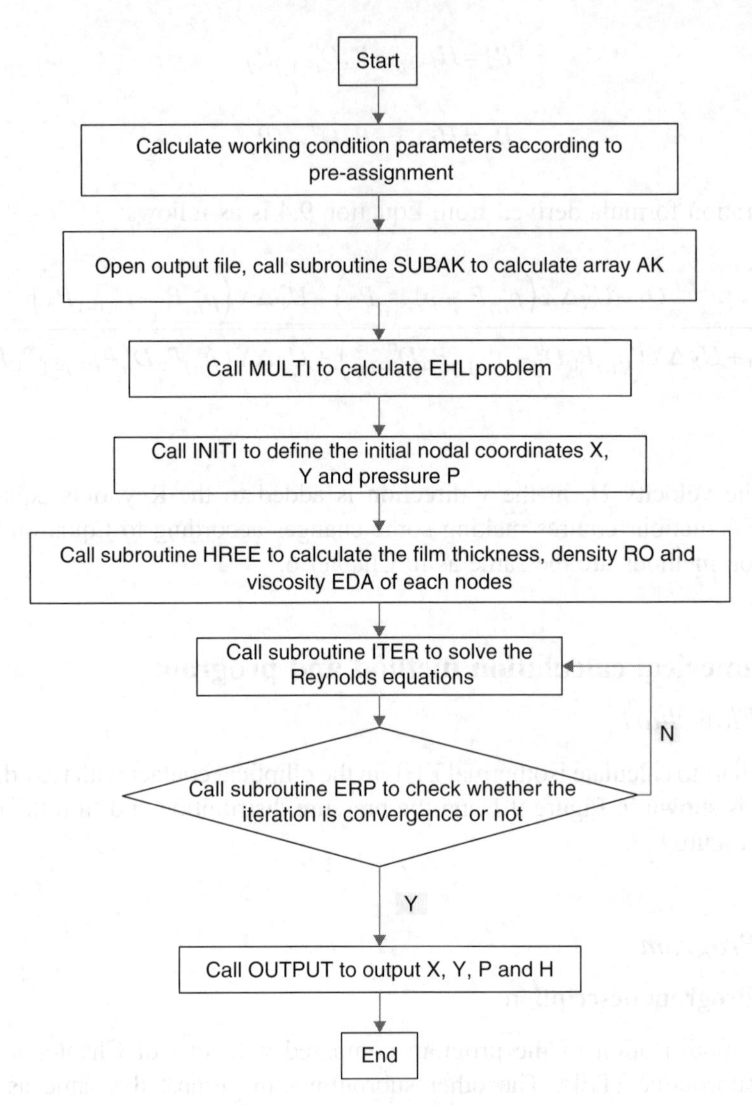

Figure 9.1 Diagram to calculate isothermal EHL in ellipse contact with two-dimension velocities

correction can be completed in the following statements based on the velocity direction.

3. Due to the change of the velocity direction, the problem is no longer symmetric. Therefore, the asymmetric computations in the *y* direction should be carried out as shown in the statement DO 70 J=2, N-1 other than DO 70 J=2, N/2+1. If not, it will lead to an error.

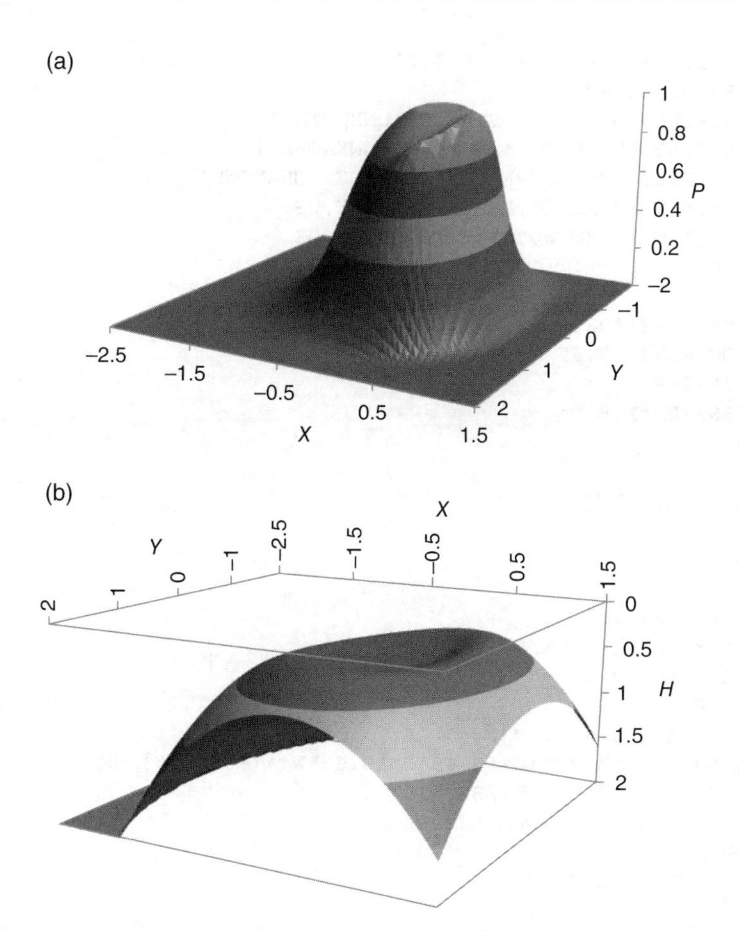

Figure 9.2 Pressure distribution and film thickness of isothermal EHL in ellipse contact with two-dimension velocities. (a) Pressure distribution and (b) film thickness

9.3.2.2 Program codes

```
PROGRAM ELLIPEHL2V
     COMMON /COM1/ENDA,A1,A2,A3,Z,HM0,DH
     COMMON /COMEK/EK,BX,BY,PTR
     REAL*8 RX,RY,KA,KB
     DATA
     N,PAI,Z,E1,EDA0,RX,RY,X0,XE,W0,US/65,3.14159265,0.68,2.21E11,0.05,
     0.3,0.6,-2.5,1.5,1000.,1.0/
     EK=RX/RY
     AA=0.5*(1./RX+1./RY)
     BB=0.5*ABS(1./RX-1./RY)
     CALL HERTZELLIPTIC(RX,RY,KA,KB)
     EA=KA*(1.5*W0/AA/E1)**(1./3.0)
```

```
      EB=KB*(1.5*W0/AA/E1)**(1./3.0)
      PH=1.5*W0/(EA*EB*PAI)
      OPEN(4,FILE='OUT.DAT',STATUS='UNKNOWN')
      OPEN(8,FILE='FILM.DAT',STATUS='UNKNOWN')
      OPEN(10,FILE='PRESSURE.DAT',STATUS='UNKNOWN')
      WRITE(*,*)N,X0,XE,W0,PH,E1,EDA0,RX,US
      WRITE(4,*)N,X0,XE,W0,PH,E1,EDA0,RX,US
      H00=0.0
      MM=N-1
      U=EDA0*US/(E1*RX)
      A1=ALOG(EDA0)+9.67
      A2=5.1E-9*PH
      A3=0.59/(PH*1.E-9)
      BX=EB
      BY=EA
      IF(RX.GT.RY) THEN
      BX=EA
      BY=EB
      ENDIF
      W=W0/(E1*RX**2)
      PTR=3*W*(RX/BY)*(RX/BX)**2/(PAI**2)
      ALFA=Z*5.1E-9*A1
      G=ALFA*E1
      AHM=1.0-EXP(-0.68*1.03)
      HM0=3.63*(RX/BX)**2*G**0.49*U**0.68*W**(-0.073)*AHM
      ENDA=12.*U*(E1/PH)*(RX/BX)**3
      WRITE(*,*)'         Wait please'
      CALL SUBAK(MM)
      CALL EHL(N,X0,XE,H00)
      STOP
      END
      SUBROUTINE EHL(N,X0,XE,H00)
     DIMENSION X(65),Y(65),H(4500),RO(4500),EPS(4500),EDA(4500),P(4500),
     POLD(4500),V(4500)
      COMMON /COM1/ENDA,A1,A2,A3,Z,HM0,DH
      COMMON /COMEK/EK,BX,BY,PTR
      DATA MK,G00/1,2.0943951/
      G0=G00*BY/BX
      NX=N
      NY=N
      CALL INITI(N,DX,X0,XE,X,Y,P,POLD)
      CALL HREE(N,DX,G0,X,Y,H,RO,EPS,EDA,P,V)
14    KK=15
      CALL ITER(N,KK,DX,G0,X,Y,H,RO,EPS,EDA,P,V)
      MK=MK+1
      CALL ERP(N,ER,P,POLD)
      WRITE(*,*)'ER=',ER
      IF(ER.GT.1.E-5.AND.DH.GT.1.E-6)THEN
      IF(MK.GE.10)THEN
      MK=1
      DH=0.5*DH
```

```
      ENDIF
      GOTO 14
      ENDIF
      IF(DH.LE.1.E-6)WRITE(*,*)'Pressures are not convergent!!!'
      CALL OUTPUT(N,DX,X,Y,H,P)
      RETURN
      END
      SUBROUTINE ERP(N,ER,P,POLD)
      DIMENSION P(N,N),POLD(N,N)
      ER=0.0
      SUM=0.0
      DO 10 I=1,N
      DO 10 J=1,N
      ER=ER+ABS(P(I,J)-POLD(I,J))
      SUM=SUM+P(I,J)
10    CONTINUE
      ER=ER/SUM
      DO I=1,N
      DO J=1,N
      POLD(I,J)=P(I,J)
      ENDDO
      ENDDO
      RETURN
      END
      SUBROUTINE INITI(N,DX,X0,XE,X,Y,P,POLD)
      DIMENSION X(N),Y(N),P(N,N),POLD(N,N)
      DX=(XE-X0)/(N-1.)
      Y0=-0.5*(XE-X0)
      DO 5 I=1,N
      X(I)=X0+(I-1)*DX
      Y(I)=Y0+(I-1)*DX
5     CONTINUE
      DO 10 I=1,N
      D=1.-X(I)*X(I)
      DO 10 J=1,N
      C=D-Y(J)*Y(J)
      IF(C.LE.0.0)P(I,J)=0.0
10    IF(C.GT.0.0)P(I,J)=SQRT(C)
      DO I=1,N
      DO J=1,N
      POLD(I,J)=P(I,J)
      ENDDO
      ENDDO
      RETURN
      END
      SUBROUTINE HREE(N,DX,G0,X,Y,H,RO,EPS,EDA,P,V)
      DIMENSION X(N),Y(N),P(N,N),H(N,N),RO(N,N),EPS(N,N),EDA(N,N),V(N,N)
      COMMON /COM1/ENDA,A1,A2,A3,Z,HM0,DH/COMAK/AK(0:65,0:65)
      COMMON /COMEK/EK,BX,BY,PTR
      DATA PAI,PAI1/3.14159265,0.2026423/
      CALL VI(N,DX,P,V)
```

```
      HMIN=1.E3
      DO 30 I=1,N
      DO 30 J=1,N
      RAD=X(I)*X(I)+EK*Y(J)*Y(J)
      W1=0.5*RAD
      H0=W1+V(I,J)
      IF(H0.LT.HMIN)HMIN=H0
30    H(I,J)=H0
      IF(KK.EQ.0)THEN
      KK=1
      DH=0.01*HM0
      H00=-HMIN+HM0
      ENDIF
      W1=0.0
      DO 32 I=1,N
      DO 32 J=1,N
32    W1=W1+P(I,J)
      W1=DX*DX*W1/G0
      DW=1.-W1
      IF(DW.LT.0.0)H00=H00+DH
      IF(DW.GT.0.0)H00=H00-DH
      DO 60 I=1,N
      DO 60 J=1,N
      H(I,J)=H00+H(I,J)
      IF(P(I,J).LT.0.0)P(I,J)=0.0
      EDA1=EXP(A1*(-1.+(1.+A2*P(I,J))**Z))
      EDA(I,J)=EDA1
      RO(I,J)=(A3+1.34*P(I,J))/(A3+P(I,J))
60    EPS(I,J)=RO(I,J)*H(I,J)**3/(ENDA*EDA1)
      RETURN
      END

      SUBROUTINE ITER(N,KK,DX,G0,X,Y,H,RO,EPS,EDA,P,V)
      DIMENSION X(N),Y(N),P(N,N),H(N,N),RO(N,N),EPS(N,N),EDA(N,N),V(N,N)
      COMMON /COMAK/AK(0:65,0:65)
      COMMON /COMEK/EK,BX,BY,PTR
      DATA KG1,PAI/0,3.14159265/
      DATA UX,UY/1.0,1.0/
      IF(KG1.NE.0)GOTO 2
      KG1=1
      AK00=AK(0,0)
      AK10=AK(1,0)
      AK01=AK(0,1)
2     DO 100 K=1,KK
      DO 70 J=2,N-1
      J0=J-1
      J1=J+1
      D2=0.5*(EPS(1,J)+EPS(2,J))
      DO 70 I=2,N-1
      I0=I-1
      I1=I+1
```

```
      D1=D2
      D2=0.5*(EPS(I1,J)+EPS(I,J))
      D4=0.5*(EPS(I,J0)+EPS(I,J))
      D5=0.5*(EPS(I,J1)+EPS(I,J))
      D6=UY**PTR*RO(I,J)*AK00
      D7=UY*PTR*RO(I,J0)*AK01
      D8=UX*PTR*RO(I,J)*AK00
      D9=UX*PTR*RO(I0,J)*AK10
      D10=D1+D2+D4+D5+(D6+D8)*DX-(D7+D9)*DX
      D11=D1*P(I0,J)+D2*P(I1,J)+D4*P(I,J0)+D5*P(I,J1)
      D12=((UX+UY)*RO(I,J)*H(I,J)-(D6+D8)*P(I,J)-UX*RO(I0,J)*H(I0,J)-
      UY*RO(I,J0)*H(I,J0)+(D7+D9)*P(I,J))*DX
      P(I,J)=(D11-D12)/D10
      IF(P(I,J).LT.0.0)P(I,J)=0.0
70    CONTINUE
      CALL HREE(N,DX,G0,X,Y,H,RO,EPS,EDA,P,V)
100   CONTINUE
      RETURN
      END

      SUBROUTINE VI(N,DX,P,V)
      DIMENSION P(N,N),V(N,N)
      COMMON /COMEK/EK,BX,BY,PTR
      COMMON /COMAK/AK(0:65,0:65)
      DO 40 I=1,N
      DO 40 J=1,N
      H0=0.0
      DO 30 K=1,N
      IK=IABS(I-K)
      DO 30 L=1,N
      JL=IABS(J-L)
30    H0=H0+AK(IK,JL)*P(K,L)
40    V(I,J)=H0*DX*PTR
      RETURN
      END
      SUBROUTINE SUBAK(MM)
      COMMON /COMAK/AK(0:65,0:65)
      S(X,Y)=X+SQRT(X**2+Y**2)
      DO 10 I=0,MM
      XP=I+0.5
      XM=I-0.5
      DO 10 J=0,I
      YP=J+0.5
      YM=J-0.5
      A1=S(YP,XP)/S(YM,XP)
      A2=S(XM,YM)/S(XP,YM)
      A3=S(YM,XM)/S(YP,XM)
      A4=S(XP,YP)/S(XM,YP)
      AK(I,J)=XP*ALOG(A1)+YM*ALOG(A2)+XM*ALOG(A3)+YP*ALOG(A4)
10    AK(J,I)=AK(I,J)
      RETURN
```

```
      END
      SUBROUTINE OUTPUT(N,DX,X,Y,H,P)
      DIMENSION X(N),Y(N),H(N,N),P(N,N)
      A=0.0
      WRITE(8,110)A,(Y(I),I=1,N)
      DO I=1,N
      WRITE(8,110)X(I),(H(I,J),J=1,N)
      ENDDO
      WRITE(10,110)A,(Y(I),I=1,N)
      DO I=1,N
      WRITE(10,110)X(I),(P(I,J),J=1,N)
      ENDDO
  110 FORMAT(66(E12.6,1X))
      RETURN
      END
       SUBROUTINE HERTZELLIPTIC(RX,RY,KA,KB)
       IMPLICIT NONE
       REAL*8, EXTERNAL :: EE,KE
       REAL*8 RX,RY,BPA,BMA,CTH,THT,PAI,E1,KA,KB
       DATA PAI/3.1415926/
       BPA=0.5*(1./RX+1./RY)
       BMA=0.5*ABS(1./RX-1./RY)
       CTH=BMA/BPA
       THT=ACOS(CTH)*180.0/PAI
       CALL CACUE(CTH,E1)
       KA=(2.*EE(E1)/(PAI*(1-E1**2)))**(1/3.)
       KB=KA*(1.-E1**2)**(1/2.)
       END SUBROUTINE
       SUBROUTINE CACUE(CTH,E1)
       IMPLICIT NONE
       REAL*8, EXTERNAL :: EE,KE
       REAL*8, EXTERNAL :: FAB
       INTEGER FLG,I
      REAL*8 PAI,CTH,E1,E11,E12,DX,A,B,A1,A2,A3,A4,A5,T1,T2,T3,T4,T5,ER0
      DATA PAI,DX,FLG,I,T1,T5,ER0/3.1415926,0.0001,1,1,1.E-30,1.,1.E-12/
       IF(CTH.LT.1.E-6)THEN
       E1=0.
       RETURN
       ENDIF
       IF(CTH.GT.0.9999999999)THEN
       E1=1.
       RETURN
       ENDIF
       A1=FAB(T1,CTH)
       A5=FAB(T5,CTH)
       DO WHILE(FLG.EQ.1)
       T3=T1+I*DX
       A3=FAB(T3,CTH)
       I=I+1
       IF((A1*A3.LT.0.).AND.(A3*A5.LT.0.)) THEN
       FLG=0
```

```
END IF
END DO
DO WHILE((T3-T1).GT.ER0)
T2=(T1+T3)/2.
A2=FAB(T2,CTH)
IF(A2.GT.0.) T1=T2
IF(A2.LT.0.) T3=T2
IF(A2.EQ.0.)THEN
 E11=T2
 EXIT
END IF
END DO
E11=T2
DO WHILE((T5-T3).GT.ER0)
T4=(T3+T5)/2.
A4=FAB(T4,CTH)
IF(A4.GT.0.) T5=T4
IF(A4.LT.0.) T3=T4
IF(A4.EQ.0.)THEN
 E12=T2
 EXIT
END IF
END DO
E12=T4
E1=E11
IF(E11.LT.E12) E1=E12
RETURN
END SUBROUTINE
REAL*8 FUNCTION FAB(E1,CTH)
IMPLICIT NONE
REAL*8 E1,CTH,T1,T2
REAL*8, EXTERNAL :: EE,KE
T1=EE(E1)
T2=KE(E1)
FAB=2*(1-E1**2)*(T1-T2)+(1.-CTH)*E1**2*T1
END FUNCTION

REAL*8 FUNCTION KE(E1)
IMPLICIT NONE
INTEGER N,I,FLG
REAL*8 E1,PAI,H,T,T1,T2,S1,S2,P,Q
PAI=3.1415926
IF(E1.EQ.1) THEN
KE=1.E10
RETURN
ENDIF
IF(E1.LT.1.E-20) THEN
KE=PAI/2.
RETURN
ENDIF
N=1
```

```
H=PAI/2.
Q=SQRT(1.-E1*E1*SIN(H)*SIN(H))
IF(Q.LT.1.E-35) Q=1.E35
Q=1./Q
T1=.5*H*(1+Q)
S1=T1
FLG=1
DO WHILE(FLG.EQ.1)
P=0.
DO I=0,N-1
T=(I+0.5)*H
Q=SQRT(1.-E1*E1*SIN(T)*SIN(T))
IF(Q.LT.1.E-35) Q=1.E35
Q=1./Q
P=P+Q
END DO
T2=(T1+H*P)/2.
S2=(4.*T2-T1)/3.
IF(ABS(S2-S1).LT.ABS(S2)*1.E-7) FLG=0
T1=T2
S1=S2
N=N+N
H=.5*H
END DO
KE=S2
RETURN
END FUNCTION
REAL*8 FUNCTION EE(E1)
IMPLICIT NONE
INTEGER N,I,FLG
REAL*8 E1,PAI,H,T,T1,T2,S1,S2,P,Q
PAI=3.1415926
N=1
H=PAI/2.
IF(E1.EQ.1) THEN
EE=1.
RETURN
ENDIF

IF(E1.LT.1.E-20) THEN
EE=PAI/2.
RETURN
ENDIF
Q=SQRT(1.-E1*E1*SIN(H)*SIN(H))
T1=.5*H*(1+Q)
S1=T1

FLG=1
DO WHILE(FLG==1)
P=0.
DO I=0,N-1
```

```
T=(I+0.5)*H
Q=SQRT(1.-E1*E1*SIN(T)*SIN(T))
P=P+Q
END DO
T2=(T1+H*P)/2.
S2=(4.*T2-T1)/3.
IF(ABS(S2-S1).LT.ABS(S2)*1.E-7) FLG=0
T1=T2
S1=S2
N=N+N
H=.5*H
END DO
EE=S2
RETURN
END FUNCTION
```

9.3.3 Example

9.3.3.1 Preassignment parameters

Node number	N = 65 × 65
Nondimensional starting point coordinate in *x* direction	X0 = −2.5
Nondimensional ending point coordinate in *x* direction	XE = 1.5
Equivalent modulus of elasticity	E1 = 2.21E11 Pa
Initial viscosity	EDA0 = 0.03 Pa·s
Radii	RX = 0.3 m, RY = 0.6 m
Load	W0 = 1000 N
Velocity	US = 1.2 m·s^{-1}
Velocity coefficient in *x* direction	UX = 1.0
Velocity coefficient in *y* direction	UY = 0.

9.3.3.2 Output parameters

The pressure array P(I,J) will be saved in the file PRESSURE.DAT;
The film thickness array H(I,J) will be saved in the file FILM.DAT;
Other data will be saved in the file OUT.DAT.

If the user needs to change the above parameters, he or she needs to recompile and relink the program before executing it.

10

Numerical calculation method and program for thermal EHL

10.1 Basic equations for thermal EHL

10.1.1 Thermal EHL in line contact

The basic equations for the thermal elastohydrodynamic lubrication (TEHL) in the line contact include all the equations of EHL and the energy equation as well as the temperature conditions of the up- and downsurfaces. The main equations in the dimensional, nondimensional, and discrete forms are as follows:

10.1.1.1 Reynolds equation

$$\frac{d}{dx}\left(\frac{\rho h^3}{\eta}\frac{dp}{dx}\right) = 12u_s\frac{d(\rho h)}{dx} \tag{10.1}$$

The nondimensional Reynolds equation is as follows:

$$\frac{d}{dX}\left(\varepsilon\frac{dP}{dX}\right) - \frac{d(\rho^* H)}{dX} = 0 \tag{10.2}$$

Numerical Calculation of Elastohydrodynamic Lubrication: Methods and Programs, First Edition.
Ping Huang.

The boundary conditions are as follows:

At the inlet $P(X_0)=0$
At the outlet $P(X_e)=0$ and $dP(X_e)/dX=0$

By using the center and the forward differential discretization equations, we can obtain the discrete differential Reynolds equation:

$$\frac{\varepsilon_{i-1/2}P_{i-1}-(\varepsilon_{i-1/2}+\varepsilon_{i+1/2})P_i+\varepsilon_{i+1/2}P_{i+1}}{\Delta X^2}=\frac{\rho_i^* H_i-\rho_{i-1}^* H_{i-1}}{\Delta X} \tag{10.3}$$

10.1.1.2 Energy equation

$$c_p\rho\left(u\frac{\partial T}{\partial x}-w\frac{\partial T}{\partial z}\right)=k\frac{\partial^2 T}{\partial z^2}-\frac{T}{\rho}\frac{\partial \rho}{\partial T}\left(u\frac{\partial p}{\partial x}\right)+\eta\left(\frac{\partial u}{\partial z}\right)^2 \tag{10.4}$$

Since $\partial^2 T/\partial z^2$ is the second derivative, the two boundary conditions should be given for solving Equation 10.4. Usually, they can be expressed as follows [7]:

$$\begin{cases} T(x,0)=\dfrac{k}{\sqrt{\pi\rho_1 c_1 k_1 u_1}}\displaystyle\int_{-\infty}^{x}\left.\frac{\partial T}{\partial z}\right|_{x,0}\frac{ds}{\sqrt{x-s}}+T_0 \\ T(x,h)=\dfrac{k}{\sqrt{\pi\rho_2 c_2 k_2 u_2}}\displaystyle\int_{-\infty}^{x}\left.\frac{\partial T}{\partial z}\right|_{x,h}\frac{ds}{\sqrt{x-s}}+T_0 \end{cases} \tag{10.5}$$

The nondimensional and discrete forms of Equations 10.4 and 10.5 will be discussed in Section 10.2.

10.1.1.3 Film thickness equation

$$h(x)=h_0+\frac{x^2}{2R}-\frac{2}{\pi E}\int_{s_1}^{s_2}p(s)\ln(s-x)^2 ds+c \tag{10.6}$$

The nondimensional film thickness equation is as follows:

$$H(X)=H_0+\frac{X^2}{2}-\frac{1}{\pi}\int_{X_0}^{X_e}\ln|X-X'|P(X')dX' \tag{10.7}$$

The discrete film thickness equation is as follows:

$$H_i=H_0+\frac{X_i^2}{2}-\frac{1}{\pi}\sum_{j=1}^{n}K_{ij}P_j \tag{10.8}$$

10.1.1.4 Roelands viscosity–pressure–temperature equation

$$\eta=\eta_0\exp\left\{(\ln\eta_0+9.67)\left[\left(1+\frac{p}{p_0}\right)^z\left(\frac{T-138}{T_0-138}\right)^{-1.1}-1\right]\right\} \tag{10.9}$$

The nondimensional viscosity–pressure–temperature equation is as follows:

$$\eta^*=\exp\left\{(\ln\eta_0+9.67)\left[\left(1+\frac{p_H P}{p_0}\right)^z\left(\frac{T-138}{T_0-138}\right)^{-1.1}-1\right]\right\} \tag{10.10}$$

10.1.1.5 Density–pressure–temperature equation

$$\rho=\rho_0\left[1+\frac{0.6p}{1+1.7p}+D(T-T_0)\right] \tag{10.11}$$

The nondimensional density–pressure–temperature equation is as follows:

$$\rho^*=1+\frac{0.6p}{1+1.7p}+D(T-T_0) \tag{10.12}$$

10.1.1.6 Load-balancing equation

The load-balancing equation can be expressed by the following:

$$\int_{x_0}^{x_e}p(x)dx=w \tag{10.13}$$

The nondimensional form is as follows:

$$\int_{X_0}^{X_e} P(X)dX = \frac{\pi}{2} \tag{10.14}$$

After discretization, we obtain the following:

$$\Delta X \sum_{i=1}^{n} \frac{P_i + P_{i+1}}{2} = \frac{\pi}{2} \tag{10.15}$$

Furthermore, in order to calculate the temperature, not only the average velocity u_s of the surfaces must be known, but the surface velocities u_1 and u_2 should also be calculated. If the slip–roll ratio s is known, we can use the following equation to calculate them:

$$\begin{aligned} u_1 &= \frac{1}{2}(2+s)u_s \\ u_2 &= \frac{1}{2}(2-s)u_s \end{aligned} \tag{10.16}$$

Because the temperature in the energy equation, the viscosity in the viscosity–pressure–temperature equation, and the elastic deformation in the film thickness equation vary with pressure, one should give an initial pressure distribution (such as the Hertz contact pressure distribution) and a temperature distribution (such as uniform temperature distribution) first. Then, use Equations 10.6 and 10.9 to calculate the new thickness and new viscosity. Substitute them into the Reynolds equation to solve a new pressure distribution iteratively and then substitute it into the energy equation to correct the temperature further. Then, use the new temperature to correct the viscosity by Equation 10.9. In addition, use Equation 10.1 to iterate the pressure again. Repeat these steps until the pressure difference between the two neighbor iterations is less than a given error. Then, one can finally obtain the final pressure distribution, the film thickness, and the temperature distribution of the TEHL in the line contact.

10.1.2 TEHL in point contact

The basic equations for TEHL in the point contact include all the equations of EHL in the point contact and the three-dimensional energy equation (in the two velocity directions and the film thickness direction). These main equations are as follows.

10.1.2.1 Reynolds equation

$$\frac{\partial}{\partial x}\left(\frac{\rho h^3}{\eta}\frac{\partial p}{\partial x}\right)+\frac{\partial}{\partial y}\left(\frac{\rho h^3}{\eta}\frac{\partial p}{\partial y}\right)=12u_s\frac{\partial(\rho h)}{\partial x} \tag{10.17}$$

The nondimensional Reynolds equation is as follows:

$$\frac{\partial}{\partial X}\left[\varepsilon\frac{\partial P}{\partial X}\right]+\frac{\partial}{\partial Y}\left[\varepsilon\frac{\partial P}{\partial Y}\right]-\frac{\partial(\rho^* H)}{\partial X}=0 \tag{10.18}$$

The boundary conditions are as follows:

At the inlet $P(X_0,Y)=0$
At the outlet $P(X_e,Y)=0$ and $\partial P(X_e,Y)/\partial X=0$
At the two sides $P|_{Y=\pm 1}=0$

By using the center and the forward differential discretization equations, we can obtain the discrete differential Reynolds equation:

$$\frac{\varepsilon_{i-1/2,j}P_{i-1,j}+\varepsilon_{i+1/2,j}P_{i+1,j}+\varepsilon_{i,j-1/2}P_{i,j-1}+\varepsilon_{i,j+1/2}P_{i,j+1}-\varepsilon_0 P_{ij}}{\Delta X^2}=\frac{\rho^*_{ij}H_{ij}-\rho^*_{i-1,j}H_{i-1,j}}{\Delta X} \tag{10.19}$$

10.1.2.2 Energy equation

$$c_p\rho\left(u\frac{\partial T}{\partial x}+v\frac{\partial T}{\partial y}-w\frac{\partial T}{\partial z}\right)=k\frac{\partial^2 T}{\partial z^2}-\frac{T}{\rho}\frac{\partial\rho}{\partial T}\left(u\frac{\partial p}{\partial x}+v\frac{\partial p}{\partial y}\right)+\eta\left[\left(\frac{\partial u}{\partial z}\right)^2+\left(\frac{\partial v}{\partial z}\right)^2\right] \tag{10.20}$$

Since $\partial^2 T/\partial z^2$ is of the second derivative, two boundary conditions should be given for solving the energy equation. Usually, they can be expressed as follows [7]:

$$\begin{cases} T(x,y,0)=\dfrac{k}{\sqrt{\pi\rho_1 c_1 k_1 u_1}}\displaystyle\int_{-\infty}^{x}\left.\frac{\partial T}{\partial z}\right|_{x,y,0}\frac{ds}{\sqrt{x-s}}+T_0 \\ T(x,y,h)=\dfrac{k}{\sqrt{\pi\rho_2 c_2 k_2 u_2}}\displaystyle\int_{-\infty}^{x}\left.\frac{\partial T}{\partial z}\right|_{x,y,h}\frac{ds}{\sqrt{x-s}}+T_0 \end{cases} \tag{10.21}$$

The nondimensional and discrete forms of Equations 10.20 and 10.21 will be discussed in Section 10.2.

10.1.2.3 Film thickness equation

$$h(x,y) = h_0 + \frac{x^2}{2R_x} + \frac{y^2}{2R_y} + \frac{2}{\pi E}\iint_{\Omega} \frac{p(s,t)}{\sqrt{(x-s)^2 + (y-t)^2}} ds dt \tag{10.22}$$

The nondimensional film thickness equation is as follows:

$$H(X,Y) = H_0 + \frac{X^2}{2} + \frac{Y^2}{2} + \frac{2}{\pi^2}\int_{X_0}^{X_e}\int_{Y_0}^{Y_e} \frac{P(S,T) dS dT}{\sqrt{(X-S)^2 + (Y-T)^2}} \tag{10.23}$$

The discrete film thickness equation is as follows:

$$H_{ij} = H_0 + \frac{X_i^2 + Y_j^2}{2} + \frac{2}{\pi^2}\sum_{k=1}^{n}\sum_{l=1}^{m} D_{ij}^{kl} P_{kl} \tag{10.24}$$

The viscosity–pressure–temperature equation and density–pressure–temperature equation in the point contact are the same as Equations 10.9–10.12.

10.1.2.4 Load-balancing equation

The load balancing equation can be expressed as follows:

$$\int_{x_0}^{x_e}\int_{y_0}^{y_e} p(x,y) dx dy = w \tag{10.25}$$

Its nondimensional form is as follows:

$$\int_{X_0}^{X_e}\int_{Y_0}^{Y_e} P(X,Y) dX dY = \frac{2\pi}{3} \tag{10.26}$$

After discretization, we obtain the following:

$$\Delta X \Delta Y \sum_{i=1}^{n} \sum_{j=1}^{m} P_{ij} = \frac{2\pi}{3} \tag{10.27}$$

In order to calculate the temperature, not only the average velocity u_s of two surfaces must be known, but velocities u_1 and u_2 of both surfaces should also be calculated. The iteration method is the same as that mentioned in the line contact.

10.2 Viscosity and temperature across film thickness

10.2.1 Calculation of fluid velocity field

In order to solve the continuity equation and the energy equation, the fluid velocity field of the film under the given pressure distribution and film thickness should be known. The fluid velocity field can be obtained by integrating the equations of motion. The flow rates (considering viscosity change in the z direction) in the x and y directions can be written as follows [7]:

$$u = u_1 + \frac{\partial p}{\partial x}\left(\int_0^z \frac{z}{\eta}dz - \frac{\int_0^h (z/\eta)dz}{\int_0^h (1/\eta)dz}\int_0^z \frac{dz}{\eta}\right) + \frac{u_2 - u_1}{\int_0^h (1/\eta)dz}\int_0^z \frac{dz}{\eta} \tag{10.28}$$

$$v = \frac{\partial p}{\partial y}\left(\int_0^z \frac{z}{\eta}dz - \frac{\int_0^h (z/\eta)dz}{\int_0^h (1/\eta)dz}\int_0^z (dz/\eta)\right) \tag{10.29}$$

The nondimensional forms of the flow rates are as follows:

$$U^* = U_1^* + \frac{p_H a^3}{R_x^3 E}\frac{\partial P}{\partial X}\left(\int_0^Z \frac{Z}{\eta^*}dZ - \frac{\int_0^H (Z/\eta^*)dZ}{\int_0^H (1/\eta^*)dZ}\int_0^Z \frac{1}{\eta^*}dZ\right) + \frac{U_2^* - U_1^*}{\int_0^H (1/\eta^*)dZ}\int_0^Z \frac{1}{\eta^*}dZ \tag{10.30}$$

$$V^* = \frac{p_H a^3}{R_x^3 E}\frac{\partial P}{\partial Y}\left(\int_0^Z \frac{Z}{\eta^*}dZ - \frac{\int_0^H (Z/\eta^*)dZ}{\int_0^H (1/\eta^*)dZ}\int_0^Z \frac{1}{\eta^*}dZ\right) \tag{10.31}$$

where, a stands for the half-width of the contact zone in the line contact or the radius of the contact circle in the point contact, R_x stands for the equivalent curvature radius of the surfaces in the x direction in the line or point contacts, w stands for the fluid velocity in the z direction, and W^* stands for the nondimensional fluid velocity.

For the convenience in calculation, the nondimensional Z should be unitized as $\bar{Z}=Z/H$. If $\eta_1(\bar{Z})=\int_0^{\bar{Z}}(1/\eta^*)d\bar{Z}$ and $\eta_2(\bar{Z})=\int_0^{\bar{Z}}(\bar{Z}/\eta^*)d\bar{Z}$, we have $\eta_1(1)=\int_0^1(1/\eta^*)d\bar{Z}$ and $\eta_2(1)=\int_0^1(\bar{Z}/\eta^*)d\bar{Z}$. Then, Equations 10.30 and 10.31 can be written as follows:

$$U^*=U_1^*+\frac{p_H a^3 H^2}{R_x^3 E}\frac{\partial P}{\partial X}\left(\eta_2(\bar{Z})-\frac{\eta_2(1)}{\eta_1(1)}\eta_1(\bar{Z})\right)+\frac{U_2^*-U_1^*}{\eta_1(1)}\eta_1(\bar{Z}) \tag{10.32}$$

$$V^*=\frac{p_H a^3 H^2}{R_x^3 E}\frac{\partial P}{\partial Y}\left(\eta_2(\bar{Z})-\frac{\eta_2(1)}{\eta_1(1)}\eta_1(\bar{Z})\right) \tag{10.33}$$

In the program of point contact, the above equations are U(K) and V(K) in Subroutine UCAL, whereas in the program of line contact, there is only U(K).

$\eta_1(\bar{Z})$ and $\eta_2(\bar{Z})$ are calculated in DO 40 loop in Subroutine EROEQ. Since $\bar{Z}$ is a unitized parameter, and the number of nodes in the thickness direction is N_Z, then the length between nodes is $1/(N_Z-1)$ and the node values are $\bar{Z}(K)=(K-1)/(N_Z-1)$, where $K=1,\ldots,N_Z$. Obviously, $\eta_1(K=1)=0$ and $\eta_2(K=1)=0$ hold. The integration between two nodes is obtained by multiplying the average values of the nodes by the node length in the z direction. When $K=2,\ \ldots,N_Z,\eta_1(\bar{Z})$ and $\eta_2(\bar{Z})$ can be discreted as follows:

$$\eta_1(\bar{Z})=\eta_1(\bar{Z}(K))=\frac{1}{N_Z-1}\sum_{i=2}^{K}\frac{1}{2}\left(\frac{1}{\eta^*(i-1)}+\frac{1}{\eta^*(i)}\right) \tag{10.34}$$

$$\eta_2(\bar{Z})=\eta_2(\bar{Z}(K))=\frac{1}{N_Z-1}\sum_{i=2}^{K}\frac{1}{2N_Z-1}\left(\frac{i\text{-}1}{\eta^*(i-1)}+\frac{i\text{-}2}{\eta^*(i)}\right) \tag{10.35}$$

The above two equations are EDA1(K) and EDA2(K) in the program of Subroutine EROEQ. We have $\eta_1(1)=\eta_1(\bar{Z}=1)=\eta_1(\bar{Z}(N_Z))$ and $\eta_2(1)=\eta_2(\bar{Z}=1)=\eta_2(\bar{Z}(N_Z))$. The value of $\eta^*(i)$ can be calculated by using Equation 10.10. The temperature value of the current node T1(K) is obtained by the previous iteration.

The partial derivatives of Equations 10.28 and 10.29 are as follows:

$$\frac{\partial u}{\partial z}=\frac{\partial p}{\partial x}\frac{1}{\eta}\left(z-\frac{\int_0^h (z/\eta)dz}{\int_0^h (1/\eta)dz}\right)+\frac{1}{\eta}\frac{u_2-u_1}{\int_0^h (1/\eta)dz} \tag{10.36}$$

$$\frac{\partial v}{\partial z}=\frac{\partial p}{\partial y}\frac{1}{\eta}\left(z-\frac{\int_0^h (z/\eta)dz}{\int_0^h (1/\eta)dz}\right) \tag{10.37}$$

By using the unitized parameter $\bar{Z}=Z/H$, the nondimensional forms of the above equations are as follows:

$$\frac{\partial U^*}{\partial Z}=\frac{p_{\mathrm{H}}a^3H}{R_x^3E}\frac{\partial P}{\partial X}\frac{1}{\eta^*}\left(\bar{Z}-\frac{\eta_2(1)}{\eta_1(1)}\right)+\frac{U_2^*-U_1^*}{H}\frac{1}{\eta_1(1)}\frac{1}{\eta^*} \tag{10.38}$$

$$\frac{\partial V^*}{\partial Z}=\frac{p_{\mathrm{H}}a^3H}{R_x^3E}\frac{\partial P}{\partial Y}\frac{1}{\eta^*}\left(\bar{Z}-\frac{\eta_2(1)}{\eta_1(1)}\right) \tag{10.39}$$

The above equations are DU(K) and DV(K) in Subroutine UCAL in the program of point contact and only DU(K) in the program of line contact.

It should be noted that $\partial U^*/\partial Z$ and $\partial V^*/\partial Z$ (will be used when solving energy equation), instead of $\partial U^*/\partial \bar{Z}$ and $\partial V^*/\partial \bar{Z}$, are required here. The introduction of $\bar{Z}$ is just for convenience of the integration calculation of $\eta_1(\bar{Z})$ and $\eta_2(\bar{Z})$.

10.2.2 Continuity equation

The continuity equation for point contact is as follows:

$$\frac{\partial(\rho u)}{\partial x}+\frac{\partial(\rho v)}{\partial y}=\frac{\partial(\rho w)}{\partial z} \tag{10.40}$$

The nondimensional form is as follows:

$$\frac{\partial(\rho^* U^*)}{\partial X}+\frac{\partial(\rho^* V^*)}{\partial Y}=\frac{R_x}{a}\frac{\partial(\rho^* W^*)}{\partial Z} \tag{10.41}$$

The discrete form is as follows:

$$\frac{\rho^*_{i,j,k}U^*_{i,j,k}-\rho^*_{i-1,j,k}U^*_{i-1,j,k}}{\Delta X}+\frac{\rho^*_{i,j,k}V^*_{i,j,k}-\rho^*_{i,j-1,k}V^*_{i,j-1,k}}{\Delta Y}=\frac{R_x}{a}\frac{\rho^*_{i,j,k}W^*_{i,j,k}-\rho^*_{i,j,k-1}W^*_{i,j,k-1}}{\Delta Z} \tag{10.42}$$

By using the aforementioned equations $\Delta Z=H/(N_Z-1)$ and $\Delta X=\Delta Y$, Equation 10.42 can be rewritten as follows:

$$W^*_{i,j,k}=\frac{1}{\rho^*_{i,j,k}}\left[\rho^*_{i,j,k-1}W^*_{i,j,k-1}+\frac{aH}{(N_Z-1)R_x\Delta X}\left(\rho^*_{i,j,k}\left(U^*_{i,j,k}+V^*_{i,j,k}\right)\right.\right.$$
$$\left.\left.-\rho^*_{i-1,j,k}U^*_{i-1,j,k}-\rho^*_{i,j-1,k}V^*_{i,j-1,k}\right)\right] \tag{10.43}$$

The above equation is W(K) in Subroutine UCAL in the program of point contact. The continuity equation for line contact is as follows:

$$\frac{\partial(\rho u)}{\partial x}=\frac{\partial(\rho w)}{\partial z} \tag{10.44}$$

The discrete form is as follows:

$$W^*_{i,k}=\frac{1}{\rho^*_{i,k}}\left[\rho^*_{i,k-1}W^*_{i,k-1}+\frac{aH}{(N_Z-1)R_x\Delta X}\left(\rho^*_{i,k}U^*_{i,k}-\rho^*_{i-1,k}U^*_{i-1,k}\right)\right] \tag{10.45}$$

The above equation is W(K) in Subroutine UCAL in the program of line contact.

10.2.3 Energy equation

The energy equation for TEHL in the point contact is as follows:

$$c_p\rho\left(u\frac{\partial T}{\partial x}+v\frac{\partial T}{\partial y}-w\frac{\partial T}{\partial z}\right)=k\frac{\partial^2 T}{\partial z^2}-\frac{T}{\rho}\frac{\partial\rho}{\partial T}\left(u\frac{\partial p}{\partial x}+v\frac{\partial p}{\partial y}\right)+\eta\left[\left(\frac{\partial u}{\partial z}\right)^2+\left(\frac{\partial v}{\partial z}\right)^2\right] \tag{10.46}$$

The nondimensional form is as follows:

$$\frac{c_p\rho_0 Ea^3}{\eta_0 kR_X}\left(\rho^* U^*\frac{\partial T^*}{\partial X}+\rho^* V^*\frac{\partial T^*}{\partial Y}\right)-\frac{c_p\rho_0 Ea^2}{\eta_0 k}\rho^* W^*\frac{\partial T^*}{\partial Z}=\frac{\partial^2 T^*}{\partial Z^2}$$
$$-\frac{Ep_{\mathrm{H}}a^3}{k\eta_0 T_0 R_x}\frac{\partial \rho^*}{\partial T^*}\frac{1}{\rho^*}\left(U^*\frac{\partial P}{\partial X}+V^*\frac{\partial P}{\partial Y}\right)T^*+\frac{E^2R_x{}^2}{k\eta_0 T_0}\left[\left(\frac{\partial U^*}{\partial Z}\right)^2+\left(\frac{\partial V^*}{\partial Z}\right)^2\right]\eta^* \tag{10.47}$$

where, $(\partial\rho^*/\partial T^*)=DT_0$ is obtained by using Equation 10.12.

Let $A_2=(-c\rho_0 Ea^3/\eta_0 kR_x)$, $A_3=(-Ep_{\mathrm{H}}a^3 D/k\eta_0 R_x)$, $A_4=\left(-E^2R_x^2/k\eta_0 T_0\right)$, and $A_5=(-c\rho_0 Ea^2/2\eta_0 k)$, which are the variables A2–A5 in Subroutine TCAL in the program of point contact. Then, Equation 10.47 can be rewritten as follows:

$$\frac{\partial^2 T^*}{\partial Z^2}-2A_5\rho^* W^*\frac{\partial T^*}{\partial Z}+A_3\left(U^*\frac{\partial P}{\partial X}+V^*\frac{\partial P}{\partial Y}\right)\frac{T^*}{\rho^*}+A_2\left(\rho^* U^*\frac{\partial T^*}{\partial X}+\rho^* V^*\frac{\partial T^*}{\partial Y}\right)$$
$$=A_4\left[\left(\frac{\partial U^*}{\partial Z}\right)^2+\left(\frac{\partial V^*}{\partial Z}\right)^2\right]\eta^* \tag{10.48}$$

The above equation can also be written as follows:

$$\frac{1}{(\Delta Z)^2}\left(T_{k-1}^*-2T_k^*+T_{k+1}^*\right)-A_5\rho_k^* W_k^*\frac{1}{\Delta Z}\left(T_{k+1}^*-T_{k-1}^*\right)$$
$$+A_3\left(U_k^*\frac{\partial P}{\partial X}+V_k^*\frac{\partial P}{\partial Y}\right)\frac{T_k^*}{\rho_k^*}+A_2\left(\rho_k^* U_k^*\frac{T_{i,j,k}^*-T_{i-1,j,k}^*}{\Delta X}+\rho_k^* V_k^*\frac{T_{i,j,k}^*-T_{i,j-1,k}^*}{\Delta Y}\right)$$
$$=A_4\left[\left(\frac{\partial U^*}{\partial Z}\right)^2+\left(\frac{\partial V^*}{\partial Z}\right)^2\right]\eta_k^* \tag{10.49}$$

For simplicity, the subscripts i and j of the current node in the above equation are omitted. It can be rewritten as follows:

$$A_{1,k}T_{k-1}^*+A_{2,k}T_k^*+A_{3,k}T_{k+1}^*=A_{4,k} \tag{10.50}$$

where, $k=2,3,\ldots,N_Z-1,$

$$A_{1,k} = \frac{1}{(\Delta Z)^2} + A_5 \rho_k^* W_k^* \frac{1}{\Delta Z}$$

$$A_{2,k} = -2\frac{1}{(\Delta Z)^2} + A_2 \rho_k^* \frac{1}{\Delta X}\left(U_k^* + V_k^*\right) + A_3 \frac{1}{\rho_k^*}\left(U_k^* \frac{\partial P}{\partial X} + V_k^* \frac{\partial P}{\partial Y}\right)$$

$$A_{3,k} = \frac{1}{(\Delta Z)^2} - A_5 \rho_k^* W_k^* \frac{1}{\Delta Z}$$

$$A_{4,k} = A_4\left[\left(\frac{\partial U^*}{\partial Z}\right)^2 + \left(\frac{\partial V^*}{\partial Z}\right)^2\right]\eta_k^* + A_2 \rho_k^* \frac{1}{\Delta X}\left(U_k^* T_{i-1,j,k}^* + V_k^* T_{i,j-1,k}^*\right)$$

A (1, K)–A (4, K) are the arrays in Subroutine TCAL in the program of point contact. The energy equation for TEHL in the line contact is as follows:

$$c\rho\left(u\frac{\partial T}{\partial x} - w\frac{\partial T}{\partial z}\right) = k\frac{\partial^2 T}{\partial z^2} - \frac{T}{\rho}\frac{\partial \rho}{\partial T}u\frac{\partial p}{\partial x} + \eta\left(\frac{\partial u}{\partial z}\right)^2 \tag{10.51}$$

By using the nondimensional and discretization parameters, the discrete form of Equation 10.51 can be expressed as follows:

$$A_{1,k}T_{k-1}^* + A_{2,k}T_k^* + A_{3,k}T_{k+1}^* = A_{4,k} \tag{10.52}$$

where

$$A_{1,k} = \frac{1}{(\Delta Z)^2} + A_5 \rho_k^* W_k^* \frac{1}{\Delta Z}$$

$$A_{2,k} = -2\frac{1}{(\Delta Z)^2} + A_2 \rho_k^* U_k^* \frac{1}{\Delta X} + A_3 \frac{1}{\rho_k^*}\frac{\partial P}{\partial X}U_k^*$$

$$A_{3,k} = \frac{1}{(\Delta Z)^2} - A_5 \rho_k^* W_k^* \frac{1}{\Delta Z}$$

$$A_{4,k} = A_4\left(\frac{\partial U^*}{\partial Z}\right)^2 \eta_k^* + A_2 \rho_k^* U_k^* T_{i-1,k}^* \frac{1}{\Delta X}$$

A (1, K)–A (4, K) are arrays in Subroutine TCAL in the program of line contact.

10.2.4 Temperature boundary conditions

Equations 10.5 and 10.21 show that the temperature boundary conditions of the up and down surfaces in the line contact and point contact are in the same form. There are two dimensions (x and y) in the point contact. Since the variables do not change with y, and calculation is along the x direction. Both the situations can be treated in the same way. The equations of the boundary conditions of the up and down surfaces are in the same forms but have different parameters. Therefore, only the equations of the boundary conditions of the down surface will be discussed. The same approach can be used to calculate the ones of the up surface. The equation of the down surface can be expressed as follows:

$$T_1 = \frac{k}{\sqrt{\pi \rho_1 c_1 k_1 u_1}} \int_{-\infty}^{x} \frac{\partial T}{\partial z} \frac{ds}{\sqrt{x-s}} + T_0 \tag{10.53}$$

The nondimensional form is as follows:

$$T_1^* = 1 + \frac{k\sqrt{\eta_0 R_x}}{\sqrt{\pi \rho_1 c_1 u_1 E k_1 a^3}} \int_{-\infty}^{X} \frac{\partial T^*}{\partial Z} \frac{dS}{\sqrt{X-S}} \tag{10.54}$$

If let $A_6 = k\sqrt{\eta_0 R_x}/\sqrt{\pi \rho_1 c_1 u_1 E k_1 a^3}$ and use the integral transformation $X = N_X \Delta X$, then we will obtain the following:

$$T_1^* = 1 + A_6 \sqrt{\Delta X} \int_{0}^{N_X} \frac{\partial T^*}{\partial Z} \frac{dN_S}{\sqrt{N_X - N_S}} \tag{10.55}$$

Another form of $(\partial T^*/\partial Z)$ is $T^*(N_S,2) - T^*(N_S,1)/\Delta Z$, and $\Delta Z = H/(N_Z - 1)$ can be put out of the integral. Therefore, we will get the following:

$$T_1^* = 1 + A_6 \sqrt{\Delta X} \frac{N_Z - 1}{H} I_1 \tag{10.56}$$

where, $I_1 = \int_0^{N_X} [T^*(N_S,2) - T^*(N_S,1)] \frac{dN_S}{\sqrt{N_X - N_S}}$. There is a singular point at $N_S = N_X$ in this integral. It can be rewritten as follows:

$$\begin{aligned} I_1 = {} & \sum_{N_S=1}^{N_X-1} [T^*(N_S,2) - T^*(N_S,1)] \frac{1}{\sqrt{N_X - N_S}} + \frac{1}{6}[T^*(N_X-1,2) - T^*(N_X-1,1)] \\ & + \frac{2}{3} \int_{N_X-1}^{N_X} [T^*(N_X,2) - T^*(N_X,1)] \frac{dN_S}{\sqrt{N_X - N_S}} \end{aligned} \tag{10.57}$$

In this equation, the trapezoid formula is used in the step between N_X-1 and N_X. Since $T^*(N_X, 1)$ is unknown, we should further assume that $T^*(N_X,2)-T^*(N_X,1)$ will not change in the current step. That is, this expression can be put out of the integral.

The singular integrals can be calculated as $\int_{N_X-1}^{N_X} \frac{dN_S}{\sqrt{N_X-N_S}}=2$. Then, we have:

$$I_1=\sum_{N_S=1}^{N_X-2}[T^*(N_S,2)-T^*(N_S,1)]\frac{1}{\sqrt{N_X-N_S}}+\frac{7}{6}[T^*(N_X-1,2)-T^*(N_X-1,1)] \\ +\frac{4}{3}[T^*{}_2-T^*{}_1] \tag{10.58}$$

Let $I_1=C_{C1}+\frac{4}{3}\left[T_2^*-T_1^*\right]$ and substitute it into Equation 10.56, we will get the following:

$$T_1^*=1+A_6\sqrt{\Delta X}\frac{N_Z-1}{H}\left[C_{C1}+\frac{4}{3}\left(T_2^*-T_1^*\right)\right] \tag{10.59}$$

It can be rewritten as follows:

$$\left(1+\frac{4}{3}A_6\sqrt{\Delta X}\frac{N_Z-1}{H}\right)T_1^*+\left(-\frac{4}{3}A_6\sqrt{\Delta X}\frac{N_Z-1}{H}\right)T_2^*=1+A_6\sqrt{\Delta X}\frac{N_Z-1}{H}C_{C1} \tag{10.60}$$

This form can also be simply expressed as follows:

$$A_{21}T_1^*+A_{31}T_2^*=A_{41} \tag{10.61}$$

where, A_{21}, A_{31}, and A_{41} are A (2,1), A (3,1), and A (4,1) in Subroutine TCAL in the programs of line contact or point contact. In the program, $C_{C3}=A_6\sqrt{\Delta X}$, $DZ1=(N_Z-1)/H$, $C_{C5}=2/3$. The value of C_{C1} is calculated in Subroutine TBOUD.

The equation of the up surface can be analyzed in the same way. Then we have the following:

$$A_{1,N_Z}T_{N_Z-1}^*+A_{2,N_Z}T_{N_Z}^*=A_{4,N_Z} \tag{10.62}$$

where, $A_{1,N_Z}=1+\frac{4}{3}A_6\sqrt{\Delta X}\frac{N_Z-1}{H}$, $A_{2,N_Z}=-\frac{4}{3}A_6\sqrt{\Delta X}\frac{N_Z-1}{H}$, and $A_{4,N_Z}=1-A_6\sqrt{\Delta X}\frac{N_Z-1}{H}C_{C1}$ are A (1, N_Z), A (3, N_Z), and A (4, N_Z) in Subroutine TCAL in the programs of line contact or point contact.

10.2.5 *Calculation of linear equation set*

Equation 10.50 or 10.52 has N_Z-2 equations. We will get a set of linear equations with N_Z equations by adding Equations 10.61 and 10.62. Following, we will give the solution of the linear equation set by setting $N_Z=5$. This linear equation set can be expressed by the matrix multiplication.

$$\begin{bmatrix} A_{21} & A_{31} & & & \\ A_{12} & A_{22} & A_{32} & & \\ & A_{13} & A_{23} & A_{33} & \\ & & A_{14} & A_{24} & A_{34} \\ & & & A_{15} & A_{25} \end{bmatrix} \begin{bmatrix} T_1^* \\ T_2^* \\ T_3^* \\ T_4^* \\ T_5^* \end{bmatrix} = \begin{bmatrix} A_{41} \\ A_{42} \\ A_{43} \\ A_{44} \\ A_{45} \end{bmatrix} \tag{10.63}$$

The square matrix can be simplified as a down triangular matrix (by DO 10 loop in Subroutine TAR3 in the programs of line contact and point contact). It can be expressed as follows:

$$\begin{bmatrix} 1 & & & & \\ -B_{12} & 1 & & & \\ & -B_{13} & 1 & & \\ & & -B_{14} & 1 & \\ & & & -B_{15} & 1 \end{bmatrix} \begin{bmatrix} T_1^* \\ T_2^* \\ T_3^* \\ T_4^* \\ T_5^* \end{bmatrix} = \begin{bmatrix} B_{21} \\ B_{22} \\ B_{23} \\ B_{24} \\ B_{25} \end{bmatrix} \tag{10.64}$$

By using the calculation of matrix multiplication (by DO 20 loop in Subroutine TAR3 in the programs of line contact and point contact), all the temperature of the nodes in the z direction will be obtained. $T_k^*(k=1,2,\ldots,N_Z)$ is stored in the array D(N) in Subroutine TAR3.

The temperature of the nodes in the current iteration can be expressed as follows:

$$\left(T_k^*\right)_{\text{now}} = (1-\text{CC})\left(T_k^*\right)_{\text{pre}} + \text{CC}\cdot T_k^* \tag{10.65}$$

where, $\left(T_k^*\right)_{\text{pre}}$ is the temperature value obtained in the previous iteration and CC is the same as CT in the main program. Finally, $\left(T_k^*\right)_{\text{now}}$ will be stored in the array $T(N_Z)$.

10.2.6 *Program to calculate temperature*

The program to calculate temperature includes Subroutine THERM (main), Subroutine TBOUD (integral part of temperature boundary conditions of the up and down surfaces),

Subroutine EROEQ (integral part of viscosity), Subroutine UCAL (velocity distribution), subroutine TCAL (temperature iteration calculation), and Subroutine ERRO (error calculation).

The diagram is shown in Figure 10.1.

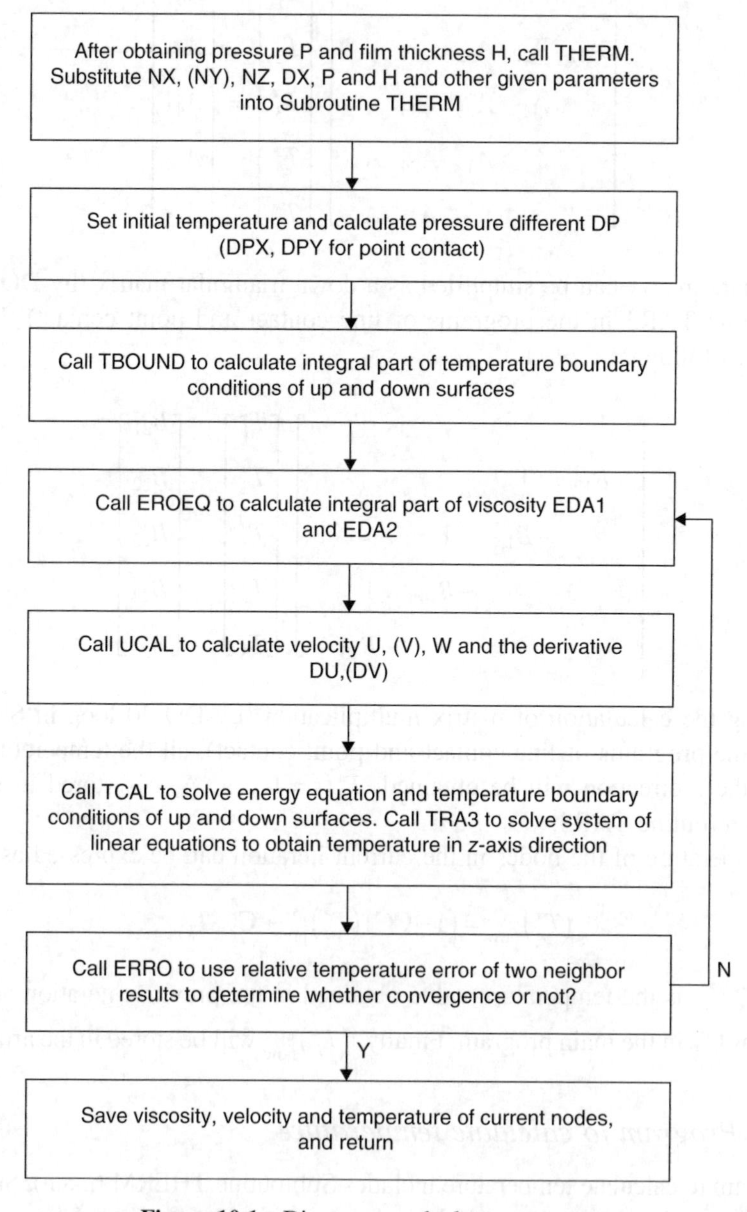

Figure 10.1 Diagram to calculate temperature

10.3 Numerical calculation method and program for TEHL in line contact

10.3.1 Flowchart

The flowchart to calculate TEHL in the line contact is shown in Figure 10.2.

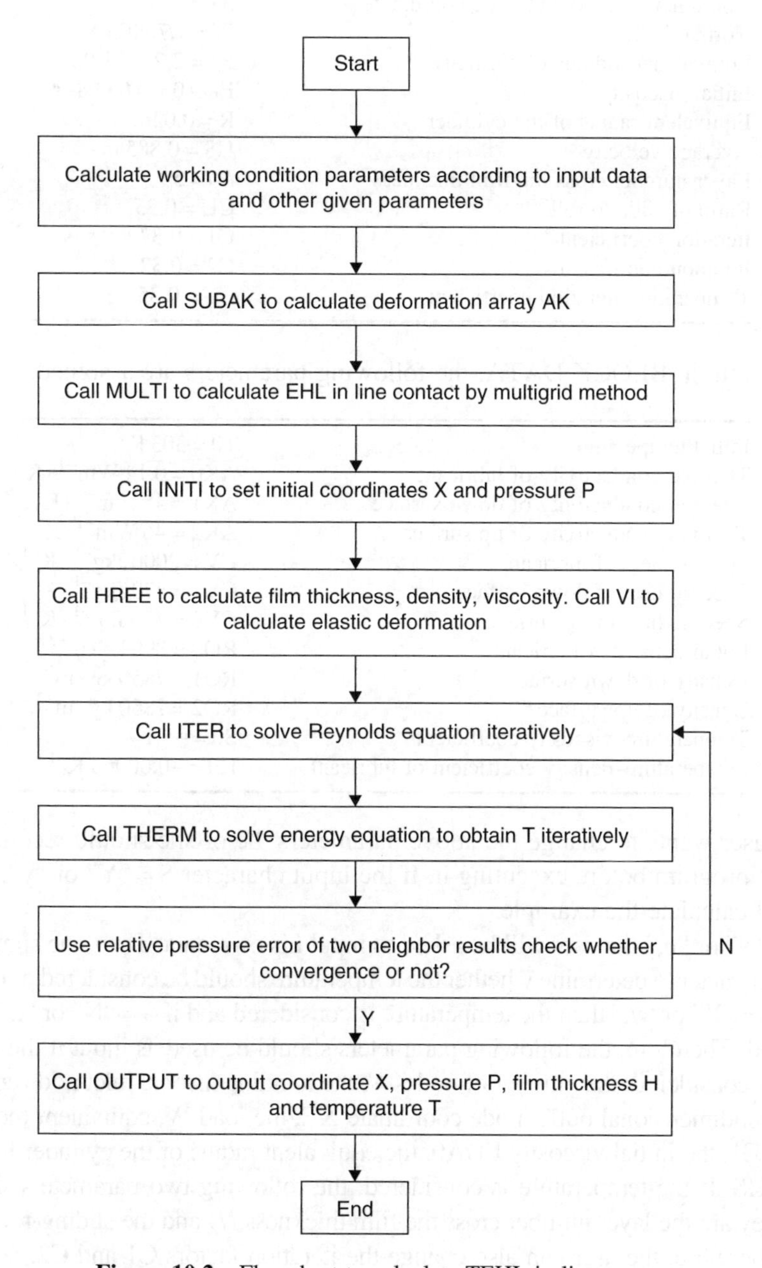

Figure 10.2 Flowchart to calculate TEHL in line contact

10.3.2 *Program*

Preassignment parameters

Number of nodes	$N = 129$
Nondimensional inlet node coordinate	X0 = −4.0
Nondimensional outlet node coordinate	XE = 1.4
Normal load	W = 1.768E5 N
Equivalent modulus of elasticity	E1 = 2.21E11 Pa
Initial viscosity	EDA0 = 0.03 Pa · s
Equivalent radius of the cylinder	R = 0.02m
Average velocity	US = 0.885 $m \cdot s^{-1}$
Layer number cross the film thickness	$N_Z = 5$
Ratio of slide to roll	CU = 0.25
Iteration coefficient	C1 = 0.37
Iteration coefficient	C2 = 0.37
Temperature iteration coefficient	CT = 0.35

In addition, in BLOCK DATA, the following parameters are assigned.

Initial temperature	T0 = 303 K
Thermal conductivity of lubricant	AK0 = 0.14 $Wm^{-1} \cdot K$
Thermal conductivity of down surface	AK1 = 46 $Wm^{-1} \cdot K$
Thermal conductivity of up surface	AK2 = 46 $Wm^{-1} \cdot K$
Specific heat of lubricant	CV = 2000 $Jkg^{-1} \cdot K^{-1}$
Specific heat of down surface	CV1 = 470 $Jkg^{-1} \cdot K^{-1}$
Specific heat of up surface	CV2 = 470 $Jkg^{-1} \cdot K^{-1}$
Initial density of lubricant	RO0 = 890 $kg \cdot m^{-3}$
Density of down surface	RO1 = 7850 $kg \cdot m^{-3}$
Density of up surface	RO2 = 7850 $kg \cdot m^{-3}$
Temperature–viscosity coefficient	S0 = −1.1
Temperature–density coefficient of lubricant	D0 = −0.00065 K^{-1}

If the user wants to change the above parameters, he or she should recompile and relink the program before executing it. If the input character S = "Y" or "y", the program will calculate the example.

If S = "N" or "n," the user will be asked to input his own data. Then one should input another character to determine whether the temperature should be considered or not. If the input is S = "Y" or "y," then the temperature is considered and if S = "N" or "n," it is not considered. Therefore, the following parameters should be used as input if the temperature is not considered: the number of nodes *N*, the nondimensional inlet node coordinate X0, the nondimensional outlet node coordinate XE, the load W, equivalent modulus of elasticity E1, the initial viscosity EDA0, the equivalent radius of the cylinder R, and the velocity US. If the temperature is considered, the following two parameters should be input. They are the layer number cross the film thickness N_Z and the sliding-rolling ratio CU. Furthermore, the user can also change the iteration factors C1 and C2.

```
PROGRAM LINEEHLT
    CHARACTER*1 S,S1,S2
    CHARACTER*16 FILEO,CDATE,CTIME
    COMMON /COM1/ENDA,A1,A2,A3,Z,C1,C2,C3,CW
    COMMON /COM2/T0,EDA0,AK0,AK1,AK2,CV,CV1,CV2,RO0,RO1,RO2,D0
    COMMON /COM3/E1,PH,B,U1,U2,R,CT/COM4/X0,XE/COM5/H2,P2,T2,ROM,HM,FM
    DATA PAI,Z,P0/3.14159265,0.68,1.96E8/,KT,S1,S2/0,1HY,1Hy/
    DATA N,X0,XE,W,E1,EDA0,R,US,C1,C2,NZ,CU,CT/129,-4.,1.4,1.768E5,
    2.21E11,0.03,0.02,0.885,0.37,0.37,5,0.25,0.35/
    OPEN(8,FILE='OUT.DAT',STATUS='UNKNOWN')
    WRITE(*,*)'Show the example or not (Y or N)?'
    READ(*,'(A)')S
    IF(S.EQ.S1.OR.S.EQ.S2)THEN
    KT=2
    GOTO 10
    ELSE
    WRITE(*,*)' Temperature is considered or not (Y or N) ?'
    READ(*,'(A)')S
    IF(S.EQ.S1.OR.S.EQ.S2)KT=2
    ENDIF
    WRITE(*,*)'N,X0,XE,W,E,EDA0,R,US='
    READ(*,*)N,X0,XE,W,E1,EDA0,R,US
    IF(KT.EQ.2)THEN
    WRITE(*,*)'NZ,CU='
    READ(*,*)NZ,CU
    ENDIF
    WRITE(*,*)' Change iteration factors or not (Y or N) ?'
    READ(*,'(A)')S
    IF(S.EQ.S1.OR.S.EQ.S2)THEN
    WRITE(*,*)'C1,C2='
    READ(*,*)C1,C2
    ENDIF
10  CW=N+0.1
    LMAX=ALOG(CW)/ALOG(2.)
    N=2**LMAX+1
    LMIN=(ALOG(CW)-ALOG(SQRT(CW)))/ALOG(2.)
    LMAX=LMIN
    H00=0.0
    W1=W/(E1*R)
    PH=E1*SQRT(0.5*W1/PAI)
    A1=(ALOG(EDA0)+9.67)
    A2=PH/P0
    A3=0.59/(PH*1.E-9)
    T2=0.0
    B=4.*R*PH/E1
    ALFA=Z*A1/P0
    G=ALFA*E1
    U=EDA0*US/(E1*R)
    CC1=SQRT(2.*U)
    AM=2.*PAI*(PH/E1)**2/CC1
```

```
      AL=G*SQRT(CC1)
      CW=(PH/E1)*(B/R)
      C3=1.6*(R/B)**2*G**0.6*U**0.7*W1**(-0.13)
      ENDA=3.*(PAI/AM)**2/8.
      U1=0.5*(2.+CU)*U
      U2=0.5*(2.-CU)*U
      CW=-1.13*C3
      WRITE(*,40)
40    FORMAT(2X,'                Wait  Please',//)
      CALL SUBAK(N)
      CALL MULTI(N,NZ,KT,LMIN,LMAX,H00)
      STOP
      END
      SUBROUTINE MULTI(N,NZ,KT,LMIN,LMAX,H00)
      DIMENSION X(1100),P(1100),H(1100),RO(1100),POLD(1100),EPS(1100),
      EDA(1100),P0(2200),F(1100),F0(2200),R(1100),R0(2200),G(10),
      T(22000)
      COMMON /COM1/ENDA,A1,A2,A3,Z,C1,C2,C3,CW
      COMMON
      /COMK/K/COMT/LT,T1(1100)/COM3/E1,PH,B,U1,U2,RR,CT/COM5/H2,P2,T2,
      RM,HM,FM
      DATA MK,IT,KH,NMAX,PAI,G0/0,0,0,1100,3.14159265,1.570796325/
      LT=LMAX
      NX=N
      K=LMIN
      N0=(N-1)/2**(LMIN-1)
      CALL KNDX(K,N,N0,N1,NMAX,DX,X)
      DO 10 I=1,N
      T1(I)=1.0
      IF(ABS(X(I)).GE.1.0)P(I)=0.0
10    IF(ABS(X(I)).LT.1.0)P(I)=SQRT(1.-X(I)*X(I))
12    CALL HREE(N,DX,H00,G0,X,P,H,RO,EPS,EDA,F,0)
      IF(KH.NE.0)GOTO 14
      KH=1
      GOTO 12
14    CALL FZ(N,P,POLD)
      DO 100 L=LMIN,LMAX
      K=L
      G(K)=PAI/2.
      DO 18 I=1,N
      R(I)=0.0
      F(I)=0.0
      R0(N1+I)=0.0
18    F0(N1+I)=0.0
20    KK=2
      CALL ITER(N,KK,DX,H00,G0,X,P,H,RO,EPS,EDA,F,R,0)
      KK=1
      CALL ITER(N,KK,DX,H00,G0,X,P,H,RO,EPS,EDA,F,R,1)
      G(K-1)=G(K)
      DO 24 I=1,N
      IF(I.LT.N)G(K-1)=G(K-1)-0.5*DX*(P(I)+P(I+1))
```

```
24  P0(N1+I)=P(I)
    N2=N
    K=K-1
    CALL KNDX(K,N,N0,N1,NMAX,DX,X)
    CALL TRANS(N,N2,P,H,RO,EPS,EDA,R)
    CALL ITER(N,KK,DX,H00,G0,X,P,H,RO,EPS,EDA,F,R,2)
    DO 26 I=1,N
    IF(I.LT.N)G(K)=G(K)+0.5*DX*(P(I)+P(I+1))
26  F(I)=H(I)
    G0=G(K)
    CALL HREE(N,DX,H00,G0,X,P,H,RO,EPS,EDA,F,1)
    DO 28 I=1,N
    R0(N1+I)=R(I)
28  F0(N1+I)=F(I)
    IF(K.NE.1)GOTO 20
    KK=19
    CALL ITER(N,KK,DX,H00,G0,X,P,H,RO,EPS,EDA,F,R,0)
40  DO 42 I=1,N
42  P0(N1+I)=P(I)
    N2=N1
    K=K+1
    CALL KNDX(K,N,N0,N1,NMAX,DX,X)
    G0=G(K)
    DO 50 I=2,N,2
    I1=N1+I
    I2=N2+I/2
    P(I-1)=P0(I2)
    P(I)=P0(I1)+0.5*(P0(I2)+P0(I2+1)-P0(I1-1)-P0(I1+1))
50  IF(P(I).LT.0.0)P(I)=0.
    DO 52 I=1,N
    R(I)=R0(N1+I)
52  F(I)=F0(N1+I)
    CALL HREE(N,DX,H00,G0,X,P,H,RO,EPS,EDA,F,0)
    KK=1
    CALL ITER(N,KK,DX,H00,G0,X,P,H,RO,EPS,EDA,F,R,0)
    IF(K.LT.L)GOTO 40
100 CONTINUE
    MK=MK+1
    CALL ERROP(N,P,POLD,ERP)
    IF(ERP.GT.0.01*C2.AND.MK.LE.12)GOTO 14
    MK=8
    IF(KT.NE.2)GOTO 105
    CALL THERM(NX,NZ,DX,T,P,H)
    CALL ERROM(NX,NZ,T1,T,KT)
    IT=IT+1
    IF(KT.EQ.2.AND.IT.LT.10)GOTO 14
    KT=2
    IF(IT.GE.10)THEN
    WRITE(*,*)'Temperature is not convergent !!!'
    READ(*,*)
    ENDIF
```

```
105 IF(MK.GE.10)THEN
    WRITE(*,*)'Pressures are not convergent !!!'
    READ(*,*)
    ENDIF
    FM=FRICT(N,DX,H,P,EDA)
    DO I=1,N
    WRITE(8,110)X(I),P(I),H(I)
    H(I)=H(I)*B*B/RR
    P(I)=P(I)*PH
    ENDDO
110 FORMAT(1X,6(E12.6,1X))
    DO I=2,N-1
    IF(P(I).GE.P(I-1).AND.P(I).GE.P(I+1))THEN
    HM=H(I)
    RM=RO(I)
    GOTO 120
    ENDIF
    ENDDO
120 DO I=1,N
    H(I)=H(I)*1.E6
    P(I)=P(I)*1.E-9
    ENDDO
    IF(KT.EQ.2)THEN
    CALL OUPT(NX,NZ,X,T)
    ENDIF
    RETURN
    END
    SUBROUTINE HREE(N,DX,H00,G0,X,P,H,RO,EPS,EDA,F0,KG)
    DIMENSION X(N),P(N),H(N),RO(N),EPS(N),EDA(N),F0(N)
    DIMENSION W(2200)
    COMMON
    /COM1/ENDA,A1,A2,A3,Z,C1,C2,C3,CW/COMK/K/COMT/LT,T1(1100)/COM2/T0,
    EDA0,AK0,AK1,AK2,CV,CV1,CV2,RO0,RO1,RO2,D0/COMAK/AK(0:1100)
    DATA KK,MK1,MK2,NW,PAI1/0,3,0,2200,0.318309886/
    IF(KK.NE.0)GOTO 3
    HM0=C3
 3  W1=0.0
    DO 4 I=1,N
 4  W1=W1+P(I)
    C3=(DX*W1)/G0
    DW=1.-C3
    IF(K.EQ.1)GOTO 6
    CALL VI(N,DX,P,W)
    GOTO 10
 6  WX=-PAI1*W1*DX*ALOG(DX)
    DO 8 I=1,N
    W(I)=WX
    DO 8 J=1,N
    IJ=IABS(I-J)
 8  W(I)=W(I)-PAI1*AK(IJ)*P(J)*DX
 10 HMIN=1.E3
```

```
      DO 30 I=1,N
      H0=0.5*X(I)*X(I)+W(I)
      IF(KG.EQ.1)GOTO 20
      IF(H0+F0(I).LT.HMIN)HMIN=H0+F0(I)
      H(I)=H0
      GOTO 30
20    F0(I)=F0(I)-H00-H0
30    CONTINUE
      IF(KG.EQ.1)RETURN
      H0=H00+HMIN
      IF(KK.NE.0)GOTO 32
      KK=1
      H00=-H0+HM0
32    IF(H0.LE.0.0)GOTO 48
      IF(K.NE.1)GOTO 50
40    MK=MK+1
      IF(MK.LE.MK1)GOTO 50
      IF(MK.GE.MK2)MK=0
      IF(H0+CW*DW.GT.0.0)HM0=H0+CW*DW
      IF(H0+CW*DW.LE.0.0)HM0=HM0*C3
48    H00=HM0-HMIN
50    DO 60 I=1,N
60    H(I)=H00+H(I)+F0(I)
      IT=2**(LT-K)
      DO 100 I=1,N
      II=IT*(I-1)+1
      CT1=-0.05*T0*(T1(II)-1.0)
      CT2=D0*T0*(T1(II)-1.)
      EDA(I)=EXP(CT1)*EXP(A1*(-1.+(1.+A2*P(I))**Z))
      RO(I)=(A3+1.34*P(I))/(A3+P(I))+CT2
      EPS(I)=RO(I)*H(I)**3/(ENDA*EDA(I))
100   CONTINUE
      RETURN
      END
      SUBROUTINE ITER(N,KK,DX,H00,G0,X,P,H,RO,EPS,EDA,F0,R0,KG)
      DIMENSION X(N),P(N),H(N),RO(N),EPS(N),EDA(N),F0(N),R0(N)
      COMMON /COM1/ENDA,A1,A2,A3,Z,C1,C2,C3/COMAK/AK(0:1100)
      DATA PAI/3.14159265/
      DO 100 K=1,KK
      DO 70 I=2,N-1
      D1=0.5*(EPS(I-1)+EPS(I))
      D2=0.5*(EPS(I)+EPS(I+1))
      D3=RO(I)*AK(0)/PAI
      D4=RO(I-1)*AK(1)/PAI
      D5=D1+D2-D3*DX+D4*DX
      D6=D1*P(I-1)+D2*P(I+1)
      D7=(RO(I)*H(I)+D3*P(I)-RO(I-1)*H(I-1)-D4*P(I))*DX
      D8=D6-D7
      P(I)=D8/D5
      IF(P(I).LT.0.0)P(I)=0.0
70    CONTINUE
```

```
      IF(KG.NE.0)GOTO 100
      CALL HREE(N,DX,H00,G0,X,P,H,RO,EPS,EDA,F0,0)
100   CONTINUE
      RETURN
      END
      SUBROUTINE VI(N,DX,P,V)
      DIMENSION P(N),V(N)
      COMMON /COMAK/AK(0:1100)
      PAI1=0.318309886
      C=ALOG(DX)
      DO 10 I=1,N
      V(I)=0.0
      DO 10 J=1,N
      IJ=IABS(I-J)
10    V(I)=V(I)+(AK(IJ)+C)*DX*P(J)
      DO I=1,N
      V(I)=-PAI1*V(I)
      ENDDO
      RETURN
      END
      SUBROUTINE SUBAK(MM)
      COMMON /COMAK/AK(0:1100)
      DO 10 I=0,MM
10    AK(I)=(I+0.5)*(ALOG(ABS(I+0.5))-1.)-(I-0.5)*(ALOG(ABS(I-0.5))-1.)
      RETURN
      END
      FUNCTION FRICT(N,DX,H,P,EDA)
      DIMENSION H(N),P(N),EDA(N)
      COMMON /COM3/E1,PH,B,U1,U2,R,CT
      DATA TAU0/4.E7/
      TP=TAU0/PH
      TE=TAU0/E1
      BR=B/R
      FRICT=0.0
      DO I=1,N
      DP=0.0
      IF(I.NE.N)DP=(P(I+1)-P(I))/DX
      TAU=0.5*H(I)*ABS(DP)*(BR/TP)+2.*ABS(U1-U2)*EDA(I)/(H(I)*BR**2*TE)
      FRICT=FRICT+TAU
      ENDDO
      FRICT=FRICT*DX*B*TAU0
      RETURN
      END
      SUBROUTINE FZ(N,P,POLD)
      DIMENSION P(N),POLD(N)
      DO 10 I=1,N
10    POLD(I)=P(I)
      RETURN
      END
      SUBROUTINE ERROP(N,P,POLD,ERP)
      DIMENSION P(N),POLD(N)
```

```
      SD=0.0
      SUM=0.0
      DO 10 I=1,N
      SD=SD+ABS(P(I)-POLD(I))
10    SUM=SUM+P(I)
      ERP=SD/SUM
      RETURN
      END
      SUBROUTINE KNDX(K,N,N0,N1,NMAX,DX,X)
      DIMENSION X(NMAX)
      COMMON /COM4/X0,XE
      N=2**(K-1)*N0
      DX=(XE-X0)/N
      N=N+1
      N1=N+K
      DO 10 I=1,N
10    X(I)=X0+(I-1)*DX
      RETURN
      END
      SUBROUTINE TRANS(N1,N2,P,H,RO,EPS,EDA,R)
      DIMENSION P(N2),H(N2),RO(N2),EPS(N2),EDA(N2),R(N2)
      DO 10 I=1,N1
      II=2*I-1
      P(I)=P(II)
      H(I)=H(II)
      R(I)=R(II)
      RO(I)=RO(II)
      EPS(I)=EPS(II)
10    EDA(I)=EDA(II)
      RETURN
      END
      SUBROUTINE OUPT(NX,NZ,X,T)
      DIMENSION X(NX),T(NX,NZ)
      DO I=1,NX
      DO K=1,NZ
      T(I,K)=303.*(T(I,K)-1.0)
      END DO
      END DO
      DO I=1,NX
      WRITE(8,30)X(I),(T(I,K),K=1,NZ)
      ENDDO
30    FORMAT(6(1X,E12.6))
      RETURN
      END
      SUBROUTINE ERROM(NX,NZ,T1,T,KT)
      DIMENSION T(NX,NZ),T1(NX)
      KT=2
      ERM=0.
      C1=1./FLOAT(NZ)
      DO 20 I=1,NX
      TT=0.
```

```
      DO 10 K=1,NZ
10    TT=TT+T(I,K)
      TT=C1*TT
      ER=ABS((TT-T1(I))/TT)
      IF(ER.GT.ERM)ERM=ER
20    T1(I)=TT
      IF(ERM.LT.0.003)KT=1
      RETURN
      END
      SUBROUTINE THERM(NX,NZ,DX,T,P,H)
      DIMENSION
      T(NX,NZ),P(NX),H(NX),T1(21),TI(21),TF(21),U(21),DU(21),W(21),
      EDA(21),RO(21),EDA1(21),EDA2(21),ROR(21),UU(21)
      DATA KK/0/
      IF(KK.NE.0)GOTO 4
      DO 2 K=1,NZ
2     T(1,K)=1.0
4     DO 30 I=2,NX
      KG=0
      DO 8 K=1,NZ
      TF(K)=T(I-1,K)
      IF(KK.NE.0)GOTO 6
      T1(K)=T(I-1,K)
      GOTO 8
6     T1(K)=T(I,K)
8     TI(K)=T1(K)
      P1=P(I)
      H1=H(I)
      DP=(P(I)-P(I-1))/DX
      CALL TBOUD(NX,NZ,I,CC1,CC2,T)
10    CALL EROEQ(NZ,T1,P1,EDA,RO,EDA1,EDA2,KG)
      CALL UCAL(NZ,DX,H1,EDA,RO,ROR,EDA1,EDA2,U,UU,DU,W,DP)
      CALL TCAL(NZ,DX,CC1,CC2,T1,TF,U,W,DU,H1,DP,EDA,RO)
      CALL ERRO(NZ,TI,T1,ETS)
      KG=KG+3
      IF(ETS.GT.1.E-4.AND.KG.LE.50)GOTO 10
      DO 20 K=1,NZ
      ROR(K)=RO(K)
      UU(K)=U(K)
20    T(I,K)=T1(K)
30    CONTINUE
      KK=1
      RETURN
      END
      SUBROUTINE TBOUD(NX,NZ,I,CC1,CC2,T)
      DIMENSION T(NX,NZ)
      CC1=0.
      CC2=0.
      DO 10 L=1,I-1
      DS=1./SQRT(FLOAT(I-L))
      IF(L.EQ.I-1)DS=1.1666667
```

```
      CC1=CC1+DS*(T(L,2)-T(L,1))
10    CC2=CC2+DS*(T(L,NZ)-T(L,NZ-1))
      RETURN
      END
      SUBROUTINE ERRO(NZ,T0,T,ETS)
      DIMENSION T0(NZ),T(NZ)
      ETS=0.0
      DO 10 K=1,NZ
      IF(T(K).LT.1.E-5)ETS0=1.
      IF(T(K).GE.1.E-5)ETS0=ABS((T(K)-T0(K))/T(K))
      IF(ETS0.GT.ETS)ETS=ETS0
10    T0(K)=T(K)
      RETURN
      END
      SUBROUTINE EROEQ(NZ,T,P,EDA,RO,EDA1,EDA2,KG)
      DIMENSION T(NZ),EDA(NZ),RO(NZ),EDA1(NZ),EDA2(NZ)
      COMMON
      /COM1/ENDA,A1,A2,A3,Z,O1,O2,O3/COM2/T0,EDA0,AK0,AK1,AK2,CV,CV1,
      CV2,RO0,RO1,RO2,D0/COM3/E1,PH,B,U1,U2,R,CC
      DATA A4,A5/0.455445545,0.544554455/
      IF(KG.NE.0)GOTO 20
      B1=(1.+A2*P)**Z
      B2=(A3+1.34*P)/(A3+P)
      B3=-0.05*T0
20    DO 30 K=1,NZ
      EDA(K)=EXP(A1*(-1.+B1))*EXP(B3*(T(K)-1.0))
30    RO(K)=B2+D0*T0*(T(K)-1.)
      CC1=0.5/(NZ-1.)
      CC2=1./(NZ-1.)
      C1=0.
      C2=0.
      DO 40 K=1,NZ
      IF(K.EQ.1)GOTO 32
      C1=C1+0.5/EDA(K)+0.5/EDA(K-1)
      C2=C2+CC1*((K-1.)/EDA(K)+(K-2.)/EDA(K-1))
32    EDA1(K)=C1*CC2
40    EDA2(K)=C2*CC2
      IF(KG.NE.2)RETURN
      C1=0.
      C2=0.
      C3=0.
      DO 50 K=1,NZ
      IF(K.EQ.1)GOTO 50
      C1=C1+0.5*(RO(K)+RO(K-1))
      C2=C2+0.5*(RO(K)*EDA1(K)+RO(K)*EDA1(K-1))
      C3=C3+0.5*(RO(K)*EDA2(K)+RO(K)*EDA2(K-1))
50    CONTINUE
      B1=12.*CC2*(C1*EDA2(NZ)/EDA1(NZ)-C2)
60    B2=2.*CC2/(U1+U2)*(C1*(U1-U2)/EDA1(NZ)+C3*U1)
      RETURN
      END
```

```
      SUBROUTINE UCAL(NZ,DX,H,EDA,RO,ROR,EDA1,EDA2,U,UU,DU,W,DP)
      DIMENSION
      U(NZ),DU(NZ),W(NZ),ROR(NZ),UU(NZ),EDA(NZ),RO(NZ),EDA1(NZ),EDA2(NZ)
      COMMON
      /COM2/T0,EDA0,AK0,AK1,AK2,CV,CV1,CV2,RO0,RO1,RO2,D0/COM3/E1,PH,B,
      U1,U2,R,CC
      DATA KK/0/
      IF(KK.NE.0)GOTO 20
      A1=U1
      A2=PH*(B/R)**3/E1
      A3=U2-U1
20    CC1=A2*DP*H
      CC2=CC1*H
      CC3=A3/H
      CC4=1./EDA1(NZ)
      DO 30 K=1,NZ
      U(K)=A1+CC2*(EDA2(K)-CC4*EDA2(NZ)*EDA1(K))+A3*CC4*EDA1(K)
      IF(U(K).LT.0.0)U(K)=0.
30    DU(K)=CC1/EDA(K)*((K-1.)/(NZ-1.)-CC4*EDA2(NZ))+CC3*CC4/EDA(K)
      A4=B/((NZ-1)*R*DX)
      C1=A4*H
      IF(KK.EQ.0)GOTO 50
      DO 40 K=2,NZ-1
      W(K)=(RO(K-1)*W(K-1)+C1*(RO(K)*U(K)-ROR(K)*UU(K)))/RO(K)
40    CONTINUE
50    KK=1
      RETURN
      END
      SUBROUTINE TCAL(NZ,DX,CC1,CC2,T,TF,U,W,DU,H,DP,EDA,RO)
      DIMENSION T(NZ),TF(NZ),U(NZ),DU(NZ),W(NZ),EDA(NZ),RO(NZ),A(4,21),
      D(21),AA(2,21)
      COMMON /COM2/T0,EDA0,AK0,AK1,AK2,CV,CV1,CV2,RO0,RO1,RO2,D0/COM3/E1,
      PH,B,U1,U2,R,CC
      DATA KK,CC5,PAI,TAU0/0,0.6666667,3.14159265,4.E7/
      IF(KK.NE.0)GOTO 5
      KK=1
      TAU=TAU0*B*B/(E1*R*R)
      A2=-CV*RO0*E1*B**3/(EDA0*AK0*R)
      A3=-E1*PH*B**3*D0/(AK0*EDA0*R)
      A4=-(E1*R)**2/(AK0*EDA0*T0)
      A5=0.5*R/B*A2
      A6=AK0*SQRT(EDA0*R/(PAI*RO1*CV1*U1*E1*AK1*B**3))
      A7=AK0*SQRT(EDA0*R/(PAI*RO2*CV2*U2*E1*AK2*B**3))
5     CC3=A6*SQRT(DX)
      CC4=A7*SQRT(DX)
      DZ=H/(NZ-1.)
      DZ1=1./DZ
      DZ2=DZ1*DZ1
      CC6=A3*DP
      DO 10 K=2,NZ-1
```

```
      A(1,K)=DZ2+DZ1*A5*RO(K)*W(K)
      A(2,K)=-2.*DZ2+A2*RO(K)*U(K)/DX+CC6*U(K)/RO(K)
      A(3,K)=DZ2-DZ1*A5*RO(K)*W(K)
      AE=ABS(EDA(K)*DU(K))
10    A(4,K)=A4*ABS(DU(K))*AE+A2*RO(K)*U(K)*TF(K)/DX
      A(1,1)=0.
      A(2,1)=1.+2.*DZ1*CC3*CC5
      A(3,1)=-2.*DZ1*CC3*CC5
      A(1,NZ)=-2.*DZ1*CC4*CC5
      A(2,NZ)=1.+2.*DZ1*CC4*CC5
      A(3,NZ)=0.
      A(4,1)=1.+CC1*CC3*DZ1
      A(4,NZ)=1.-CC2*CC4*DZ1
      CALL TRA3(NZ,D,A,AA)
      DO 20 K=1,NZ
20    T(K)=(1.-CC)*T(K)+CC*D(K)
30    CONTINUE
      RETURN
      END
      SUBROUTINE TRA3(N,D,A,B)
      DIMENSION D(N),A(4,N),B(2,N)
      C=1./A(2,N)
      B(1,N)=-A(1,N)*C
      B(2,N)=A(4,N)*C
      DO 10 I=1,N-1
      IN=N-I
      IN1=IN+1
      C=1./(A(2,IN)+A(3,IN)*B(1,IN1))
      B(1,IN)=-A(1,IN)*C
10    B(2,IN)=(A(4,IN)-A(3,IN)*B(2,IN1))*C
      D(1)=B(2,1)
      DO 20 I=2,N
20    D(I)=B(1,I)*D(I-1)+B(2,I)
      RETURN
      END
      BLOCK DATA
      COMMON /COM2/T0,EDA0,AK0,AK1,AK2,CV,CV1,CV2,RO0,RO1,RO2,D0
      DATA T0,AK0,AK1,AK2,CV,CV1,CV2,RO0,RO1,RO2,D0/
      303.,0.14,46.,46.,2000.,470.,470.,890.,7850.,7850.,-0.00065/
      END
```

10.3.3 Example

The obtained film thickness and pressure distributions under the given working conditions are shown in Figure 10.3. The film thickness is divided into five layers across the film thickness direction. The calculated temperature distribution curves are shown in Figure 10.4. We see that the temperature in the middle layer is the highest, while the temperature in the fast moving surface is the lowest.

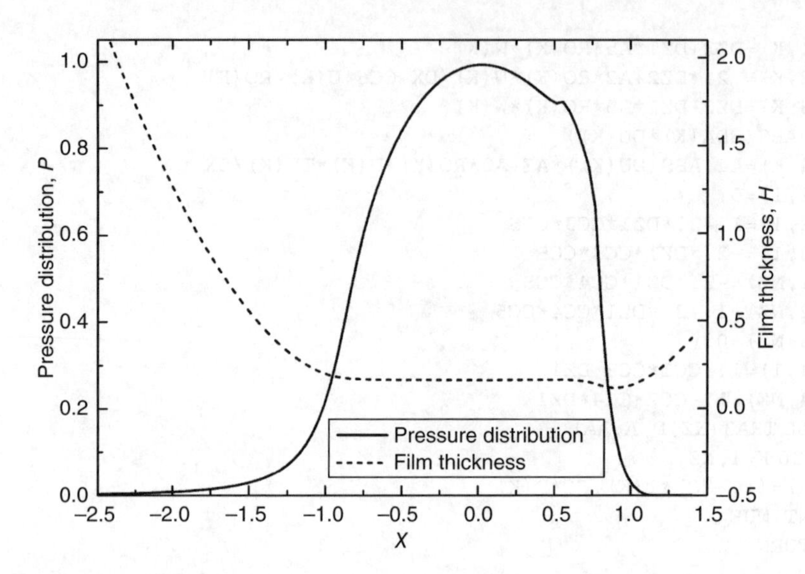

Figure 10.3 Film thickness and pressure distribution of TEHL in line contact

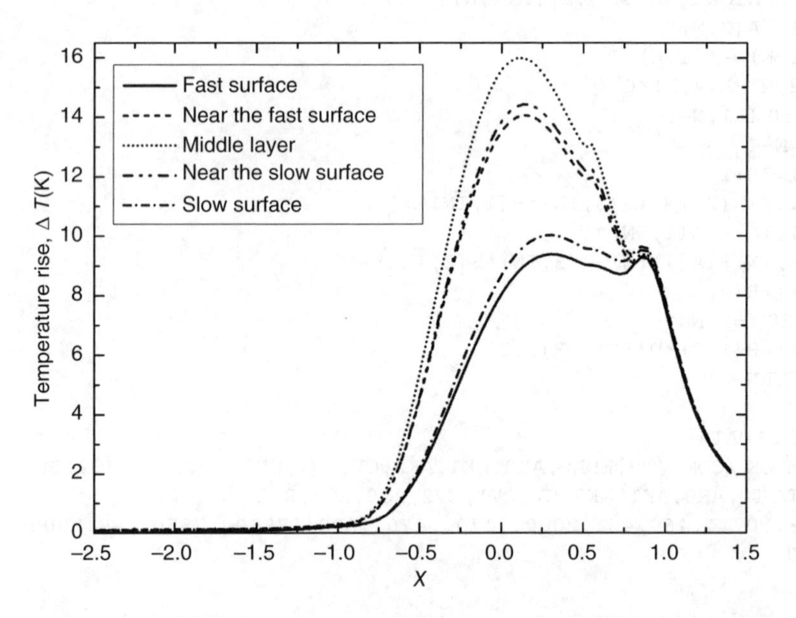

Figure 10.4 Temperature distribution of each layer for TEHL in line contact

10.4 Numerical calculation method and program for TEHL in point contact

10.4.1 Flowchart

The flowchart to calculate TEHL in the point contact is shown in Figure 10.5.

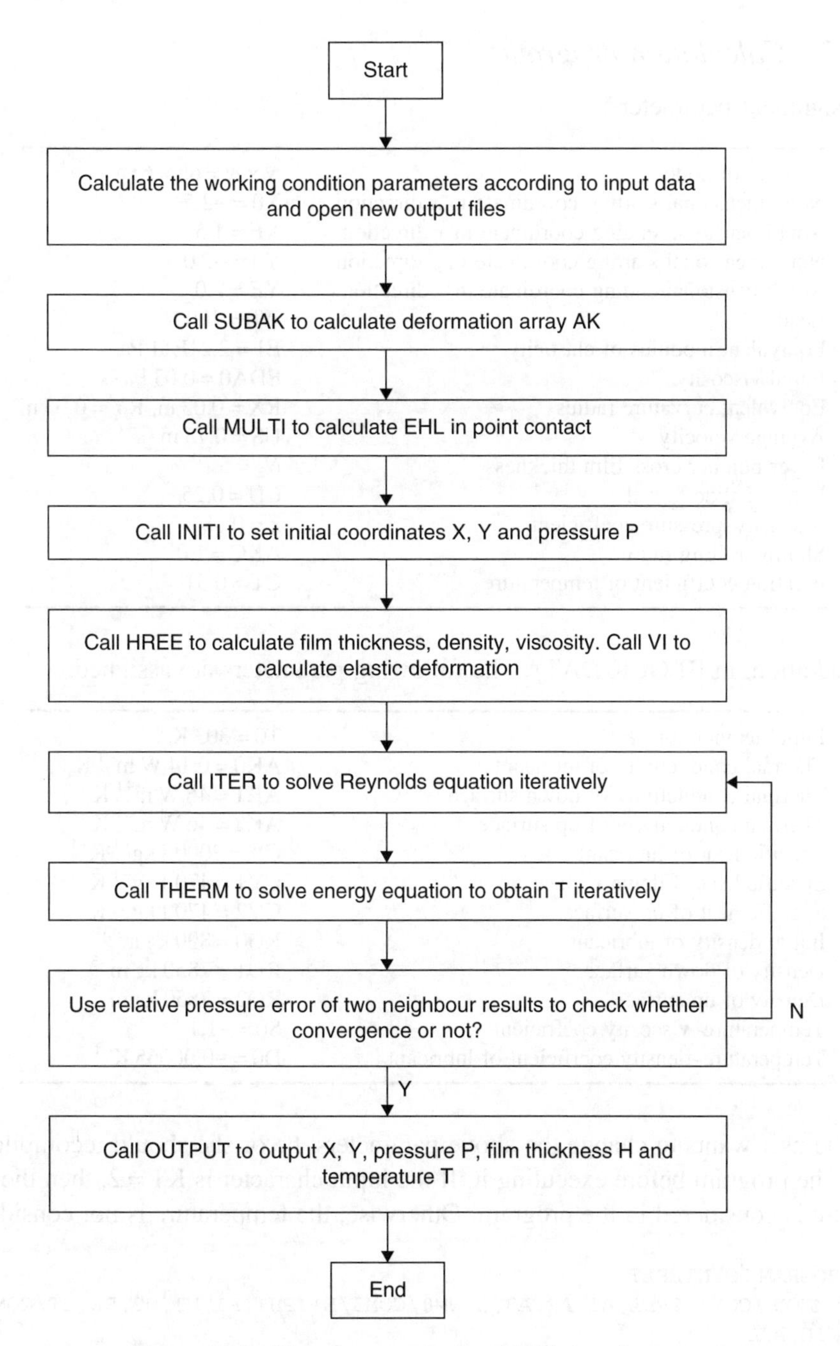

Figure 10.5 Flowchart to calculate TEHL in point contact

10.4.2 *Calculation program*

Preassignment parameters:

Number of nodes	$N \times N = 65 \times 65$
Nondimensional starting coordinate in x direction	X0 = −2.5
Nondimensional ending coordinate in x direction	XE = 1.5
Nondimensional starting coordinate in y direction	Y0 = −2.0
Nondimensional ending coordinate in y direction	YE = 2.0
Load	W0 = 100 N
Equivalent modulus of elasticity	E1 = 2.21E11 Pa
Initial viscosity	EDA0 = 0.03 Pa · s
Equivalent curvature radius	RX = 0.03 m, RY = 0.03 m
Average velocity	US = 0.75 $\text{m} \cdot \text{s}^{-1}$
Layer number cross film thickness	$N_Z = 5$
Ratio of slide to roll	CU = 0.25
Viscosity–pressure coefficient	$Z = 0.68$
Sliding–rolling ratio	AKC = 1.0
Iteration coefficient of temperature	CT = 0.31

In addition, in BLOCK DATA, the following parameters are assigned.

Initial temperature	T0 = 303 K
Thermal conductivity of lubricant	AK0 = 0.14 W m^{-1} K
Thermal conductivity of down surface	AK1 = 46 W m^{-1} K
Thermal conductivity of up surface	AK2 = 46 W m^{-1} K
Specific heat of lubricant	CV = 2000 J kg^{-1} K^{-1}
Specific heat of down surface	CV1 = 470 J kg^{-1} K^{-1}
Specific heat of up surface	CV2 = 470 J kg^{-1} K^{-1}
Initial density of lubricant	RO0 = 890 kg m^{-3}
Density of down surface	RO1 = 7850 kg m^{-3}
Density of up surface	RO2 = 7850 kg m^{-3}
Temperature–viscosity coefficient	S0 = −1.1
Temperature–density coefficient of lubricant	D0 = −0.00065 K^{-1}

If the user wants to change the above parameters, he or she should recompile and relink the program before executing it. If the input character is KT = 2, then the temperature is considered in the program. Otherwise, the temperature is not considered.

```
PROGRAM POINTEHLT
COMMON /COM1/ENDA,A1,A2,A3,Z,HM0/COM3/E1,PH1,B1,U1,U2,RE,CT/COMK/
LMIN,AKC
COMMON /COM2/T0,AK0,AK1,AK2,CV,CV1,CV2,RO0,RO1,RO2,S0,D0
COMMON /COMW/W0,T1,T2,RX,B,PH/COMC/KT,NF/COMD/AD,AD1,KK1,KK2,KK3,
KK4/COME/US,EDA0/COMH/HMC
DATA PAI,Z,N,W0,E1,EDA0,RX,RY,US,X0,XE/
3.14159265,0.68,65,100.,2.21E11,0.03,0.03,0.03,.75,-2.5,1.5/
DATA NZ,CT,AKC/5,0.31,1.0/
```

```
      DATA KK1,KK2,KK3,KK4,NF,AD,AD1/0,0,0,0,0,0.0,0.0/
      WRITE(*,*)'KT='
      READ(*,*)KT
     OPEN(8,FILE='FILM.DAT',STATUS='UNKNOWN')
      OPEN(9,FILE='PRESS.DAT',STATUS='UNKNOWN')
      OPEN(10,FILE='OUT.DAT',STATUS='UNKNOWN')
      B=(1.5*W0*RX/E1)**(1./3.0)
      PH=1.5*W0/(B**2*PAI)
      H00=0.0
      MM=N-1
      U=EDA0*US/(E1*RX)
      U1=0.5*(2.+AKC)*U
      U2=0.5*(2.-AKC)*U
      A1=ALOG(EDA0)+9.67
      A2=5.1E-9*PH
      A3=0.59/(PH*1.E-9)
      PH1=PH
      B1=B
      RE=RX
      W=2.*PAI*PH/(3.*E1)*(B/RX)**2
      ALFA=Z*5.1E-9*A1
      G=ALFA*E1
      AHM=1.0-EXP(-0.68*1.03)
      AHC=1.0-0.61*EXP(-0.73*1.03)
      HM0=3.63*(RX/B)**2*G**0.49*U**0.68*W**(-0.073)*AHM
      HMC=2.69*(RX/B)**2*G**0.53*U**0.67*W**(-0.067)*AHC
      ENDA=12.*U*(E1/PH)*(RX/B)**3
      T1=PH*B/RX
      T2=EDA0*US*RX/(B*B)
      WRITE(*,*)"N,X0,XE,PH,E1,EDA0,RX,US=",N,X0,XE,PH,E1,EDA0,RX,US
      WRITE(*,*)'          Wait please'
      CALL SUBAK(MM)
      CALL MULTI(N,NZ,X0,XE,H00)
      STOP
      END
      SUBROUTINE MULTI(N,NZ,X0,XE,H00)
      DIMENSION X(65),Y(65),H(4500),RO(4500),EPS(4500),EDA(4500),P(4500),
      POLD(4500),T(65,65,5)
      COMMON /COMT/T1(65,65)/COMC/KT,NF
      DATA MK,KTK,G00/200,1,2.0943951/
      G0=G00
      NX=N
      NY=N
      NN=(N+1)/2
      DO I=1,N
      DO J=1,N
      T1(I,J)=1.0
      DO K=1,5
      T(I,J,K)=1.0
      ENDDO
      ENDDO
```

```
      ENDDO
      CALL INITI(N,DX,X0,XE,X,Y,P,POLD)
      CALL HREE(N,DX,H00,G0,X,Y,H,RO,EPS,EDA,P)
      M=0
      KTK=0
14    KK=15
15    CALL ITER(N,KK,DX,H00,G0,X,Y,H,RO,EPS,EDA,P)
      M=M+1
      CALL ERP(N,ER,P,POLD)
      ER=ER/KK
      WRITE(*,*)'ER=',ER
      IF(KT.NE.0)GOTO 17
      IF(M.LT.MK.AND.ER.GT.1.E-6)GOTO 14
      GOTO 120
17    KT1=0
18    CALL THERM(NX,NY,NZ,DX,P,H,T)
      CALL ERROM(NX,NY,NZ,T,ERM)
      IF(ER.LT.1.0E-6)GOTO 120
      IF(KT1.LT.1)THEN
      KT1=KT1+1
      GOTO 18
      ENDIF
      IF(KTK.LT.MK)THEN
      KTK=KTK+1
      GOTO 14
      ENDIF
120   CONTINUE
      OPEN(11,FILE='TEM.DAT',STATUS='UNKNOWN')
      WRITE(11,110)X0,(Y(I),I=1,N)
      TMAX=0.0
      DO I=1,N
      WRITE(11,110)X(I),(303.0*(T1(I,JJ)-1.),JJ=1,N)
      DO J=1,N
      IF(TMAX.LT.303.0*(T1(I,J)-1.))TMAX=303.*(T1(I,J)-1.)
      ENDDO
      ENDDO
110   FORMAT(66(E12.6,1X))
130   CALL OUPT(N,DX,X,Y,H,P,EDA,TMAX)
      RETURN
      END
      SUBROUTINE INITI(N,DX,X0,XE,X,Y,P,POLD)
      DIMENSION X(N),Y(N),P(N,N),POLD(N,N)
      NN=(N+1)/2
      DX=(XE-X0)/(N-1.)
      Y0=-0.5*(XE-X0)
      DO 5 I=1,N
      X(I)=X0+(I-1)*DX
      Y(I)=Y0+(I-1)*DX
5     CONTINUE
      DO 10 I=1,N
      D=1.-X(I)*X(I)
```

```
      DO 10 J=1,NN
      C=D-Y(J)*Y(J)
      IF(C.LE.0.0)P(I,J)=0.0
10    IF(C.GT.0.0)P(I,J)=SQRT(C)
      DO 20 I=1,N
      DO 20 J=NN+1,N
      JJ=N-J+1
20    P(I,J)=P(I,JJ)
      DO I=1,N
      DO J=1,N
      POLD(I,J)=P(I,J)
      ENDDO
      ENDDO
      RETURN
      END
      SUBROUTINE HREE(N,DX,H00,G0,X,Y,H,RO,EPS,EDA,P)
      DIMENSION X(N),Y(N),P(N,N),H(N,N),RO(N,N),EPS(N,N),EDA(N,N)
      DIMENSION W(150,150),P0(150,150),ROU(65,65)
      COMMON /COM1/ENDA,A1,A2,A3,Z,HM0/COMAK/AK(0:65,0:65)
      COMMON /COM2/T0,EAK,EAK1,EAK2,CV,CV1,CV2,RO0,RO1,RO2,S0,D0
      COMMON /COMT/T1(65,65)/COMK/LMIN,AKC/COMD/AD,AD1,KK,KK2,KK3,KK4/
      COMC/KT,NF
      DATA KR,NW,PAI,PAI1,DELTA/0,150,3.14159265,0.2026423,0.0/
      NN=(N+1)/2
      CALL VI(NW,N,DX,P,W)
      HMIN=1.E3
      IF(KR.EQ.0)THEN
      OPEN(12,FILE='ROUGH2.DAT',STATUS='UNKNOWN')
      DO I=1,N
      DO J=NN+1,N
      ROU(I,J)=ROU(I,N+1-J)
      ENDDO
100   FORMAT(33(1X,F10.6))
      ENDDO
      CLOSE(12)
      KR=1
      ENDIF
      DO 30 I=1,N
      DO 30 J=1,NN
      RAD=X(I)*X(I)+Y(J)*Y(J)
      W1=0.5*RAD+DELTA
      ZZ=0.5*AD1*AD1+X(I)*ATAN(AD*PAI/180.0)
      IF(W1.LE.ZZ)W1=ZZ
      H0=W1+W(I,J)
      IF(H0.LT.HMIN)HMIN=H0
30    H(I,J)=H0
      IF(KK.EQ.0)THEN
      KG1=0
      H01=-HMIN+HM0
      DH=0.005*HM0
      H02=-HMIN
```

```
      H00=0.5*(H01+H02)
      ENDIF
      W1=0.0
      DO 32 I=1,N
      DO 32 J=1,N
   32 W1=W1+P(I,J)
      W1=DX*DX*W1/G0
      DW=1.-W1
      IF(KK.EQ.0)THEN
      KK=1
      GOTO 50
      ENDIF
      IF(DW.LT.0.0)THEN
      KG1=1
      H00=AMIN1(H01,H00+DH)
      ENDIF
      IF(DW.GT.0.0)THEN
      KG2=2
      H00=AMAX1(H02,H00-DH)
      ENDIF
   50 DO 60 I=1,N
      DO 60 J=1,NN
      H(I,J)=H00+H(I,J)
      CT1=((T1(I,J)-0.455445545)/0.544554455)**S0
      CT2=D0*T0*(T1(I,J)-1.)
      IF(P(I,J).LT.0.0)P(I,J)=0.0
      EDA1=EXP(A1*(-1.+(1.+A2*P(I,J))**Z*CT1))
      EDA(I,J)=EDA1
      IF(NF.EQ.0)GOTO 55
      IF(I.NE.1.AND.J.NE.1)THEN
      DPDX=(P(I,J)-P(I-1,J))/DX
      DPDY=(P(I,J)-P(I,J-1))/DX
      EDA(I,J)=EQEDA(DPDX,DPDY,P(I,J),H(I,J),EDA1)
      ENDIF
      EDA1=EDA(I,J)
   55 RO(I,J)=(A3+1.34*P(I,J))/(A3+P(I,J))+CT2
   60 EPS(I,J)=RO(I,J)*H(I,J)**3/(ENDA*EDA1)
      DO 70 J=NN+1,N
      JJ=N-J+1
      DO 70 I=1,N
      H(I,J)=H(I,JJ)
      RO(I,J)=RO(I,JJ)
      EDA(I,J)=EDA(I,JJ)
   70 EPS(I,J)=EPS(I,JJ)
      RETURN
      END
      SUBROUTINE ITER(N,KK,DX,H00,G0,X,Y,H,RO,EPS,EDA,P)
      DIMENSION X(N),Y(N),P(N,N),H(N,N),RO(N,N),EPS(N,N),EDA(N,N)
      COMMON /COMAK/AK(0:65,0:65)
      DATA PAI/3.14159265/
      NN=(N+1)/2
```

```
      DO 100 K=1,KK
      DO 70 J=2,NN
      J0=J-1
      J1=J+1
      DO 70 I=2,N-1
      I0=I-1
      I1=I+1
      D1=0.5*(EPS(I0,J)+EPS(I,J))
      D2=0.5*(EPS(I1,J)+EPS(I,J))
      D3=0.5*(EPS(I,J0)+EPS(I,J))
      D4=0.5*(EPS(I,J1)+EPS(I,J))
      D5=2.0*RO(I,J)*AK(0,0)/PAI**2
      D6=2.0*RO(I0,J)*AK(1,0)/PAI**2
      D7=D1+D2+D3+D4+D5*DX-D6*DX
      D8=D1*P(I0,J)+D2*P(I1,J)+D3*P(I,J0)+D4*P(I,J1)
      D9=(RO(I,J)*H(I,J)-D5*P(I,J)-RO(I0,J)*H(I0,J)+D6*P(I,J))*DX
      P(I,J)=(D8-D9)/D7
      IF(P(I,J).LT.0.0)P(I,J)=0.0
70    CONTINUE
      DO 80 J=1,NN
      JJ=N+1-J
      DO 80 I=1,N
80    P(I,JJ)=P(I,J)
      CALL HREE(N,DX,H00,G0,X,Y,H,RO,EPS,EDA,P)
100   CONTINUE
      RETURN
      END
      SUBROUTINE ERP(N,ER,P,POLD)
      DIMENSION P(N,N),POLD(N,N)
      ER=0.0
      SUM=0.0
      NN=(N+1)/2
      DO 10 I=1,N
      DO 10 J=1,NN
      ER=ER+ABS(P(I,J)-POLD(I,J))
      SUM=SUM+P(I,J)
10    CONTINUE
      ER=ER/SUM
      DO I=1,N
      DO J=1,N
      POLD(I,J)=P(I,J)
      ENDDO
      ENDDO
      RETURN
      END
      SUBROUTINE VI(NW,N,DX,P,V)
      DIMENSION P(N,N),V(NW,NW)
      COMMON /COMAK/AK(0:65,0:65)
      PAI1=0.2026423
      DO 40 I=1,N
      DO 40 J=1,N
```

```
      H0=0.0
      DO 30 K=1,N
      IK=IABS(I-K)
      DO 30 L=1,N
      JL=IABS(J-L)
30    H0=H0+AK(IK,JL)*P(K,L)
40    V(I,J)=H0*DX*PAI1
      RETURN
      END
      SUBROUTINE SUBAK(MM)
      COMMON /COMAK/AK(0:65,0:65)
      S(X,Y)=X+SQRT(X**2+Y**2)
      DO 10 I=0,MM
      XP=I+0.5
      XM=I-0.5
      DO 10 J=0,I
      YP=J+0.5
      YM=J-0.5
      A1=S(YP,XP)/S(YM,XP)
      A2=S(XM,YM)/S(XP,YM)
      A3=S(YM,XM)/S(YP,XM)
      A4=S(XP,YP)/S(XM,YP)
      AK(I,J)=XP*ALOG(A1)+YM*ALOG(A2)+XM*ALOG(A3)+YP*ALOG(A4)
10    AK(J,I)=AK(I,J)
      RETURN
      END
      SUBROUTINE THERM(NX,NY,NZ,DX,P,H,T)
      DIMENSION
      T(NX,NY,NZ),T1(21),TI(21),U(21),DU(21),UU(65,65,21),
      V(21),DV(21),VV(65,65,21),W(21),EDA(21),RO(21),EDA1(21),EDA2(21),
      ROR(65,65,21),P(NX,NX),H(NX,NX),TFX(21),TFY(21)
      COMMON /COMD/AD,AD1,KK1,KK,KK3,KK4
      IF(KK.NE.0)GOTO 4
      DO 2 K=1,NZ
      DO 1 J=1,NY
1     T(1,J,K)=1.0
      DO 2 I=1,NX
2     T(I,1,K)=1.0
4     DO 30 J=2,NY
      DO 30 I=2,NX
      KG=0
      DO 6 K=1,NZ
      TFX(K)=T(I-1,J,K)
      TFY(K)=T(I,J-1,K)
      IF(KK.NE.0)GOTO 5
      T1(K)=T(I-1,J,K)
      GOTO 6
5     T1(K)=T(I,J,K)
6     TI(K)=T1(K)
      P1=P(I,J)
      H1=H(I,J)
```

```
      DPX=(P(I,J)-P(I-1,J))/DX
      DPY=(P(I,J)-P(I,J-1))/DX
      CALL TBOUD(NX,NY,NZ,I,J,CC1,CC2,T)
10    CALL EROEQ(NZ,T1,P1,H1,DPX,DPY,EDA,RO,EDA1,EDA2,KG)
      CALL UCAL(I,J,NX,NY,NZ,DX,H1,EDA,RO,ROR,EDA1,EDA2,U,UU,DU,V,VV,
      DV,W,DPX,DPY)
      CALL TCAL(NZ,DX,CC1,CC2,T1,TFX,TFY,U,V,W,DU,DV,H1,DPX,DPY,EDA,RO)
      CALL ERRO(NZ,TI,T1,ETS)
      KG=KG+3
      IF(ETS.GT.1.E-4.AND.KG.LE.50)GOTO 10
      DO 20 K=1,NZ
      ROR(I,J,K)=RO(K)
      UU(I,J,K)=U(K)
      VV(I,J,K)=V(K)
20    T(I,J,K)=T1(K)
30    CONTINUE
      KK=1
      RETURN
      END
      SUBROUTINE TBOUD(NX,NY,NZ,I,J,CC1,CC2,T)
      DIMENSION T(NX,NY,NZ)
      CC1=0.
      CC2=0.
      DO 10 L=1,I-1
      DS=1./SQRT(FLOAT(I-L))
      IF(L.EQ.I-1)DS=1.1666667
      CC1=CC1+DS*(T(L,J,2)-T(L,J,1))
10    CC2=CC2+DS*(T(L,J,NZ)-T(L,J,NZ-1))
      RETURN
      END
      SUBROUTINE ERRO(NZ,T0,T,ETS)
      DIMENSION T0(NZ),T(NZ)
      ETS=0.0
      DO 10 K=1,NZ
      IF(T(K).LT.1.E-5)ETS0=1.
      IF(T(K).GE.1.E-5)ETS0=ABS((T(K)-T0(K))/T(K))
      IF(ETS0.GT.ETS)ETS=ETS0
10    T0(K)=T(K)
      RETURN
      END
      SUBROUTINE EROEQ(NZ,T,P,H,DPX,DPY,EDA,RO,EDA1,EDA2,KG)
      DIMENSION T(NZ),EDA(NZ),RO(NZ),EDA1(NZ),EDA2(NZ)
      COMMON /COM1/ENDA,A1,A2,A3,Z,C3/COM2/T0,AK0,AK1,AK2,CV,CV1,CV2,RO0,
 RO1,RO2,S0,D0/COM3/E1,PH,B,U1,U2,R,CC/COMC/KT,NF
      DATA A4,A5/0.455445545,0.544554455/
      IF(KG.NE.0)GOTO 20
      B1=(1.+A2*P)**Z
      B2=(A3+1.34*P)/(A3+P)
20    DO 30 K=1,NZ
      EDA3=EXP(A1*(-1.+B1*((T(K)-A4)/A5)**S0))
      EDA(K)=EDA3
```

```
      IF(NF.NE.0)EDA(K)=EQEDA(DPX,DPY,P,H,EDA3)
30    RO(K)=B2+D0*T0*(T(K)-1.)
      CC1=0.5/(NZ-1.)
      CC2=1./(NZ-1.)
      C1=0.
      C2=0.
      DO 40 K=1,NZ
      IF(K.EQ.1)GOTO 32
      C1=C1+0.5/EDA(K)+0.5/EDA(K-1)
      C2=C2+CC1*((K-1.)/EDA(K)+(K-2.)/EDA(K-1))
32    EDA1(K)=C1*CC2
40    EDA2(K)=C2*CC2
      RETURN
      END
      SUBROUTINE UCAL(I,J,NX,NY,NZ,DX,H,EDA,RO,ROR,EDA1,EDA2,U,UU,DU,V,
      VV,DV,W,DPX,DPY)
      DIMENSION U(NZ),UU(NX,NY,NZ),DU(NZ),V(NZ),VV(NX,NY,NZ),DV(NZ),
      W(NZ),ROR(NX,NY,NZ),EDA(NZ),RO(NZ),EDA1(NZ),EDA2(NZ)
      COMMON /COM2/T0,AK0,AK1,AK2,CV,CV1,CV2,RO0,RO1,RO2,S0,D0/COM3/E1,
      PH,B,U1,U2,R,CC
      COMMON /COMD/AD,AD1,KK1,KK2,KK,KK4
      IF(KK.NE.0)GOTO 20
      A1=U1
      A2=PH*(B/R)**3/E1
      A3=U2-U1
20    CUA=A2*DPX*H
      CUB=CUA*H
      CVA=A2*DPY*H
      CVB=CVA*H
      CC3=A3/H
      CC4=1./EDA1(NZ)
      DO 30 K=1,NZ
      U(K)=A1+CUB*(EDA2(K)-CC4*EDA2(NZ)*EDA1(K))+A3*CC4*EDA1(K)
      V(K)=CVB*(EDA2(K)-CC4*EDA2(NZ)*EDA1(K))
      DU(K)=CUA/EDA(K)*((K-1.)/(NZ-1.)-CC4*EDA2(NZ))+CC3*CC4/EDA(K)
30    DV(K)=CVA/EDA(K)*((K-1.)/(NZ-1.)-CC4*EDA2(NZ))
      A4=B/((NZ-1)*R*DX)
      C1=A4*H
      IF(KK.EQ.0)GOTO 50
      DO 40 K=2,NZ-1
      W(K)=(RO(K-1)*W(K-1)+C1*(RO(K)*(U(K)+V(K))-ROR(I-1,J,K)*UU(I-1,
      J,K)-ROR(I,J-1,K)*VV(I,J-1,K)))/RO(K)
40    CONTINUE
50    KK=1
      RETURN
      END
      SUBROUTINE TCAL(NZ,DX,CC1,CC2,T,TFX,TFY,U,V,W,DU,DV,H,DPX,DPY,
      EDA,RO)
      DIMENSION T(NZ),U(NZ),DU(NZ),V(NZ),DV(NZ),W(NZ),EDA(NZ),RO(NZ),
      A(4,21),D(21),AA(2,21),TFX(NZ),TFY(NZ)
      COMMON /COM2/T0,AK0,AK1,AK2,CV,CV1,CV2,RO0,RO1,RO2,S0,D0/COM3/E1,
      PH,B,U1,U2,R,CC
```

```
      COMMON /COMD/AD,AD1,KK1,KK2,KK3,KK/COME/US,EDA0
      DATA CC5,PAI/0.6666667,3.14159265/
      IF(KK.NE.0)GOTO 5
      KK=1
      A2=-CV*RO0*E1*B**3/(EDA0*AK0*R)
      A3=-E1*PH*B**3*D0/(AK0*EDA0*R)
      A4=-(E1*R)**2/(AK0*EDA0*T0)
      A5=0.5*R/B*A2
      A6=AK0*SQRT(EDA0*R/(PAI*RO1*CV1*U1*E1*AK1*B**3))
      A7=AK0*SQRT(EDA0*R/(PAI*RO2*CV2*U2*E1*AK2*B**3))
5     CC3=A6*SQRT(DX)
      CC4=A7*SQRT(DX)
      DZ=H/(NZ-1.)
      DZ1=1./DZ
      DZ2=DZ1*DZ1
      CC6=A3*DPX
      CC7=A3*DPY
      DO 10 K=2,NZ-1
      A(1,K)=DZ2+DZ1*A5*RO(K)*W(K)
      A(2,K)=-2.*DZ2+A2*RO(K)*(U(K)+V(K))/DX+(CC6*U(K)+CC7*V(K))/RO(K)
      A(3,K)=DZ2-DZ1*A5*RO(K)*W(K)
10    A(4,K)=A4*EDA(K)*(DU(K)**2+DV(K)**2)+A2*RO(K)*(U(K)*TFX(K)+V(K)
      *TFY(K))/DX
      A(1,1)=0.
      A(2,1)=1.+2.*DZ1*CC3*CC5
      A(3,1)=-2.*DZ1*CC3*CC5
      A(1,NZ)=-2.*DZ1*CC4*CC5
      A(2,NZ)=1.+2.*DZ1*CC4*CC5
      A(3,NZ)=0.
      A(4,1)=1.+CC1*CC3*DZ1
      A(4,NZ)=1.-CC2*CC4*DZ1
      CALL TRA3(NZ,D,A,AA)
      DO 20 K=1,NZ
      T(K)=(1.-CC)*T(K)+CC*D(K)
20    IF(T(K).LT.1.)T(K)=1.
30    CONTINUE
      RETURN
      END
      SUBROUTINE TRA3(N,D,A,B)
      DIMENSION D(N),A(4,N),B(2,N)
      C=1./A(2,N)
      B(1,N)=-A(1,N)*C
      B(2,N)=A(4,N)*C
      DO 10 I=1,N-1
      IN=N-I
      IN1=IN+1
      C=1./(A(2,IN)+A(3,IN)*B(1,IN1))
      B(1,IN)=-A(1,IN)*C
10    B(2,IN)=(A(4,IN)-A(3,IN)*B(2,IN1))*C
      D(1)=B(2,1)
      DO 20 I=2,N
20    D(I)=B(1,I)*D(I-1)+B(2,I)
```

```
      RETURN
      END
      SUBROUTINE ERROM(NX,NY,NZ,T,ERM)
      DIMENSION T(NX,NY,NZ)
      COMMON /COMT/T1(65,65)
      ERM=0.
      C1=1./FLOAT(NZ)
      DO 20 I=2,NX
      DO 20 J=2,NY
      TT=0.
      DO 10 K=1,NZ
10    TT=TT+T(I,J,K)
      TT=C1*TT
      ER=ABS((TT-T1(I,J))/TT)
      IF(ER.GT.ERM)ERM=ER
20    T1(I,J)=TT
      RETURN
      END
      SUBROUTINE OUPT(N,DX,X,Y,H,P,EDA,TMAX)
      DIMENSION X(N),Y(N),H(N,N),P(N,N),EDA(N,N)
      COMMON /COM1/ENDA,A1,A2,A3,Z,HM0/COMH/HMC
      COMMON /COMW/W0,T1,T2,RX,B,PH/COMK/LMIN,AKC/COMD/AD,AD1,KK1,KK2,
      KK3,KK4
      NN=(N+1)/2
      A=0.0
      WRITE(8,110)A,(Y(I),I=1,N)
      DO I=1,N
      WRITE(8,110)X(I),(H(I,J),J=1,N)
      ENDDO
      WRITE(9,110)A,(Y(I),I=1,N)
      DO I=1,N
      WRITE(9,110)X(I),(P(I,J),J=1,N)
      ENDDO
110   FORMAT(66(E12.6,1X))
      F=0.0
      HMIN=H(1,1)
      PMAX=0.0
      HM=0.0
      NCOUN=0
      NPA=0
      DO I=2,N
      DO J=1,N
      IF(X(I).LE.0.0.AND.X(I+1).GE.0.0)THEN
      IF(Y(J).LE.0.0.AND.Y(I+1).GE.0.0)HC=H(I,J)
      ENDIF
      DPDX=(P(I,J)-P(I-1,J))/DX
      TAU=T1*DPDX*H(I,J)+0.5*AKC*T2*EDA(I,J)/H(I,J)
      F=F+TAU
      IF(H(I,J).LT.HMIN)HMIN=H(I,J)
      IF(P(I,J).GT.PMAX)PMAX=P(I,J)
      RAD=SQRT(X(I)*X(I)+Y(J)*Y(J))
      IF(RAD.LE.0.5)THEN
```

```
      NCOUN=NCOUN+1
      HM=HM+H(I,J)
      ENDIF
      IF(P(I,J).GT.1.E-6)THEN
      PA=PA+P(I,J)
      NPA=NPA+1
      ENDIF
      ENDDO
      ENDDO
      PA=PA/FLOAT(NPA)
      HM=HM/FLOAT(NCOUN)
      F=B*B*F*DX*DX/W0
      HMIN=HMIN*B*B/RX
      HM=HM*B*B/RX
      HC=HC*B*B/RX
      PMAX=PMAX*PH
      HDM=HM0*B*B/RX
      HDC=HMC*B*B/RX
      WRITE(10,*)'W0,F,HMIN,HC,HDM,HDC,PMAX,HM,TMAX,PA'
      WRITE(10,120)W0,F,HMIN,HC,HDM,HDC,PMAX,HM,TMAX,PA
  120 FORMAT(10(1X,E12.6))
      RETURN
      END
      FUNCTION EQEDA(DPDX,DPDY,P,H,EDA)
      COMMON /COME/U0,EDA0/COMW/W0,T1,T2,R,B,PH/COMK/LMIN,AKC
      DATA TAU0/2.E7/
      DPDX1=DPDX*PH/B
      DPDY1=DPDY*PH/B
      P1=P*PH
      H1=H*B*B/R
      EDA1=EDA*EDA0
      TAUL=TAU0+0.036*P1
      C1=-0.5*EDA1*AKC*U0/H1-0.5*H1*DPDX1
      TAU1=DPDX1*H1+C1
      TAU2=C1
      TAUY=0.5*DPDY1*H1
      TAUX=AMAX1(ABS(TAU1),ABS(TAU2))
      TAU=SQRT(TAUX**2+TAUY**2)
      X=TAUL/TAU
      EQEDA=EDA
      IF(X.LT.1)THEN
      EQEDA=EDA*X
      ENDIF
      IF(EQEDA.LT.1)EQEDA=1.
      RETURN
      END
      BLOCK DATA
      COMMON /COM2/T0,AK0,AK1,AK2,CV,CV1,CV2,RO0,RO1,RO2,S0,D0
      DATA T0,AK0,AK1,AK2,CV,CV1,CV2,RO0,RO1,RO2,S0,D0/303.,0.14,46.,
      46.,2000.,470.,470.,890.,7850.,7850.,-1.1,-0.00065/
      END
```

10.4.3 Example

By using the conditions given in the program, the film thickness, pressure distribution, and average temperature rise of TEHL in the point contact can be obtained and are shown in Figure 10.6.

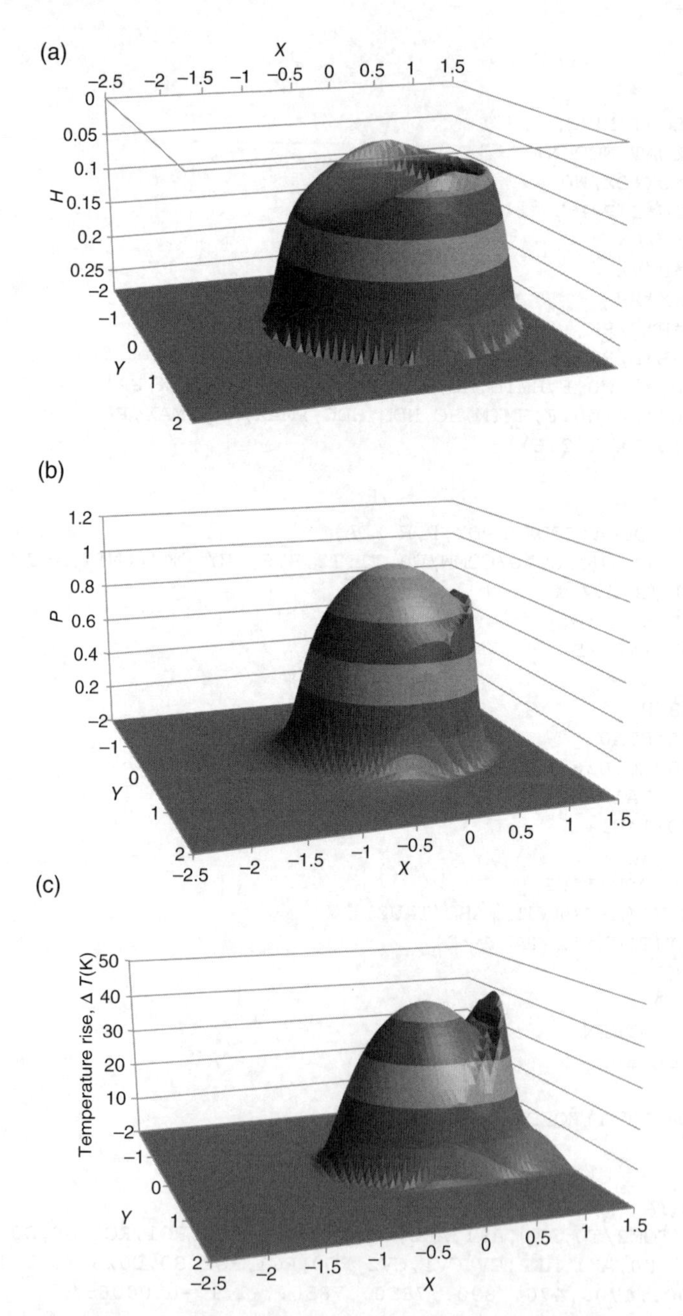

Figure 10.6 Film thickness, pressure distribution, and average temperature rise distribution of TEHL in point contact. (a) Film thickness distribution, (b) pressure distribution, and (c) average temperature rise distribution

11

Numerical calculation method and program for grease EHL

11.1 Basic equations of grease EHL

Grease is a kind of lubricant that is obtained by adding thickening agents to oil to form a semisolid or jelly-like substance. The thickening agents added are commonly metallic soaps, whose fibers form mesh frameworks for storing oil. As grease is of a three-dimensional frame structure composed of fiber, it is not laminar, and hence the grease lubrication process shows complex mechanical properties, for example, time-dependent and viscoplastic. Figure 11.1 shows the rheological behavior of grease. The main characteristics of grease can be summarized as follows.

1. The viscosity of grease usually increases with decrease in the shear strain rate. Thus, the relationship between the shear stress and shear rate is nonlinear.
2. As shown in Figure 11.1, grease has a yield shear stress τ_s. Only when the shear stress τ is larger than τ_s, grease presents the properties of a fluid. When $\tau \le \tau_s$, grease behaves like a solid. It may have a certain amount of elastic deformation under a load. Because grease has a yield shear stress, the grease lubricant film appears a non-flow layer when $\tau \le \tau_s$. In the flowing layer, the velocities perpendicular to the layer are the same.
3. Grease has a thixotropic property. This means that when grease flows under a certain shear stress, both the shear strain and the viscosity decrease gradually with increase of time. Furthermore, its viscosity will partially recover after the shear

Numerical Calculation of Elastohydrodynamic Lubrication: Methods and Programs, First Edition. Ping Huang.

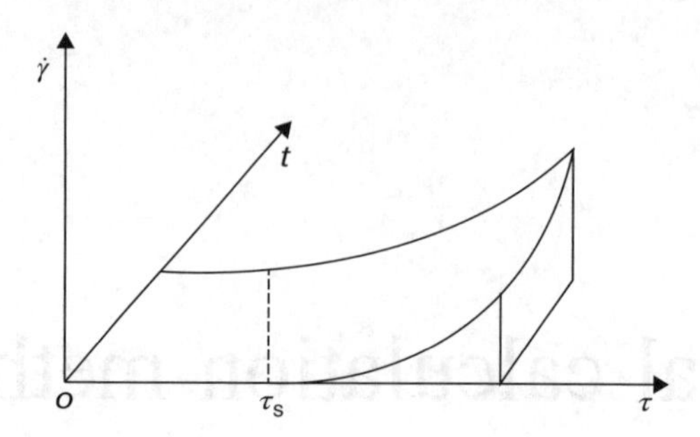

Figure 11.1 Rheological property of grease

process ceases. Thus, grease lubrication is time-dependent. If grease lubrication is "steady," it is assumed not to be time-dependent.

Three kinds of constitutive equations are commonly used to describe the rheological property of grease. They are as follows:

1. Ostwald model

$$\tau = \phi\dot{\gamma}^n \tag{11.1}$$

2. Bingham model

$$\tau = \tau_s + \phi\dot{\gamma} \tag{11.2}$$

3. Herschel–Bulkley model

$$\tau = \tau_s + \phi\dot{\gamma}^n \tag{11.3}$$

The Herschel–Bulkley model is much closer to the experimental results and is especially more accurate at a low velocity. In addition, when $n = 1$, the Herschel–Bulkley model will turn into the Bingham model, but when $\tau_s = 0$, it turns into the Ostwald model. Therefore, the Herschel–Bulkley model is universal. Strictly speaking, the three rheological parameters τ_s, ϕ, and n should be the functions of temperature and pressure. For isothermal lubrication, we do not need to consider the influence of temperature. Furthermore, the rheological index n is usually assumed not to be related to pressure p, but the yield shear stress τ_s and the plastic viscosity ϕ vary with pressure p according to the following relationships:

$$\tau_s = \tau_{s0} e^{ap}$$
$$\phi = \phi_0 e^{ap} \tag{11.4}$$

The equations of grease lubrication are similar to those of oil lubrication. The one-dimensional Reynolds equation of grease lubrication can also be derived from the constitutive equation, the equilibrium equation, and the continuity equation. However, because the constitutive equation of grease contains the yield shear stress τ_s, the lubricant film will be divided into two parts, known as the nonshearing flow part and the shearing flow part. The two parts must be dealt with separately so that the deriving process is complicated. For example, the one-dimensional Reynolds equation based on the Herschel–Bulkley model is as follows:

$$\frac{dp}{dx} = 2\phi\left[2\left(2+\frac{1}{n}\right)\right]^n \frac{U^n(h-\bar{h})^n}{h^{2n+1}}\left[1-\frac{2\tau_s}{(dp/dx)h}\right]^{-(n+1)} \left[1+\frac{n}{n+1}\frac{2\tau_s}{(dp/dx)h}\right]^{-n} \tag{11.5}$$

where, $\bar{h}$ is the film thickness at $dp/dx = 0$.

If $\tau_s = 0$, then Equation 11.5 will become the Reynolds equation of the Ostwald model and is as follows:

$$\frac{dp}{dx} = \text{sign}(h-\bar{h})\frac{2\phi}{h^{2n+1}}U^n\left[2\left(2+\frac{1}{n}\right)\right]^n |h-\bar{h}|^n \tag{11.6}$$

where, $\text{sign}(h-\bar{h})$ is the function; it takes the same sign of $h-\bar{h}$ and its value is 1.

If $\tau_s = 0$ and $n = 1$, Equation 11.5 turns into the Reynolds equation of the Newtonian fluid.

$$\frac{dp}{dx} = 12\phi U\frac{h-\bar{h}}{h^3} \tag{11.7}$$

It should be noted that the Reynolds equation can be derived according to different rheological models, but its applications are not exactly the same. In order to obtain the practical solutions to meet the requirements, one must carefully choose the rheological model according to the working conditions.

11.2 Numerical calculation and program for isothermal grease EHL

11.2.1 Line contact problem

11.2.1.1 Basic equations and calculation method

Reynolds equation

According to Equation 11.5 of the Ostwald model, the Reynolds equation for grease elastohydrodynamic lubrication (GEHL) in the line contact is as follows:

$$\frac{n}{2n+1}\cdot\left(\frac{1}{2}\right)^{\frac{n+1}{n}}\left\{\frac{d}{dx}\left[ph^{\frac{2n+1}{n}}\left(\frac{1}{\phi}\frac{dp}{dx}\right)^{\frac{1}{n}}\right]\right\}=U\frac{d(ph)}{dx} \tag{11.8}$$

The nondimensional Reynolds equation for GEHL in the line contact is as follows:

$$\frac{d}{dX}\left[\varepsilon\cdot\left(\frac{dp}{dX}\right)^{1/n}\right]-\frac{dH}{dX}=0 \tag{11.9}$$

The boundary conditions are given as follows:

In the inlet $P(X_0)=0$
In the outlet $P(X_e)=0$ and $dP(X_e)/dX=0$

The nondimensional Reynolds equation in the discrete form can be written as follows:

$$\frac{\varepsilon_{i+1/2}(P_{i+1}-P_i)^{1/n}-\varepsilon_{i-1/2}(P_i-P_{i-1})^{1/n}}{\Delta X^{1+1/n}}-\frac{H_i-H_{i-1}}{\Delta X}=0 \tag{11.10}$$

At the inlet, the boundary condition is $P(X_0)=0$ and the boundary of the outlet is determined by setting the negative pressure zero.

Film thickness equation

$$h(x)=h_c+\frac{x^2}{2R}-\frac{2}{\pi E}\int_{x_0}^{x_e}p(s)\ln(x-s)^2ds+c \tag{11.11}$$

The nondimensional film thickness equation

$$H(X)=H_0+\frac{X^2}{2}-\frac{1}{\pi}\int_{X_0}^{X_e}\ln|X-X'|p(X')dX' \tag{11.12}$$

Its discrete form is as follows

$$H_i = H_0 + \frac{x_i^2}{2} - \frac{1}{\pi}\sum_{j=1}^{n} K_{ij} P_j \tag{11.13}$$

Viscosity–pressure equation and density–pressure equation

Because there is no widely recognized viscosity–pressure equation and the density–pressure equation of grease, here we have used the equations of oil but changed the initial viscosity. If one has his/her own formulas for grease, he/she can replace that in the equation for calculation without affecting the program execution.

Viscosity–pressure equation

$$\phi = \phi_0 \ \exp\{(\ln\phi_0 + 9.67)\left[\left(1 + 5.1\times10^{-9}p\right)^z - 1\right]\} \tag{11.14}$$

The nondimensional viscosity–pressure equation

$$\phi^* = \exp\left\{[\ln(\phi_0) + 9.67]\left[-1 + \left(1 + 5.1\times10^{-9}P\cdot p_{\mathrm{H}}\right)^{0.68}\right]\right\} \tag{11.15}$$

The density of grease is considered as a constant, that is, $\rho = \text{const}$. The nondimensional density $\rho^* = 1$.

Load equation

$$\int_{x_0}^{x_e} p(x)dx = w \tag{11.16}$$

$$W = \int_{X_0}^{X_e} Pdx = \frac{\pi}{2} \tag{11.17}$$

Its discrete form is as follows:

$$\Delta X \sum_{i=1}^{n} \frac{(P_i + P_{i+1})}{2} = \frac{\pi}{2} \tag{11.18}$$

The iterative methods of GEHL program are similar to those of EHL in the line contact, and hence are not repeated here.

11.2.1.2 Calculation program and results

Calculation diagram (Figure 11.2)

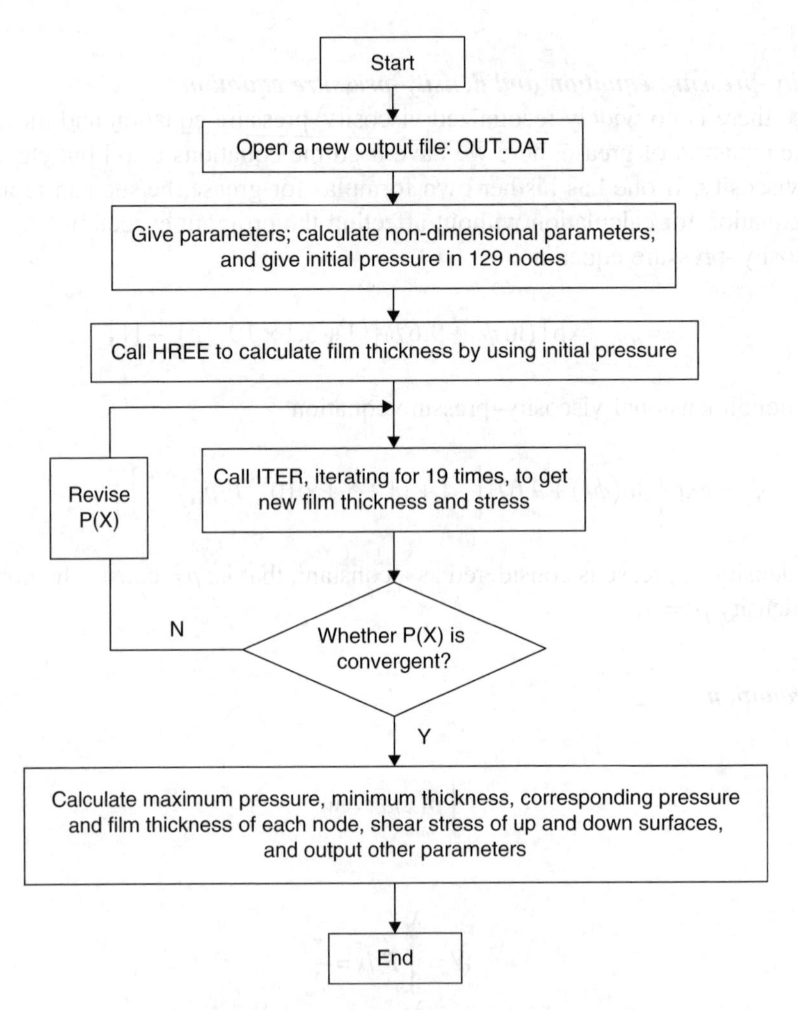

Figure 11.2 Diagram of GEHL in line contact

Functions of main program

Parameters setting

Number of nodes	N = 129
Nondimensional coordinates of inlet	X0 = −4.0
Nondimensional coordinates of outlet	XE = 1.4
Load	W = 1.0E5 N
Equivalent elastic modulus	E1 = 2.26E11 Pa
Initial viscosity of grease	EDA0 = 0.4 Pa · s
Equivalent radius	R = 0.012183 m
Average velocity	US = 0.87 m · s^{-1}
Ratio of sliding to rolling	CU = 0.67
Iterative coefficient	C1 = 0.5
Rheology parameter	FN = 0.8

Input parameters

When the characteristic coefficient S = "Y" or "y," the program calculates the example and gives the results. If the input character is not equal to "Y" or "y," the program does not give the example results. One should input the maximum Hertz contact pressure PH.

Program

```
PROGRAM GREASELINEEHL
CHARACTER*1 S, S1, S2
CHARACTER*16 CDATE, CTIME
COMMON /COM1/ENDA, A1, A2, A3, Z, C1, C3, CW, LMAX, FF/COM2/EDA0/COM4/X0, XE/
COM3/E1, PH, B, U1, U2, R
DATA PAI, Z, P0/3.14159265, 0.68, 1.96E8/S1, S2/1HY, 1Hy/
DATA N, X0, XE, W, E1, EDA0, R, Us, CU, C1, FN/129, -4., 1.4, 1.E5, 2.26E11,
0.4, 0.012183, 0.87, 0.67, 0.5, 0.8/CDATE, CTIME/'The date is', 'The
time is'/
OPEN(8, FILE='OUT.DAT', STATUS='UNKNOWN')
1 FORMAT(20X, A12, I2.2, ':', I2.2, ':', I4.4)
2 FORMAT(20X, A12, I2.2, ':', I2.2, ':', I2.2, '.', I2.2)
WRITE(*, *)'Show the example or not (Y or N)?'
READ(*, '(A)')S
IF(S.EQ.S1.OR.S.EQ.S2)THEN
GOTO 10
ENDIF
WRITE(*, *)'PH='
READ(*, *)PH
W=2.*PAI*R*PH*(PH/E1)
WRITE(*, *)'W=', W
10 CW=N+0.1
FF=1./FN
LMAX=ALOG(CW)/ALOG(2.)
N=2**LMAX+1
```

```
LMIN=(ALOG(CW)-ALOG(SQRT(CW)))/ALOG(2.)
LMAX=LMIN
W1=W/(E1*R)
PH=E1*SQRT(0.5*W1/PAI)
A1=(ALOG(EDA0)+9.67)
A2=PH/P0
A3=0.59/(PH*1.E-9)
B=4.*R*PH/E1
ALFA=Z*A1/P0
G=ALFA*E1
U=EDA0*US/(2.*E1*R)
C3=1.6*(R/B)**2*G**0.6*U**0.7*W1**(-0.13)
ENDA=B**(2.+FF)*(PH/2/EDA0)**FF/R**(1+FF)/US/(2.+FF)
U1=0.5*(2.+CU)*U
U2=0.5*(2.-CU)*U
WRITE(*, *)'B, PH, G, U=', B, PH, G, U
CW=-1.13*C3
WRITE(*, *)N, X0, XE, W, E1, EDA0, R, US, PH
WRITE(8, *)N, W, E1, EDA0, R, US, B, PH, FF
WRITE(*, 40)
40 FORMAT(2X, 'Wait Please', //)
CALL SUBAK(N)
CALL MULTI(N)
STOP
END
SUBROUTINE MULTI(N)
REAL*8 X(1100), P(1100), H(1100), RO(1100), POLD(1100), EPS(1100), EDA
(1100), R(1100), K(1100), E(1100)
COMMON /COM1/ENDA, A1, A2, A3, Z, C1, C3, CW, LMAX, FF/COM4/X0, XE/COM3/E1,
PH, B, U1, U2, RR
DATA MK, G0/1, 1.570796325/
NX=N
DX=(XE-X0)/(N-1.0)
DO 10 I=1, N
X(I)=X0+(I-1)*DX
IF(ABS(X(I)).GE.1.0)P(I)=0.0
IF(ABS(X(I)).LT.1.0)P(I)=SQRT(1.-X(I)*X(I))
10 CONTINUE
CALL HREE(N, DX, H00, G0, X, P, H, RO, EPS, EDA)
CALL FZ(N, P, POLD)
14 KK=19
CALL ITER(N, KK, DX, H00, G0, X, P, H, RO, EPS, EDA, R)
MK=MK+1
CALL ERROP(N, P, POLD, ERP)
IF(ERP.GT.1.E-4.AND.MK.LE.800)THEN
GOTO 14
ENDIF
WRITE(*, *)PH, RR, B
105 IF(MK.GE.800)THEN
WRITE(*, *)'Pressures are not convergent !!!'
READ(*, *)
```

```
ENDIF
FM=FRICT(N, DX, X, H, P, EDA)
H2=1.E3
P2=0.0
DO 106 I=1, N
IF(H(I).LT.H2)H2=H(I)
IF(P(I).GT.P2)P2=P(I)
106 CONTINUE
DO 108 I=1, N
K(I)=P(I)*PH/1.E9
E(I)=H(I)*B*B*1.E6/RR
108 CONTINUE
H3=H2*B*B/RR
P3=P2*PH
110 FORMAT(6(1X, E12.6))
120 CONTINUE
WRITE(8, *)'P2, H2, P3, H3=', P2, H2, P3, H3
CALL OUTHP(N, X, K, E)
RETURN
END
SUBROUTINE OUTHP(N, X, K, E)
REAL*8 X(N), K(N), E(N)
DX=X(2)-X(1)
DO 10 I=1, N
WRITE(8, 20)X(I), K(I), E(I)
10 CONTINUE
20 FORMAT(1X, 6(F20.6, 1X))
RETURN
END
SUBROUTINE HREE(N, DX, H00, G0, X, P, H, RO, EPS, EDA)
REAL*8 X(N), P(N), H(N), RO(N), EPS(N), EDA(2200)
REAL*8 W(2200)
COMMON /COM1/ENDA, A1, A2, A3, Z, C1, C3, CW, K, FF/COM2/EDA0/COMAK/
AK(0:1100)
DATA KK, NW, PAI1/0, 2200, 0.318309886/
IF(KK.NE.0)GOTO 3
HM0=C3
H00=0.0
3  W1=0.0
DO 4 I=1, N
4  W1=W1+P(I)
C3=(DX*W1)/G0
DW=1.-C3
CALL DISP(N, NW, K, DX, P, W)
HMIN=1.E3
DO 30 I=1, N
H0=0.5*X(I)*X(I)-PAI1*W(I)
IF(H0.LT.HMIN)HMIN=H0
H(I)=H0
30  CONTINUE
IF(KK.NE.0)GOTO 32
```

```
KK=1
H00=-HMIN+HM0
32   H0=H00+HMIN
IF(H0.LE.0.0)GOTO 48
IF(H0+0.3*CW*DW.GT.0.0)HM0=H0+0.3*CW*DW
IF(H0+0.3*CW*DW.LE.0.0)HM0=HM0*C3
48   H00=HM0-HMIN
50   DO 60 I=1, N
60   H(I)=H00+H(I)
DO 100 I=1, N
EDA(I)=EXP(A1*(-1.+(1.+A2*P(I))**Z))
RO(I)=(A3+1.34*P(I))/(A3+P(I))
EPS(I)=RO(I)*H(I)**(2+FF)*ENDA/EDA(I)**FF
100  CONTINUE
RETURN
END
SUBROUTINE ITER(N, KK, DX, H00, G0, X, P, H, RO, EPS, EDA, R)
REAL*8 X(N), P(N), H(N), RO(N), EPS(N), EDA(N), R(N)
COMMON /COM1/ENDA, A1, A2, A3, Z, C1, C3, CW, LMAX, FF/COMAK/AK(0:1100)
DATA KG1, PAI/0, 3.14159265/
IF(KG1.NE.0)GOTO 5
KG1=1
DX1=1./DX
DX2=DX*DX
DX3=1./DX2
DX4=DX1/PAI
DX5=DX1**(1+FF)
DXL=DX*ALOG(DX)
AK0=DX*AK(0)+DXL
AK1=DX*AK(1)+DXL5DO 100 K=1, KK
D2=0.5*(EPS(1)+EPS(2))
D3=0.5*(EPS(2)+EPS(3))
D5=DX1*(RO(2)*H(2)-RO(1)*H(1))
D7=DX4*(RO(2)*AK0-RO(1)*AK1)
PP=0.
DO 70 I=2, N-1
D1=D2
D2=D3
D4=D5
D6=D7
IF(I+2.LE.N)D3=0.5*(EPS(I+1)+EPS(I+2))
D5=DX1*(RO(I+1)*H(I+1)-RO(I)*H(I))
D7=DX4*(RO(I+1)*AK0-RO(I)*AK1)
DD=(D1+D2)*DX3
IF(0.05*DD.LT.ABS(D6))GOTO 10
RI=-DX5*(D2*SIGN(1.0, (P(I+1)-P(I)))*ABS((P(I+1)-P(I)))**(FF)-D1*SIGN
(1.0, (P(I)-P(I-1)))*ABS((P(I)-P(I-1)))**(FF))+D4
R(I)=RI
DLDP=-FF*DX5*(D1*ABS((P(I)-P(I-1)))**(FF-1)+D2*ABS((P(I+1)-P(I)))**
(FF-1))+D6
RI=RI/DLDP
```

```
RI=RI/C1
GOTO 20
10  RI=-DX5*(D2*SIGN(1.0, (P(I+1)-P(I)))*ABS((P(I+1)-P(I)))**
(FF)-D1*SIGN(1.0, (P(I)-PP))*ABS((P(I)-PP))**(FF))+D4
R(I)=RI
DLDP=-FF*DX5*(2*D1*ABS((P(I)-PP))**(FF-1)+D2*ABS((P(I+1)-P(I)))**
(FF-1))+2.*D6
RI=RI/DLDP
RI=0.5*RI
IF(I.GT.2.AND.P(I-1)-C1*RI.GT.0)P(I-1)=P(I-1)-C1*RI
20  PP=P(I)
P(I)=P(I)+C1*RI
IF(P(I).LT.0.0)P(I)=0.0
IF(P(I).LE.0.0)R(I)=0.0
70  CONTINUE
CALL HREE(N, DX, H00, G0, X, P, H, RO, EPS, EDA)
100  CONTINUE
RETURN
END
SUBROUTINE DISP(N, NW, KMAX, DX, P1, W)
REAL*8 P1(N), W(NW), P(2200), AK1(0:50), AK2(0:50)
COMMON /COMAK/AK(0:1100)
DATA NMAX, KMIN/2200, 1/
N2=N
M=3+2*ALOG(FLOAT(N))
K1=N+KMAX
DO 10 I=1, N
10  P(K1+I)=P1(I)
DO 20 KK=KMIN, KMAX-1
K=KMAX+KMIN-KK
N1=(N2+1)/2
CALL DOWNP(NMAX, N1, N2, K, P)
20  N2=N1
DX1=DX*2**(KMAX-KMIN)
CALL WI(NMAX, N1, KMIN, KMAX, DX, DX1, P, W)
DO 30 K=KMIN+1, KMAX
N2=2*N1-1
DX1=DX1/2.
CALL AKCO(M+5, KMAX, K, AK1)
CALL AKIN(M+6, AK1, AK2)
CALL WCOS(NMAX, N1, N2, K, W)
CALL CORR(NMAX, N2, K, M, 1, DX1, P, W, AK1)
CALL WINT(NMAX, N2, K, W)
CALL CORR(NMAX, N2, K, M, 2, DX1, P, W, AK2)
30  N1=N2
DO 40 I=1, N
40  W(I)=W(K1+I)
RETURN
END
SUBROUTINE DOWNP(NMAX, N1, N2, K, P)
REAL*8 P(NMAX)
```

```
K1=N1+K-1
K2=N2+K-1
DO 10 I=3, N1-2
I2=2*I+K2
10  P(K1+I)=(16.*P(I2)+9.*(P(I2-1)+P(I2+1))-(P(I2-3)+P(I2+3)))/32.
P(K1+2)=0.25*(P(K2+3)+P(K2+5))+0.5*P(K2+4)
P(K1+N1-1)=0.25*(P(K2+N2-2)+P(K2+N2))+0.5*P(K2+N2-1)
RETURN
END

SUBROUTINE WCOS(NMAX, N1, N2, K, W)
REAL*8 W(NMAX)
K1=N1+K-1
K2=N2+K
DO 10 I=1, N1
II=2*I-1
10  W(K2+II)=W(K1+I)
RETURN
END
SUBROUTINE WINT(NMAX, N, K, W)
REAL*8 W(NMAX)
K2=N+K
DO 10 I=4, N-3, 2
II=K2+I
10  W(II)=(9.*(W(II-1)+W(II+1))-(W(II-3)+W(II+3)))/16.
I1=K2+2
I2=K2+N-1
W(I1)=0.5*(W(I1-1)+W(I1+1))
W(I2)=0.5*(W(I2-1)+W(I2+1))
RETURN
END
SUBROUTINE CORR(NMAX, N, K, M, I1, DX, P, W, AK)
REAL*8 P(NMAX), W(NMAX), AK(0:M)
K1=N+K
IF(I1.EQ.2)GOTO 20
DO 10 I=1, N, 2
II=K1+I
J1=MAX0(1, I-M)
J2=MIN0(N, I+M)
DO 10 J=J1, J2
IJ=IABS(I-J)
10 W(II)=W(II)+AK(IJ)*DX*P(K1+J)
RETURN
20  DO 30 I=2, N, 2
II=K1+I
J1=MAX0(1, I-M)
J2=MIN0(N, I+M)
DO 30 J=J1, J2
IJ=IABS(I-J)
30  W(II)=W(II)+AK(IJ)*DX*P(K1+J)
RETURN
```

```
END
SUBROUTINE WI(NMAX, N, KMIN, KMAX, DX, DX1, P, W)
REAL*8 P(NMAX), W(NMAX)
COMMON /COMAK/AK(0:1100)
K1=N+1
K=2**(KMAX-KMIN)
C=ALOG(DX)
DO 10 I=1, N
II=K1+I
W(II)=0.0
DO 10 J=1, N
IJ=K*IABS(I-J)
10  W(II)=W(II)+(AK(IJ)+C)*DX1*P(K1+J)
RETURN
END
SUBROUTINE AKCO(KA, KMAX, K, AK1)
REAL*8 AK1(0:KA)
COMMON /COMAK/AK(0:1100)
J=2**(KMAX-K)
DO 10 I=0, KA
II=J*I
10  AK1(I)=AK(II)
RETURN
END
SUBROUTINE AKIN(KA, AK1, AK2)
REAL*8 AK1(KA), AK2(KA)
DO 10 I=4, KA-3
10  AK2(I)=(9.*(AK1(I-1)+AK1(I+1))-(AK1(I-3)+AK1(I+3)))/16.
AK2(1)=(9.*AK1(2)-AK1(4))/8.
AK2(2)=(9.*(AK1(1)+AK1(3))-(AK1(3)+AK1(5)))/16.
AK2(3)=(9.*(AK1(2)+AK1(4))-(AK1(2)+AK1(6)))/16.
DO 20 I=1, KA
20  AK2(I)=AK1(I)-AK2(I)
DO 30 I=1, KA-1, 2
I1=I+1
AK1(I)=0.0
30  AK1(I1)=AK2(I1)
RETURN
END
SUBROUTINE SUBAK(MM)
COMMON /COMAK/AK(0:1100)
DO 10 I=0, MM
10  AK(I)=(I+0.5)*(ALOG(ABS(I+0.5))-1.)-(I-0.5)*(ALOG(ABS(I-0.5))-1.)
RETURN
END
FUNCTION FRICT(N, DX, X, H, P, EDA)
REAL*8 X(N), H(N), P(N), EDA(N)
COMMON /COM3/E1, PH, B, U1, U2, R
DATA TAU0, AT/1.98E7, 0.078/
OPEN (10, FILE='TAU.DAT')
TP=TAU0/PH
```

```
TE=TAU0/E1
A=AT/TAU0
BR=B/R
FRICT=0.0
DO 10 I=1, N
DP=0.0
A=1.0+AT*P(I)/TAU0
IF(I.NE.N)DP=(P(I+1)-P(I))/DX
TAU1=0.5*H(I)*DP*(BR/TP/A)
TAU2=2.*U1*EDA(I)/(H(I)*BR**2*TE*A)
FRICT=FRICT+ABS(TAU1)+TAU2
TAU2=0.05*TAU2
WRITE(10, 5)X(I), TAU1, TAU2, P(I), H(I)
5  FORMAT(5(E12.6, 1X))
10 CONTINUE
FRICT=FRICT*DX*B*TAU0
RETURN
END
SUBROUTINE FZ(N, P, POLD)
REAL*8 P(N), POLD(N)
DO 10 I=1, N
10  POLD(I)=P(I)
RETURN
END
SUBROUTINE ERROP(N, P, POLD, ERP)
REAL*8 P(N), POLD(N)
SD=0.0
SUM=0.0
DO 10 I=1, N
SD=SD+ABS(P(I)-POLD(I))
POLD(I)=P(I)
10  SUM=SUM+P(I)
ERP=SD/SUM
RETURN
END
```

Results

The characteristics of GEHL in the line contact can be analyzed according to the operation results. A typical GEHL film is shown in Figure 11.3, which has the following main features:

1. The film thickness shape and the pressure distribution of GEHL are similar to that of oil lubrication.
2. In the central contact area, the lubrication pressure is close to the Hertz stress and the shape of the film thickness is approximately parallel.
3. At the outlet, the pressure appears as an obvious secondary pressure peak, but its width is extremely narrow. Then, the pressure falls sharply to the environmental pressure;

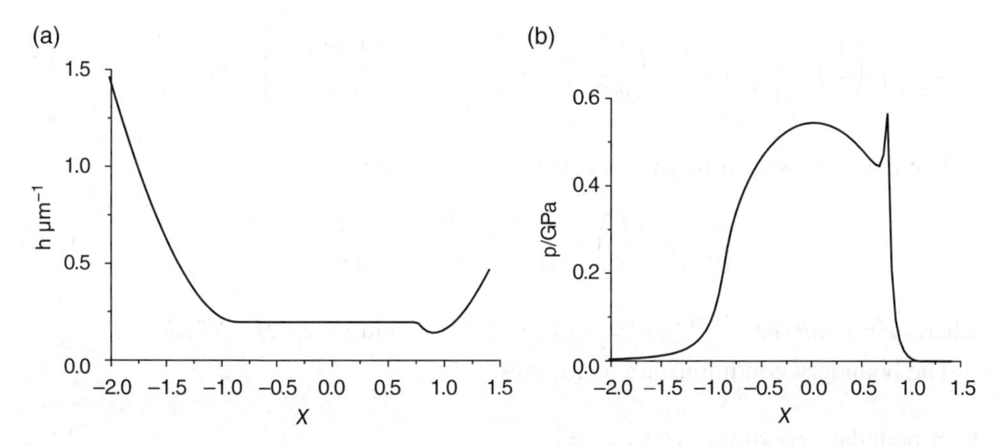

Figure 11.3 Film thickness and pressure distribution of GEHL ($w = 100$ kN, $u_s = 0.87$ m·s^{-1}, $n = 0.8$). (a) Film thickness and (b) pressure

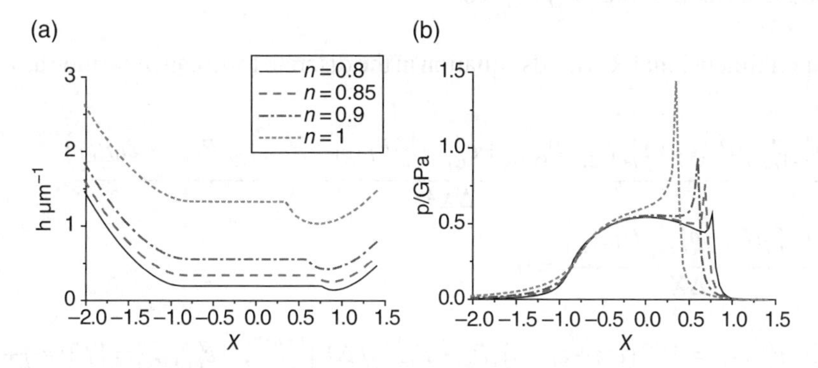

Figure 11.4 Influence of n on GEHL ($w = 100$ kN, $u_s = 0.87$ m · s^{-1}). (a) Film thickness and (b) pressure

4. At the second pressure peak, the film begins to shrink and appears the necking. The film thickness at the necking place is the minimum.

Figure 11.4 shows the influence of the rheology parameter n on the lubrication film. From the figure, we can see that as n increases, the pressure peak increases and its position moves toward the inlet and the average film thickness also increases.

11.2.2 Point contact problem

11.2.2.1 Basic equations and calculation method

Reynolds equation

Based on the Ostwald rheological equation, the Reynolds equation for GEHL in the point contact can be written as follows:

$$\frac{n}{2n+1}\cdot\left(\frac{1}{2}\right)^{\frac{n+1}{n}}\left\{\frac{\partial}{\partial x}\left[\rho h^{\frac{2n+1}{n}}\cdot\left(\frac{1}{\phi}\frac{\partial p}{\partial x}\right)^{\frac{1}{n}}\right]+\frac{\partial}{\partial y}\left[\rho h^{\frac{2n+1}{n}}\cdot\left(\frac{1}{\phi}\frac{\partial p}{\partial y}\right)^{\frac{1}{n}}\right]\right\}=u_s\frac{\partial(\rho h)}{\partial x} \tag{11.19}$$

The nondimensional Reynolds equation is as follows:

$$\frac{\partial}{\partial X}\left(\varepsilon'\cdot\frac{\partial P}{\partial X}\right)+\frac{\partial}{\partial Y}\left(\varepsilon''\cdot\frac{\partial P}{\partial Y}\right)-\frac{\partial(\rho^* H)}{\partial X}=0 \tag{11.20}$$

where, $\varepsilon'=\varepsilon\cdot|\partial P/\partial X|^{(1-n)/n}$, $\varepsilon''=\varepsilon\cdot|\partial P/\partial Y|^{(1-n)/n}$, and $\varepsilon=\lambda\rho^* H^{2+1/n}/\phi^{*1/n}$.

The boundary conditions are as follows:

Inlet boundary condition: $P(X_0,Y)=0$

Outlet boundary conditions: $P(X_e,Y)=0$ and $\dfrac{\partial P(X_e,Y)}{\partial X}=0$

Side boundary conditions: $P|_{Y=\pm 1}=0$

The nondimensional Reynolds equation in the discrete form can be written as follows:

$$\begin{aligned}&\frac{\left(\varepsilon'_{i-1/2,j}\cdot P_{i-1,j}+\varepsilon'_{i+1/2,j}\cdot P_{i+1,j}+\varepsilon''_{i,j-1/2}\cdot P_{i,j-1}+\varepsilon''_{i,j+1/2}\cdot P_{i,j+1}-\varepsilon_0 P_{i,j}\right)}{\Delta X^2}\\&-\frac{\left(\rho^*_{i,j}H_{i,j}-\rho^*_{i-1,j}H_{i-1,j}\right)}{\Delta X}=0\end{aligned} \tag{11.21}$$

where, $\varepsilon'_{i-1/2,j}=1/2\left(\varepsilon_{i,j}+\varepsilon_{i-1,j}\right)\left|P_{i,j}-P_{i-1,j}/\Delta X\right|^{1-n/n}$, $\varepsilon'_{i+1/2,j}=1/2\left(\varepsilon_{i,j}+\varepsilon_{i+1,j}\right)\left|P_{i+1,j}-P_{i,j}/\Delta X\right|^{1-n/n}$, $\varepsilon''_{i,j-1/2}=1/2\left(\varepsilon_{i,j}+\varepsilon_{i,j-1}\right)\left|P_{i,j}-P_{i,j-1}/\Delta Y\right|^{1-n/n}$, $\varepsilon''_{i,j+1/2}=1/2\left(\varepsilon_{i,j}+\varepsilon_{i,j+1}\right)\left|P_{i,j}-P_{i,j+1}/\Delta Y\right|^{1-n/n}$, and $\varepsilon_0=\varepsilon'_{i-1/2,j}+\varepsilon'_{i+1/2,j}+\varepsilon''_{i,j-1/2}+\varepsilon''_{i,j+1/2}$. Since the grids of the mesh are uniform, the nondimensional node spaces are the same, that is, $\Delta Y=\Delta X$.

Film thickness equation

$$h(x,y)=h_0+\frac{x^2}{2R_x}+\frac{y^2}{2R_y}+v(x,y) \tag{11.22}$$

where, $v(x, y)$ is the elastic deformation

$$v(x,y)=\frac{2}{\pi E}\iint_{\Omega}\frac{p(s,t)}{\sqrt{(x-s)^2+(y-t)^2}}dsdt \tag{11.23}$$

The nondimensional film thickness equation

$$H(X,Y)=H_0+\frac{X^2+Y^2}{2}+\frac{2}{\pi^2}\int_{X_0}^{X_e}\int_{Y_0}^{Y_e}\frac{P(S,T)dSdT}{\sqrt{(X-S)^2+(Y-T)^2}} \tag{11.24}$$

Its discrete form is as follows:

$$H_{ij}=H_0+\frac{X_i^2+Y_j^2}{2}+\frac{2}{\pi^2}\sum_{k=1}^{n}\sum_{l=1}^{n}D_{ij}^{kl}P_{kl} \tag{11.25}$$

The viscosity–pressure equation and the density–pressure equation are same as those of isothermal EHL in the line contact.

Load equation

$$\int_{x_0}^{x_e}\int_{y_0}^{y_e}p(x,x)dxdy=w \tag{11.26}$$

The nondimensional load equation

$$\int_{X_0}^{X_e}\int_{Y_0}^{Y_e}P(X,Y)dXdY=\frac{2}{3}\pi \tag{11.27}$$

Its discrete form is as follows

$$\Delta X\Delta Y\sum_{i=1}^{n}\sum_{j=1}^{n}P_{ij}=\frac{2\pi}{3} \tag{11.28}$$

When calculating the pressure distribution of oil EHL in the point contact, the Gauss–Seidel iteration method is used for the low-pressure region and the Jacobi Bipolar iteration method is used for the high-pressure region. In Equation 11.21, the nonlinear part of the pressure gradient is calculated in the parameter. The Jacobi Bipolar iteration method used in the high-pressure region will result easily in iteration divergence and calculation failure. Therefore, in this chapter, we use the Gauss–Seidel iteration method in the whole region.

11.2.2.2 Flowchart and results

Flowchart (Figure 11.5)

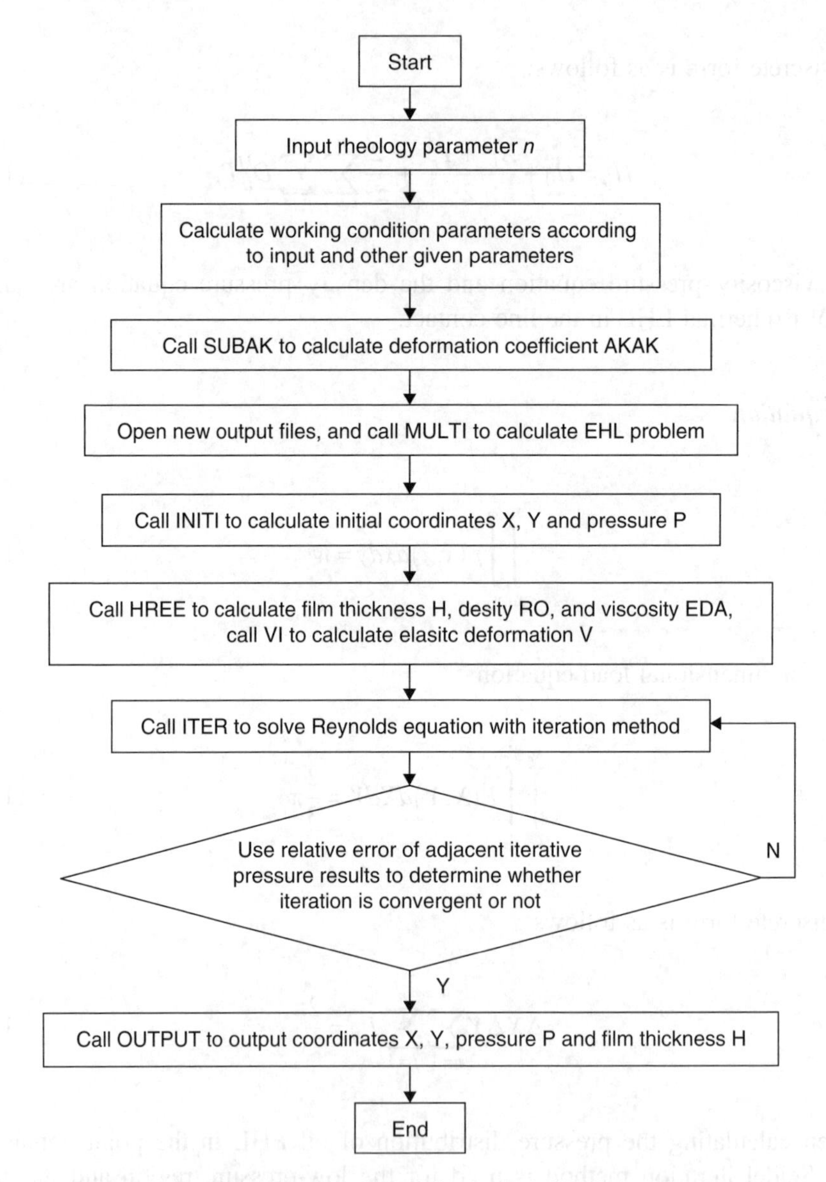

Figure 11.5 Flowchart to calculate isothermal GEHL in point contact

Calculation program

The parameters used in this program are listed as follow:

Number of nodes	N = 65 × 65
Nondimensional starting coordinate in *x* direction	X0 = −2.5
Nondimensional ending coordinate in *x* direction	XE = 1.5
Nondimensional starting coordinate in *y* direction	Y0 = −2.0
Nondimensional ending coordinate in *y* direction	YE = 2.0
Equivalent modulus of elasticity	E1 = 2.21E11 Pa
Initial plastic viscosity	EDA0 = 0.058 Pa · s
Equivalent curvature radius	RX = RY = 0.05 m

Input parameters:

Rheology parameter, *n*

Output data:

The pressure array P(I,J) will be saved in the file PRESSURE.DAT;
The film thickness array H(I,J) will be saved in the file FILM.DAT;
Other data will be saved in the file OUT.DAT.

```
   PROGRAM GREASEPOINTEHL
   DIMENSION THETA(15), EALFA(15), EBETA(15)
   COMMON /COM1/Z, ENDA, AKC, HM0, HMC, EK, EAL, EBE, AD, AD1, KK1, KK2, KK3,
KK4, FN, FN1, FF
   COMMON /COM2/W0, E1, RX, B, PH, US, U1, U2, EDA0
   COMMON /COM3/A1, A2, A3, LMIN
   DATA PAI, Z, AKC, AD, AD1/3.14159265, 0.68, 1.0, 0.0, 0.0/
  DATA T0, EDA0, AK, AK1, AK2, CV, CV1, CV2, RO0, RO1, RO2, S0, D0/303., 0.058,
0.14, 46., 46., 2000., 470., 470., 890., 7850., 7850., -1.1, -0.00065/
   DATA N, NZ, RX, RY, X0, XE,  E1, US, CT, W0/65, 5, 0.05, 0.05, -2.5, 1.5,
2.21E11, 1.5, 0.31, 39.24/
   DATA THETA/10., 20., 30., 35., 40., 45., 50., 55., 60., 65., 70., 75., 80.,
85., 90./
  DATA EALFA/6.612, 3.778, 2.731, 2.397, 2.136, 1.926, 1.754, 1.611, 1.486,
1.378, 1.284, 1.202, 1.128, 1.061, 1.0/
   DATA EBETA/0.319, 0.408, 0.493, 0.53, 0.567, 0.604, 0.641, 0.678, 0.717,
0.759, 0.802, 0.846, 0.893, 0.944, 1.0/
   DATA KK1, KK2, KK3, KK4/0, 0, 0, 0/
   WRITE(*, *)'n<=1 INPUT n=?'
   READ(*, *)FN
   FN1=1.0/FN
   FF=1.0/FN-1.0
   WRITE(*, *)"FF=", FF
   EK=RX/RY
   AA=0.5*(1./RX+1./RY)
```

```
BB=0.5*ABS(1./RX-1./RY)
CC=ACOS(BB/AA)*180.0/PAI
EAL=1.0
EBE=1.0
DO I=1, 15
IF(CC.LT.THETA(I))THEN
WRITE(*, *)I
EAL=EALFA(I-1)+(CC-THETA(I))*(EALFA(I)-EALFA(I-1))/(THETA(I)-THETA
(I-1))
EBE=EBETA(I-1)+(CC-THETA(I))*(EBETA(I)-EBETA(I-1))/(THETA(I)-THETA
(I-1))
GOTO 10
ENDIF
ENDDO
10  EA=EAL*(1.5*W0/AA/E1)**(1./3.0)
EB=EBE*(1.5*W0/AA/E1)**(1./3.0)
PH=1.5*W0/(EA*EB*PAI)
OPEN(8, FILE='FILM.DAT', STATUS='UNKNOWN')
OPEN(9, FILE='PRESS.DAT', STATUS='UNKNOWN')
OPEN(10, FILE='OUT.DAT', STATUS='UNKNOWN')
WRITE(*, *)"N, X0, XE, PH, E1, EDA0, RX, US"
WRITE(*, *)N, X0, XE, PH, E1, EDA0, RX, US
WRITE(16, *)"N, X0, XE, PH, E1, EDA0, RX, US"
WRITE(16, *)N, X0, XE, PH, E1, EDA0, RX, US
H00=0.0
MM=N-1
LMIN=ALOG(N-1.)/ALOG(2.)-1.99
U=EDA0*(US/2.)**FN/(E1*RX**FN)
WRITE(*, *)"U=", U
U1=0.5*(2.+AKC)*U
U2=0.5*(2.-AKC)*U
A1=ALOG(EDA0)+9.67
A2=5.1E-9*PH
A3=0.59/(PH*1.E-9)
B=PAI*PH*RX/E1
W=2.*PAI*PH/(3.*E1)*(B/RX)**2
ALFA=Z*5.1E-9*A1
G=ALFA*E1
AHM=1.0-EXP(-0.68*1.03)
AHC=1.0-0.61*EXP(-0.73*1.03)
HM0=3.63*(RX/B)**2*G**0.49*U**0.68*W**(-0.073)*AHM
HMC=2.69*(RX/B)**2*G**0.53*U**0.67*W**(-0.067)*AHC
ENDA=2.*U*(3.+FF)*2.0**(1.0+FF)*(E1/PH)**(1.0+FF)*(RX/B)**(3.0+FF)
WRITE(*, *)"ENDA=", ENDA
UTL=EDA0*US*RX/(B*B*2.E7)
W0=2.0*PAI*EA*EB*PH/3.0
T1=PH*B/RX
T2=EDA0*US*RX/(B*B)
WRITE(*, *)'          Wait please'
CALL SUBAK(MM)
CALL MULTI(N, NZ, X0, XE, H00)
STOP
```

```
END
SUBROUTINE MULTI(N, NZ, X0, XE, H00)
DIMENSION X(65), Y(65), H(4500), RO(4500), EPS(4500), EDA(4500), P(4500),
POLD(4500), T(65, 65, 5)
COMMON /COM1/Z, ENDA, AKC, HM0, HMC, EK, EAL, EBE, AD, AD1, KK1, KK2, KK3,
KK4, FN, FN1, FF
DATA MK, KTK, G00/200, 1, 2.0943951/
G0=G00*EAL*EBE
NX=N
NY=N
NN=(N+1)/2
CALL INITI(N, DX, X0, XE, X, Y, P, POLD)
CALL HREE(N, DX, H00, G0, X, Y, H, RO, EPS, EDA, P)
M=0
14  KK=15
CALL ITER(N, KK, DX, H00, G0, X, Y, H, RO, EPS, EDA, P)
CALL ERP(N, ER, P, POLD)
ER=ER/KK
WRITE(*, *)'ER=', ER
M=M+1
IF(M.LT.MK.AND.ER.GT.1.E-4)GOTO 14
CALL OUPT(N, DX, X, Y, H, P, EDA, TMAX)
RETURN
END
SUBROUTINE INITI(N, DX, X0, XE, X, Y, P, POLD)
DIMENSION X(N), Y(N), P(N, N), POLD(N, N)
NN=(N+1)/2
DX=(XE-X0)/(N-1.)
Y0=-0.5*(XE-X0)
DO 5 I=1, N
X(I)=X0+(I-1)*DX
Y(I)=Y0+(I-1)*DX
5  CONTINUE
DO 10 I=1, N
D=1.-X(I)*X(I)
DO 10 J=1, NN
C=D-Y(J)*Y(J)
IF(C.LE.0.0)P(I, J)=0.0
10  IF(C.GT.0.0)P(I, J)=SQRT(C)
DO 20 I=1, N
DO 20 J=NN+1, N
JJ=N-J+1
20  P(I, J)=P(I, JJ)
DO I=1, N
DO J=1, N
POLD(I, J)=P(I, J)
ENDDO
ENDDO
RETURN
END
SUBROUTINE HREE(N, DX, H00, G0, X, Y, H, RO, EPS, EDA, P)
DIMENSION X(N), Y(N), P(N, N), H(N, N), RO(N, N), EPS(N, N), EDA(N, N)
```

```
DIMENSION W(150, 150), P0(150, 150), ROU(65, 65)
COMMON /COM1/Z, ENDA, AKC, HM0, HMC, EK, EAL, EBE, AD, AD1, KK1, KK2, KK3,
KK4, FN, FN1, FF
COMMON /COM2/W0, E1, RX, B, PH, US, U1, U2, EDA0
COMMON /COM3/A1, A2, A3, LMIN
DATA KR, NW, PAI, PAI1, DELTA/0, 150, 3.14159265, 0.2026423, 0.0/
NN=(N+1)/2
CALL VI(NW, N, DX, P, W)
HMIN=1.E3
DO 30 I=1, N
DO 30 J=1, NN
RAD=X(I)*X(I)+EK*Y(J)*Y(J)
W1=0.5*RAD+DELTA
ZZ=0.5*AD1*AD1+X(I)*ATAN(AD*PAI/180.0)
IF(W1.LE.ZZ)W1=ZZ
H0=W1+W(I, J)
IF(H0.LT.HMIN)HMIN=H0
30  H(I, J)=H0
IF(KK.EQ.0)THEN
KG1=0
H01=-HMIN+HM0
DH=0.005*HM0
H02=-HMIN
H00=0.5*(H01+H02)
ENDIF
W1=0.0
DO 32 I=1, N
DO 32 J=1, N
32  W1=W1+P(I, J)
W1=DX*DX*W1/G0
DW=1.-W1
IF(KK.EQ.0)THEN
KK=1
GOTO 50
ENDIF
IF(DW.LT.0.0)THEN
KG1=1
H00=AMIN1(H01, H00+DH)
ENDIF
IF(DW.GT.0.0)THEN
KG2=2
H00=AMAX1(H02, H00-DH)
ENDIF
50  DO 60 I=1, N
DO 60 J=1, NN
H(I, J)=H00+H(I, J)
IF(P(I, J).LT.0.0)P(I, J)=0.0
EDA1=EXP(A1*(-1.+(1.+A2*P(I, J))**Z))
EDA(I, J)=EDA1
RO(I, J)=1.
EPS(I, J)=ENDA*RO(I, J)*H(I, J)**(2.+FN1)/(EDA(I, J)**FN1)
60  CONTINUE
```

```
DO 70 J=NN+1, N
JJ=N-J+1
DO 70 I=1, N
H(I, J)=H(I, JJ)
RO(I, J)=RO(I, JJ)
EDA(I, J)=EDA(I, JJ)
70  EPS(I, J)=EPS(I, JJ)
RETURN
END


SUBROUTINE ITER(N, KK, DX, H00, G0, X, Y, H, RO, EPS, EDA, P)
DIMENSION X(N), Y(N), P(N, N), H(N, N), RO(N, N), EPS(N, N), EDA(N, N)
DIMENSION D(70), A(350), B(210), ID(70)
COMMON /COM1/Z, ENDA, AKC, HM0, HMC, EK, EAL, EBE, AD, AD1, KK1, KK2, KK3,
KK4, FN, FN1, FF
COMMON /COMAK/AK(0:65, 0:65)
DATA KG1, PAI1, C1, C2/0, 0.2026423, 0.27, 0.27/
IF(KG1.NE.0)GOTO 2
KG1=1
AK00=AK(0, 0)
AK10=AK(1, 0)
AK20=AK(2, 0)
BK00=AK00-AK10
BK10=AK10-0.25*(AK00+2.*AK(1, 1)+AK(2, 0))
BK20=AK20-0.25*(AK10+2.*AK(2, 1)+AK(3, 0))
2  NN=(N+1)/2
MM=N-1
DX1=1./DX
DX2=DX*DX
DX3=1./DX2
DO 100 K=1, KK
PMAX=0.0
DO 70 J=2, NN
J0=J-1
J1=J+1
IA=1
8  MM=N-IA
IF(P(MM, J0).GT.1.E-6)GOTO 20
IF(P(MM, J).GT.1.E-6)GOTO 20
IF(P(MM, J1).GT.1.E-6)GOTO 20
IA=IA+1
IF(IA.LT.N)GOTO 8
GOTO 70
20  IF(MM.LT.N-1)MM=MM+1
DPDX1=ABS((P(2, J)-P(1, J))*DX1)**(FF)
D2=0.5*(EPS(1, J)+EPS(2, J))*DPDX1
DO 50 I=2, MM
I0=I-1
I1=I+1
II=5*I0
DPDX2=ABS((P(I1, J)-P(I, J))*DX1)**(FF)
```

```
DPDY1=ABS((P(I, J)-P(I, J0))*DX1)**(FF)
DPDY2=ABS((P(I, J1)-P(I, J))*DX1)**(FF)
D1=D2
D2=0.5*(EPS(I1, J)+EPS(I, J))*DPDX2
D4=0.5*(EPS(I, J0)+EPS(I, J))*DPDY1
D5=0.5*(EPS(I, J1)+EPS(I, J))*DPDY2
P1=P(I0, J)
P2=P(I1, J)
P3=P(I, J)
P4=P(I, J0)
P5=P(I, J1)
D3=D1+D2+D4+D5
IF(H(I, J).LE.0.0)THEN
ID(I)=0
A(II+1)=0.0
A(II+2)=0.0
A(II+3)=1.0
A(II+4)=0.0
A(II+5)=1.0
A(II-4)=0.0
GOTO 50
ENDIF
ID(I)=1
IF(J.EQ.NN)P5=P4
A(II+1)=PAI1*(RO(I0, J)*AK10-RO(I, J)*AK20)
A(II+2)=DX3*D1+PAI1*(RO(I0, J)*AK00-RO(I, J)*AK10)
A(II+3)=-DX3*D3+PAI1*(RO(I0, J)*AK10-RO(I, J)*AK00)
A(II+4)=DX3*D2+PAI1*(RO(I0, J)*AK20-RO(I, J)*AK10)
A(II+5)=-DX3*(D1*P1+D2*P2+D4*P4+D5*P5-D3*P3)+DX1*(RO(I, J)*H(I, J)-
RO(I0, J)*H(I0, J))
50  CONTINUE
CALL TRA4(MM, D, A, B)
DO 60 I=2, MM
IF(ID(I).EQ.1)P(I, J)=P(I, J)+C1*D(I)
IF(P(I, J).LT.0.0)P(I, J)=0.0
IF(PMAX.LT.P(I, J))PMAX=P(I, J)
60  CONTINUE
70  CONTINUE
DO 80 J=1, NN
JJ=N+1-J
DO 80 I=1, N
80  P(I, JJ)=P(I, J)
CALL HREE(N, DX, H00, G0, X, Y, H, RO, EPS, EDA, P)
100  CONTINUE
RETURN
END

SUBROUTINE TRA4(N, D, A, B)
DIMENSION D(N), A(5, N), B(3, N)
C=1./A(3, N)
```

```
B(1, N)=-A(1, N)*C
B(2, N)=-A(2, N)*C
B(3, N)=A(5, N)*C
DO 10 I=1, N-2
IN=N-I
IN1=IN+1
C=1./(A(3, IN)+A(4, IN)*B(2, IN1))
B(1, IN)=-A(1, IN)*C
B(2, IN)=-(A(2, IN)+A(4, IN)*B(1, IN1))*C
10  B(3, IN)=(A(5, IN)-A(4, IN)*B(3, IN1))*C
D(1)=0.0
D(2)=B(3, 2)
DO 20 I=3, N
20  D(I)=B(1, I)*D(I-2)+B(2, I)*D(I-1)+B(3, I)
RETURN
END

SUBROUTINE VI(NW, N, DX, P, V)
DIMENSION P(N, N), V(NW, NW)
COMMON /COMAK/AK(0:65, 0:65)
PAI1=0.2026423
DO 40 I=1, N
DO 40 J=1, N
H0=0.0
DO 30 K=1, N
IK=IABS(I-K)
DO 30 L=1, N
JL=IABS(J-L)
30  H0=H0+AK(IK, JL)*P(K, L)
40  V(I, J)=H0*DX*PAI1
RETURN
END

SUBROUTINE SUBAK(MM)
COMMON /COMAK/AK(0:65, 0:65)
S(X, Y)=X+SQRT(X**2+Y**2)
DO 10 I=0, MM
XP=I+0.5
XM=I-0.5
DO 10 J=0, I
YP=J+0.5
YM=J-0.5
A1=S(YP, XP)/S(YM, XP)
A2=S(XM, YM)/S(XP, YM)
A3=S(YM, XM)/S(YP, XM)
A4=S(XP, YP)/S(XM, YP)
AK(I, J)=XP*ALOG(A1)+YM*ALOG(A2)+XM*ALOG(A3)+YP*ALOG(A4)
10  AK(J, I)=AK(I, J)
RETURN
END
```

```
SUBROUTINE ERP(N, ER, P, POLD)
DIMENSION P(N, N), POLD(N, N)
ER=0.0
SUM=0.0
NN=(N+1)/2
DO 10 I=1, N
DO 10 J=1, NN
ER=ER+ABS(P(I, J)-POLD(I, J))
SUM=SUM+P(I, J)
10  CONTINUE
ER=ER/SUM
DO I=1, N
DO J=1, N
POLD(I, J)=P(I, J)
ENDDO
ENDDO
RETURN
END

SUBROUTINE OUPT(N, DX, X, Y, H, P, EDA, TMAX)
DIMENSION X(N), Y(N), H(N, N), P(N, N), EDA(N, N)
COMMON /COM1/Z, ENDA, AKC, HM0, HMC, EK, EAL, EBE, AD, AD1, KK1, KK2, KK3,
KK4, FN, FN1, FF
COMMON /COM2/W0, E1, RX, B, PH, US, U1, U2, EDA0
A=0.0
WRITE(8, 40)A, (Y(I), I=1, N)
DO I=1, N
WRITE(8, 40)X(I), (H(I, J), J=1, N)
ENDDO
WRITE(9, 40)A, (Y(I), I=1, N)
DO I=1, N
WRITE(9, 40)X(I), (P(I, J), J=1, N)
ENDDO
40  FORMAT(66(E12.6, 1X))
HMIN=1.E3
PMAX=0.0
DO J=1, N
DO I=2, N
IF(H(I, J).LT.HMIN)HMIN=H(I, J)
IF(P(I, J).GT.PMAX)PMAX=P(I, J)
ENDDO
ENDDO
HMIN=HMIN*B*B/RX
PMAX=PMAX*PH
WRITE(10, *)'HMIN, PMAX, TMAX', HMIN, PMAX, TMAX
RETURN
END
```

Results

Based on the program, the results obtained for the film thickness and the pressure distribution are shown in Figure 11.6.

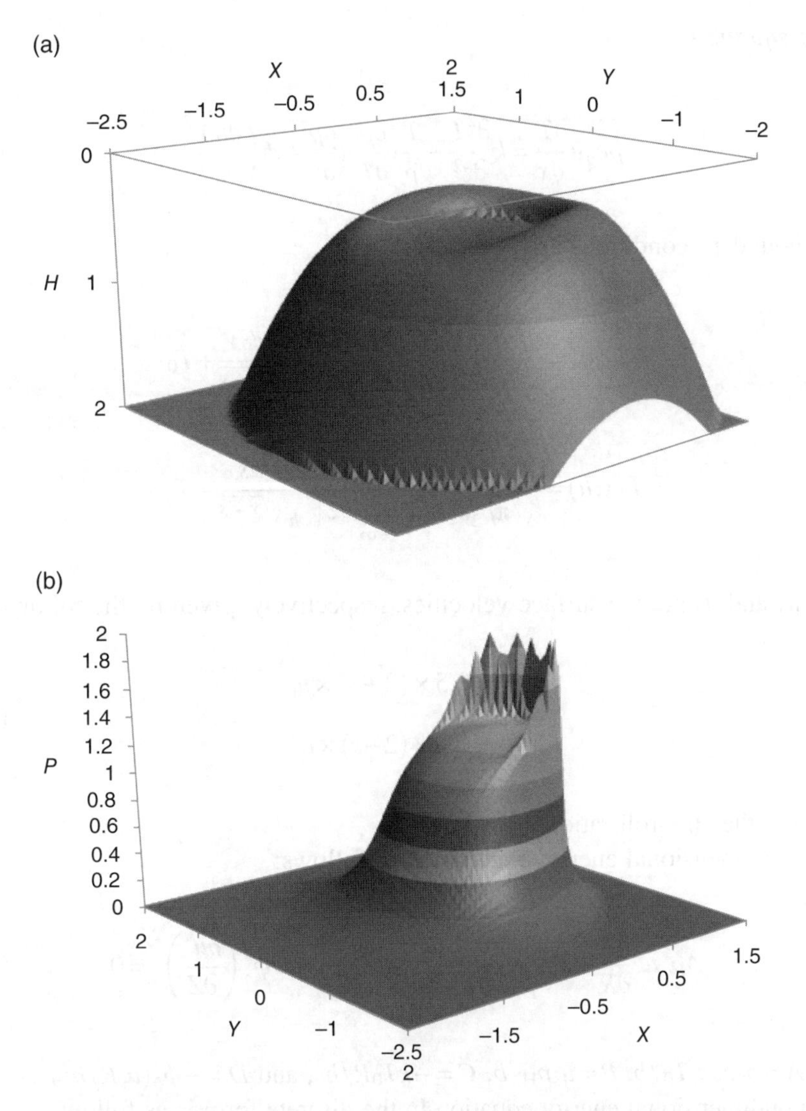

Figure 11.6 Film thickness and pressure distribution of GEHL in point contact. (a) Film thickness and (b) pressure

11.3 Numerical calculation method and program for thermal GEHL

11.3.1 Line contact problem

11.3.1.1 Basic equations and calculation method

The Reynolds equation and film thickness equation for the thermal GEHL in the line contact are the same as that of the isothermal GEHL in the line contact. Compared with the calculation of the isothermal GEHL in the line contact, the calculation of the thermal GEHL will need the following energy equation.

Energy equation

$$\rho c_p u \frac{\partial T}{\partial x} = k \frac{\partial^2 T}{\partial z^2} - \frac{T}{\rho} \cdot \frac{\partial \rho}{\partial T} u \frac{\partial p}{\partial x} + \phi \left(\frac{\partial u}{\partial z} \right)^2 \tag{11.29}$$

The boundary conditions are as follows:

$$\begin{aligned} T(x,0) &= \frac{k}{\sqrt{\pi \rho_1 c_1 k_1 u_1}} \int_{-\infty}^{x} \left. \frac{\partial T}{\partial z} \right|_{x,0} \frac{ds}{\sqrt{x-s}} + T_0 \\ T(x,h) &= \frac{k}{\sqrt{\pi \rho_2 c_2 k_2 u_2}} \int_{-\infty}^{x} \left. \frac{\partial T}{\partial z} \right|_{x,h} \frac{ds}{\sqrt{x-s}} + T_0 \end{aligned} \tag{11.30}$$

where, u_1 and u_2 are the surface velocities, respectively, given by the following:

$$\begin{aligned} u_1 &= 0.5 \times (2+s) \times u_s \\ u_2 &= 0.5 \times (2-s) \times u_s \end{aligned} \tag{11.31}$$

where, s is the slip–roll ratio.

The nondimensional energy equation is as follows:

$$A\rho^* u^* \frac{\partial T}{\partial X} + B \frac{u^* T^*}{\rho^*} \frac{\partial \rho^*}{\partial T^*} \frac{\partial P}{\partial X} + C \frac{\partial^2 T}{\partial Z^2} + D\phi^* \left(\frac{\partial u^*}{\partial Z} \right)^2 = 0 \tag{11.32}$$

where, $A = \rho_0 c_p u_s T_0 / b$, $B = u_s p_H / b$, $C = -kT_0 R / b^2$, and $D = -\phi_0 (u_s R / b^2)^2$.

The nondimensional energy equation in the discrete form is as follows:

$$\begin{aligned} &A\rho^* u^* \left(\frac{T_{i,k} - T_{i-1,k}}{\Delta X} \right) + B \frac{T^*}{\rho^*} \frac{\partial \rho^*}{\partial T^*} \left(u^* \frac{P_{i,j} - P_{i-1,j}}{\Delta X} \right) + C \frac{T_{i,k+1} - 2T_{i,k} + T_{i,k-1}}{\Delta Z^2} \\ &+ D\phi^* \left(\frac{u^*_{i,k+1} - u^*_{i,k}}{\Delta Z} \right)^2 = 0 \end{aligned} \tag{11.33}$$

where, the subscript k indicates the node number in the thickness direction; ρ^*, ϕ^*, and $\partial \rho^* / \partial T^*$ can be obtained by the analysis formulas so that they do not need to be discreted.

Viscosity–temperature–pressure equation

$$\phi=\phi_0\ \exp\left\{(\ln\phi_0+9.67)\left[(1+p/p_0)^z\times\left(\frac{T-138}{T_0-138}\right)^{-1.1}-1\right]\right\} \quad (11.34)$$

The nondimensional viscosity–temperature–pressure equation is as follows:

$$\phi^*=\exp\left\{(\ln\phi_0+9.67)\left[(1+p/p_0)^z\times\left(\frac{T-138}{T_0-138}\right)^{-1.1}-1\right]\right\} \quad (11.35)$$

Density–temperature–pressure equation

$$\rho=\rho_0\left[1+\frac{0.6p}{1+1.7p}+D(T-T_0)\right] \quad (11.36)$$

The nondimensional density–temperature–pressure equation is as follows:

$$\rho^*=\left[1+\frac{0.6p}{1+1.7p}+D(T-T_0)\right] \quad (11.37)$$

The load equation is the same as that of the isothermal GEHL in the line contact.

Similar to the EHL of oil, because the temperature of the energy equation, the viscosity of the viscosity–temperature–pressure equation, and the elastic deformation in the film thickness equation all change with pressure, we should give an initial pressure distribution (such as the Hertz contact stress) and temperature distribution (such as the uniform temperature field) first and then calculate the pressure and temperature with an iteration method.

11.3.1.2 Program and results

Program

The program to calculate the thermal GEHL in the line contact contains a main program and many subroutines. The role of the main program is to preassign initial parameters, calculate nondimensional parameters, and set output parameters. Then, call the subroutines to solve the Reynolds equation and to calculate the lubrication film thickness. Subroutine THERM is used to calculate temperature. It gets the pressure, film thickness, viscosity, etc. from the previous subroutines. After substituting them into the energy equation and with the thermal boundary conditions, the temperature can be calculated. Subroutine ERRO is to calculate the error and if the error meets the accuracy requirement, the calculation is finished. The diagram is shown in Figure 11.7.

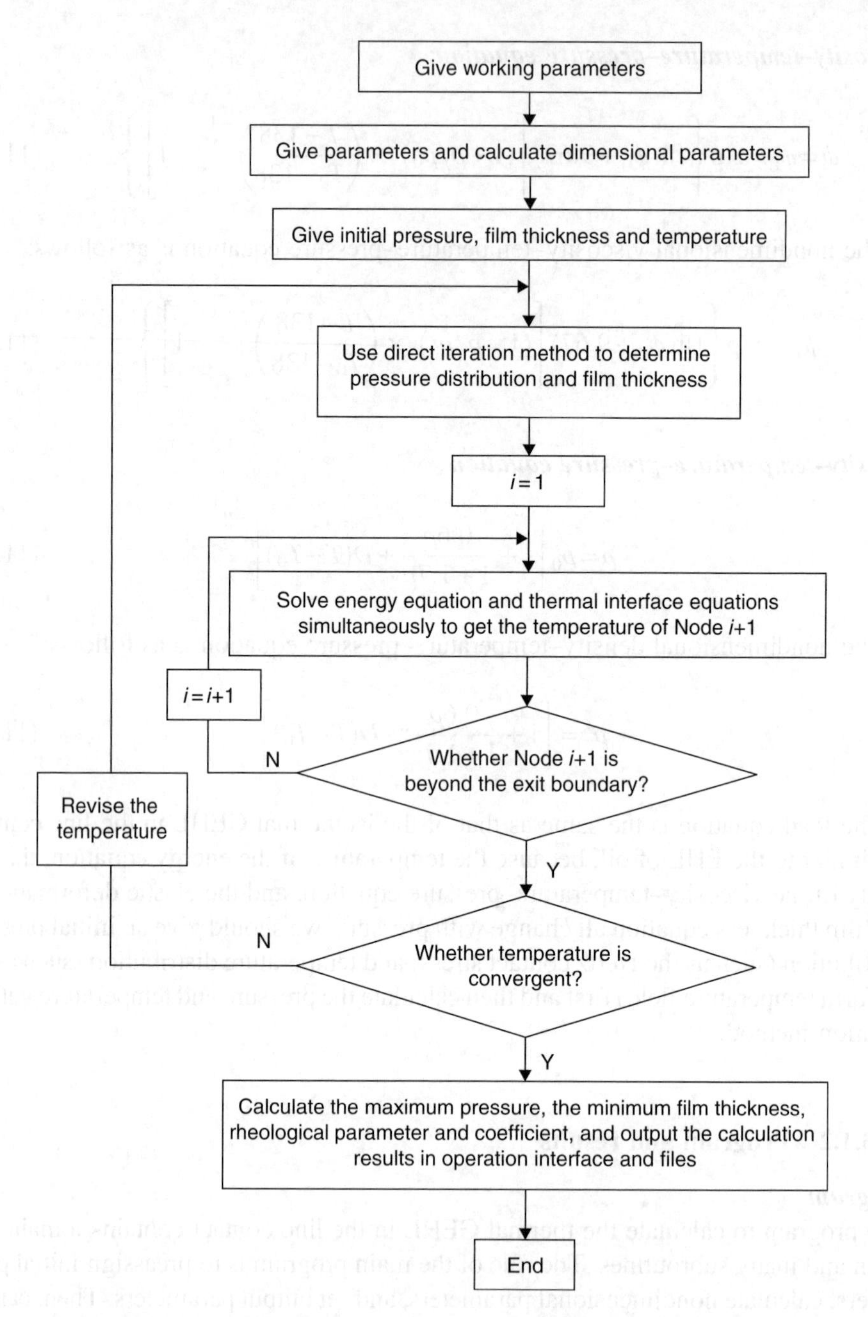

Figure 11.7 Diagram to calculate thermal GEHL in the line contact

Program

```
      PROGRAM GREASELINETHERMEHL
      CHARACTER*1 S, S1, S2
      CHARACTER*16 FILEO
      COMMON /COM1/ENDA, A1, A2, A3, Z, C1, C2, C3, CW
      COMMON /COM2/T0, EDA0/COM3/E1, PH, B, U1, U2, R, CT/COM4/X0, XE
      COMMON /COM5/H2, P2, T2, ROM, HM, FM/COM6/FF
      DATA KT, S1, S2/0, 1HY, 1Hy/
      DATA FILEO/4HDATA/
      PAI=3.14159265
      Z=0.68
      P0=1.96E8
      N=129
      X0=-4.
      XE=1.4
      W=1. E5
      E1=2.21E11
      R=0.02
      Us=0.87
      C1=0.37
      C2=0.37
      NZ=5
      CU=0.25
      CT=0.35
      T2=0.0
      FF=0.85
      OPEN(8, FILE=FILEO, STATUS='UNKNOWN')
      WRITE(*, *)'Show the example or not (Y or N)?'
      READ(*, '(A)')S
      IF(S.EQ.S1.OR.S.EQ.S2)THEN
      KT=2
      GOTO 10
      ELSE
      WRITE(*, *)' Temperature is considered or not (Y or N) ?'
      READ(*, '(A)')S
      IF(S.EQ.S1.OR.S.EQ.S2)KT=2
      ENDIF
      WRITE(*, *)'W, US, FF='
      READ(*, *)W, US, FF
      IF(KT.EQ.2)THEN
      WRITE(*, *)'NZ, CU='
      READ(*, *)NZ, CU
      ENDIF
      WRITE(*, *)' Change iteration factors or not (Y or N) ?'
      READ(*, '(A)')S
      IF(S.EQ.S1.OR.S.EQ.S2)THEN
      WRITE(*, *)'C1, C2='
      READ(*, *)C1, C2
      ENDIF
 10   CW=N+0.1
```

```
      LMAX=ALOG(CW)/ALOG(2.)
      N=2**LMAX+1
      LMIN=(ALOG(CW)-ALOG(SQRT(CW)))/ALOG(2.)
      LMAX=LMIN
      H00=0.0
      W1=W/(E1*R)
      PH=E1*SQRT(0.5*W1/PAI)
      A1=(ALOG(EDA0)+9.67)
      A2=PH/P0
      A3=0.59/(PH*1.E-9)
      T2=0.0
      B=4.*R*PH/E1
      ALFA=Z*A1/P0
      G=ALFA*E1
      U=EDA0*US/(2.*E1*R)
      CC1=SQRT(2.*U)
      AM=2.*PAI*(PH/E1)**2/CC1
      AL=G*SQRT(CC1)
      CW=(PH/E1)*(B/R)
      C3=1.6*(R/B)**2*G**0.6*U**0.7*W1**(-0.13)
      ENDA=B**(2.+1/FF)*(PH/2/EDA0)**(1/FF)/R**(1+1/FF)/Us/(2.+1/FF)
      ENDA=1./ENDA
      U1=0.5*(2.+CU)*U
      U2=0.5*(2.-CU)*U
      CW=-1.13*C3
      WRITE(*, 40)
40    FORMAT(2X, 'Wait   Please', //)
      CALL SUBAK(N)
      CALL MULTI(N, NZ, KT, LMIN, LMAX, H00)
      H2=H2*1.E-6
      P2=P2*1.E6
      Q=2.*ROM*HM*US
      FM=FM/W
      STOP
      END
      SUBROUTINE MULTI(N, NZ, KT, LMIN, LMAX, H00)
      DIMENSION X(1100), P(1100), H(1100), RO(1100), POLD(1100), EPS(1100),
      EDA(1100), P0(2200), F(1100), F0(2200), R(1100), R0(2200), G(10),
      T(22000)
      COMMON /COM1/ENDA, A1, A2, A3, Z, C1, C2, C3, CW/COM6/FF
      COMMON /COMK/K/COMT/LT, T1(1100)/COM3/E1, PH, B, U1, U2, RR, CT
      COMMON /COM5/H2, P2, T2, RM, HM, FM
      DATA MK, IT, KH, NMAX, PAI, G0/0, 0, 0, 1100, 3.14159265, 1.570796325/
      LT=LMAX
      NX=N
      K=LMIN
      N0=(N-1)/2**(LMIN-1)
      CALL KNDX(K, N, N0, N1, NMAX, DX, X)
      DO 10 I=1, N
      T1(I)=1.0
      IF(ABS(X(I)).GE.1.0)P(I)=0.0
```

```
10  IF(ABS(X(I)).LT.1.0)P(I)=SQRT(1.-X(I)*X(I))
12  CALL HREE(N, DX, H00, G0, X, P, H, RO, EPS, EDA, F, 0)
    IF(KH.NE.0)GOTO 14
    KH=1
    GOTO 12
14  CALL FZ(N, P, POLD)
    DO 100 L=LMIN, LMAX
    K=L
    G(K)=PAI/2.
    DO 18 I=1, N
    R(I)=0.0
    F(I)=0.0
    R0(N1+I)=0.0
18  F0(N1+I)=0.0
20  KK=2
    CALL ITER(N, KK, DX, H00, G0, X, P, H, RO, EPS, EDA, F, R, 0)
    KK=1
    CALL ITER(N, KK, DX, H00, G0, X, P, H, RO, EPS, EDA, F, R, 1)
    G(K-1)=G(K)
    DO 24 I=1, N
    IF(I.LT.N)G(K-1)=G(K-1)-0.5*DX*(P(I)+P(I+1))
24  P0(N1+I)=P(I)
    N2=N
    K=K-1
    CALL KNDX(K, N, N0, N1, NMAX, DX, X)
    CALL TRANS(N, N2, P, H, RO, EPS, EDA, R)
    CALL ITER(N, KK, DX, H00, G0, X, P, H, RO, EPS, EDA, F, R, 2)
    DO 26 I=1, N
    IF(I.LT.N)G(K)=G(K)+0.5*DX*(P(I)+P(I+1))
26  F(I)=H(I)
    G0=G(K)
    CALL HREE(N, DX, H00, G0, X, P, H, RO, EPS, EDA, F, 1)
    DO 28 I=1, N
    R0(N1+I)=R(I)
28  F0(N1+I)=F(I)
    IF(K.NE.1)GOTO 20
    KK=19
    CALL ITER(N, KK, DX, H00, G0, X, P, H, RO, EPS, EDA, F, R, 0)
40  DO 42 I=1, N
42  P0(N1+I)=P(I)
    N2=N1
    K=K+1
    CALL KNDX(K, N, N0, N1, NMAX, DX, X)
    G0=G(K)
    DO 50 I=2, N, 2
    I1=N1+I
    I2=N2+I/2
    P(I-1)=P0(I2)
    P(I)=P0(I1)+0.5*(P0(I2)+P0(I2+1)-P0(I1-1)-P0(I1+1))
50  IF(P(I).LT.0.0)P(I)=0.
    DO 52 I=1, N
```

```
    R(I)=R0(N1+I)
52  F(I)=F0(N1+I)
    CALL HREE(N, DX, H00, G0, X, P, H, RO, EPS, EDA, F, 0)
    KK=1
    CALL ITER(N, KK, DX, H00, G0, X, P, H, RO, EPS, EDA, F, R, 0)
    IF(K.LT.L)GOTO 40
100 CONTINUE
    MK=MK+1
    CALL ERROP(N, P, POLD, ERP)
    IF(ERP.GT.0.01*C2.AND.MK.LE.12)GOTO 14
    MK=8
    IF(KT.NE.2)GOTO 105
    CALL THERM(NX, NZ, DX, T, P, H)
    CALL ERROM(NX, NZ, T1, T, KT)
    IT=IT+1
    IF(KT.EQ.2.AND.IT.LT.25)GOTO 14
    KT=2
    IF(IT.GE.25)THEN
    WRITE(*, *)'Temperature is not convergent !!!'
    READ(*, *)
    ENDIF
105 IF(MK.GE.10)THEN
    WRITE(*, *)'Pressures are not convergent !!!'
    READ(*, *)
    ENDIF
    FM=FRICT(N, DX, H, P, EDA)
110 FORMAT(6(1X, E12.6))
    DO I=2, N-1
    IF(P(I).GE.P(I-1).AND.P(I).GE.P(I+1))THEN
    HM=H(I)*B*B/RR
    RM=RO(I)
    GOTO 120
    ENDIF
    ENDDO
120 H2=1.E5
    P2=1.E-10
    DO I=1, N
    WRITE(8, 110)X(I), P(I), H(I)
    ENDDO
    WW=0
    DO I=1, N
    WW=WW+P(I)
    H(I)=H(I)*B*B*1.E6/RR
    P(I)=P(I)*PH/1.E6
    IF(H(I).LT.H2)H2=H(I)
    IF(P(I).GT.P2)P2=P(I)
    ENDDO
    IF(KT.EQ.2)THEN
    CALL OUPT(NX, NZ, X, T, T2)
    ENDIF
    RETURN
```

```
    END
    SUBROUTINE HREE(N, DX, H00, G0, X, P, H, RO, EPS, EDA, F0, KG)
    DIMENSION X(N), P(N), H(N), RO(N), EPS(N), EDA(N), F0(N)
    DIMENSION W(2200)
   COMMON /COM1/ENDA, A1, A2, A3, Z, C1, C2, C3, CW/COMK/K/COMT/LT, T1(1100)
   COMMON /COM2/T0, EDA0, AK0, AK1, AK2, CV, CV1, CV2, RO0, RO1, RO2, S0, D0/
COMAK/AK(0:1100)/COM6/FF
    DATA KK, MK1, MK2, NW, PAI1/0, 3, 0, 2200, 0.318309886/
    IF(KK.NE.0)GOTO 3
    HM0=C3
3   W1=0.0
    DO 4 I=1, N
4   W1=W1+P(I)
    C3=(DX*W1)/G0
    DW=1.-C3
    IF(K.EQ.1)GOTO 6
    CALL DISP(N, NW, K, DX, P, W)
    GOTO 10
6   WX=W1*DX*ALOG(DX)
    DO 8 I=1, N
    W(I)=WX
    DO 8 J=1, N
    IJ=IABS(I-J)
8   W(I)=W(I)+AK(IJ)*P(J)*DX
10  HMIN=1.E3
    DO 30 I=1, N
    H0=0.5*X(I)*X(I)-PAI1*W(I)
    IF(KG.EQ.1)GOTO 20
    IF(H0+F0(I).LT.HMIN)HMIN=H0+F0(I)
    H(I)=H0
    GOTO 30
20  F0(I)=F0(I)-H00-H0
30  CONTINUE
    IF(KG.EQ.1)RETURN
    H0=H00+HMIN
    IF(KK.NE.0)GOTO 32
    KK=1
    H00=-H0+HM0
32  IF(H0.LE.0.0)GOTO 48
    IF(K.NE.1)GOTO 50
40  MK=MK+1
    IF(MK.LE.MK1)GOTO 50
    IF(MK.GE.MK2)MK=0
    IF(H0+CW*DW.GT.0.0)HM0=H0+CW*DW
    IF(H0+CW*DW.LE.0.0)HM0=HM0*C3
48  H00=HM0-HMIN
50  DO 60 I=1, N
60  H(I)=H00+H(I)+F0(I)
    IT=2**(LT-K)
    DO 100 I=1, N
    II=IT*(I-1)+1
```

```
      CT1=((T1(II)-0.455445545)/0.544554455)**S0
      CT2=D0*T0*(T1(II)-1.)
      EDA(I)=EXP(A1*(-1.+(1.+A2*P(I))**Z*CT1))
      RO(I)=(A3+1.34*P(I))/(A3+P(I))+CT2
      EPS(I)=RO(I)*H(I)**(2+1/FF)/ENDA/EDA(I)**(1/FF)
100   CONTINUE
      RETURN
      END
      SUBROUTINE ITER(N, KK, DX, H00, G0, X, P, H, RO, EPS, EDA, F0, R0, KG)
      DIMENSION X(N), P(N), H(N), RO(N), EPS(N), EDA(N), F0(N), R0(N)
      COMMON /COM1/ENDA, A1, A2, A3, Z, C1, C2, C3/COM6/FF/COMAK/AK(0:1100)
      DATA PAI/3.14159265/
      DX1=1./DX
      DX2=DX*DX
      DX3=1./DX2
      DX4=DX1/PAI
      DX5=DX1**(1.0+1/FF)
      DXL=DX*ALOG(DX)
      AK0=DX*AK(0)+DXL
      AK1=DX*AK(1)+DXL
      DO 100 K=1, KK
      RMAX=0.0
      D2=0.5*(EPS(1)+EPS(2))
      D3=0.5*(EPS(2)+EPS(3))
      D5=DX1*(RO(2)*H(2)-RO(1)*H(1))
      D7=DX4*(RO(2)*AK0-RO(1)*AK1)
      PP=0.
      DO 70 I=2, N-1
      D1=D2
      D2=D3
      D4=D5
      D6=D7
      IF(I+2.LE.N)D3=0.5*(EPS(I+1)+EPS(I+2))
      D5=DX1*(RO(I+1)*H(I+1)-RO(I)*H(I))
      D7=DX4*(RO(I+1)*AK0-RO(I)*AK1)
      AB1=(ABS(P(I+1)-P(I)))**(1/FF-1.0)
      AB2=(ABS(P(I)-P(I-1)))**(1/FF-1.0)
      IF(KG.NE.0)GOTO 30
      DD=(D1+D2)*DX3
      IF(DD.LT.0.1*ABS(D6))GOTO 10
      RI=-DX5*(D2*(P(I+1)-P(I))*AB1-D1*(P(I)-P(I-1))*AB2)+D4+R0(I)
      DLDP=-DX3*(D1+D2)+D6
      RI=C1*RI/DLDP
      GOTO 20
10    RI=-DX5*(D2*(P(I+1)-P(I))*AB1-D1*(P(I)-PP)*AB2)+D4+R0(I)
      DLDP=-1/FF*DX5*(2*D1*AB1+D2*AB2)+2.*D6
      RI=C2*RI/DLDP
      IF(I.GT.2.AND.P(I-1)-RI.GT.0.0)P(I-1)=P(I-1)-RI
20    PP=P(I)
      P(I)=P(I)+RI
      IF(P(I).LT.0.0)P(I)=0.0
```

```
      IF(K.NE.KK)GOTO 70
      IF(RMAX.LT.ABS(RI).AND.P(I).GT.0.0)RMAX=ABS(RI)
      GOTO 70
30    IF(KG.EQ.2)GOTO 40
      R0(I)=-DX5*(D2*(P(I+1)-P(I))*AB1-D1*(P(I)-P(I-1))*AB2)+D4+R0(I)
      GOTO 70
40    R0(I)=DX5*(D2*(P(I+1)-P(I))*AB1-D1*(P(I)-P(I-1))*AB2)-D4+R0(I)
70    CONTINUE
      IF(KG.NE.0)GOTO 100
      CALL HREE(N, DX, H00, G0, X, P, H, RO, EPS, EDA, F0, 0)
100   CONTINUE
      RETURN
      END
      SUBROUTINE DISP(N, NW, KMAX, DX, P1, W)
      DIMENSION P1(N), W(NW), P(2200), AK1(0:50), AK2(0:50)
      COMMON /COMAK/AK(0:1100)
      DATA NMAX, KMIN/2200, 1/
      N2=N
      M=3+2*ALOG(FLOAT(N))
      K1=N+KMAX
      DO 10 I=1, N
10    P(K1+I)=P1(I)
      DO 20 KK=KMIN, KMAX-1
      K=KMAX+KMIN-KK
      N1=(N2+1)/2
      CALL DOWNP(NMAX, N1, N2, K, P)
20    N2=N1
      DX1=DX*2**(KMAX-KMIN)
      CALL WI(NMAX, N1, KMIN, KMAX, DX, DX1, P, W)
      DO 30 K=KMIN+1, KMAX
      N2=2*N1-1
      DX1=DX1/2.
      CALL AKCO(M+5, KMAX, K, AK1)
      CALL AKIN(M+6, AK1, AK2)
      CALL WCOS(NMAX, N1, N2, K, W)
      CALL CORR(NMAX, N2, K, M, 1, DX1, P, W, AK1)
      CALL WINT(NMAX, N2, K, W)
      CALL CORR(NMAX, N2, K, M, 2, DX1, P, W, AK2)
30    N1=N2
      DO 40 I=1, N
40    W(I)=W(K1+I)
      RETURN
      END
      SUBROUTINE DOWNP(NMAX, N1, N2, K, P)
      DIMENSION P(NMAX)
      K1=N1+K-1
      K2=N2+K-1
      DO 10 I=3, N1-2
      I2=2*I+K2
10    P(K1+I)=(16.*P(I2)+9.*(P(I2-1)+P(I2+1))-(P(I2-3)+P(I2+3)))/32.
      P(K1+2)=0.25*(P(K2+3)+P(K2+5))+0.5*P(K2+4)
```

```
      P(K1+N1-1)=0.25*(P(K2+N2-2)+P(K2+N2))+ 0.5*P(K2+N2-1)
      RETURN
      END
      SUBROUTINE WCOS(NMAX, N1, N2, K, W)
      DIMENSION W(NMAX)
      K1=N1+K-1
      K2=N2+K
      DO 10 I=1, N1
      II=2*I-1
10    W(K2+II)=W(K1+I)
      RETURN
      END
      SUBROUTINE WINT(NMAX, N, K, W)
      DIMENSION W(NMAX)
      K2=N+K
      DO 10 I=4, N-3, 2
      II=K2+I
10    W(II)=(9.*(W(II-1)+W(II+1))-(W(II-3)+W(II+3)))/16.
      I1=K2+2
      I2=K2+N-1
      W(I1)=0.5*(W(I1-1)+W(I1+1))
      W(I2)=0.5*(W(I2-1)+W(I2+1))
      RETURN
      END
      SUBROUTINE CORR(NMAX, N, K, M, I1, DX, P, W, AK)
      DIMENSION P(NMAX), W(NMAX), AK(0:M)
      K1=N+K
      IF(I1.EQ.2)GOTO 20
      DO 10 I=1, N, 2
      II=K1+I
      J1=MAX0(1, I-M)
      J2=MIN0(N, I+M)
      DO 10 J=J1, J2
      IJ=IABS(I-J)
10    W(II)=W(II)+AK(IJ)*DX*P(K1+J)
      RETURN
20    DO 30 I=2, N, 2
      II=K1+I
      J1=MAX0(1, I-M)
      J2=MIN0(N, I+M)
      DO 30 J=J1, J2
      IJ=IABS(I-J)
30    W(II)=W(II)+AK(IJ)*DX*P(K1+J)
      RETURN
      END
      SUBROUTINE WI(NMAX, N, KMIN, KMAX, DX, DX1, P, W)
      DIMENSION P(NMAX), W(NMAX)
      COMMON /COMAK/AK(0:1100)
      K1=N+1
      K=2**(KMAX-KMIN)
      C=ALOG(DX)
```

```
      DO 10 I=1, N
      II=K1+I
      W(II)=0.0
      DO 10 J=1, N
      IJ=K*IABS(I-J)
10    W(II)=W(II)+(AK(IJ)+C)*DX1*P(K1+J)
      RETURN
      END
      SUBROUTINE AKCO(KA, KMAX, K, AK1)
      DIMENSION AK1(0:KA)
      COMMON /COMAK/AK(0:1100)
      J=2**(KMAX-K)
      DO 10 I=0, KA
      II=J*I
10    AK1(I)=AK(II)
      RETURN
      END
      SUBROUTINE AKIN(KA, AK1, AK2)
      DIMENSION AK1(KA), AK2(KA)
      DO 10 I=4, KA-3
10    AK2(I)=(9.*(AK1(I-1)+AK1(I+1))-(AK1(I-3)+AK1(I+3)))/16.
      AK2(1)=(9.*AK1(2)-AK1(4))/8.
      AK2(2)=(9.*(AK1(1)+AK1(3))-(AK1(3)+AK1(5)))/16.
      AK2(3)=(9.*(AK1(2)+AK1(4))-(AK1(2)+AK1(6)))/16.
      DO 20 I=1, KA
20    AK2(I)=AK1(I)-AK2(I)
      DO 30 I=1, KA, 2
      I1=I+1
      AK1(I)=0.0
30    IF(I1.LE.KA) AK1(I1)=AK2(I1)
      RETURN
      END
      SUBROUTINE SUBAK(MM)
      COMMON /COMAK/AK(0:1100)
      DO 10 I=0, MM
10    AK(I)=(I+0.5)*(ALOG(ABS(I+0.5))-1.)-(I-0.5)*(ALOG(ABS(I-0.5))-1.)
      RETURN
      END
      FUNCTION FRICT(N, DX, H, P, EDA)
      DIMENSION H(N), P(N), EDA(N)
      COMMON /COM3/E1, PH, B, U1, U2, R, CT
      DATA TAU0/4.E7/
      TP=TAU0/PH
      TE=TAU0/E1
      BR=B/R
      FRICT=0.0
      DO I=1, N
      DP=0.0
      IF(I.LT.N)THEN
      DP=(P(I+1)-P(I))/DX
      TAU=0.5*H(I)*ABS(DP)*(BR/TP)+2.*ABS(U1-U2)*EDA(I)/(H(I)*BR**2*TE)
```

```
      FRICT=FRICT+TAU
      ENDIF
      ENDDO
      FRICT=FRICT*DX*B*TAU0
      RETURN
      END
      SUBROUTINE FZ(N, P, POLD)
      DIMENSION P(N), POLD(N)
      DO 10 I=1, N
10    POLD(I)=P(I)
      RETURN
      END
      SUBROUTINE ERROP(N, P, POLD, ERP)
      DIMENSION P(N), POLD(N)
      SD=0.0
      SUM=0.0
      DO 10 I=1, N
      SD=SD+ABS(P(I)-POLD(I))
      POLD(I)=P(I)
10    SUM=SUM+P(I)
      ERP=SD/SUM
      RETURN
      END
      SUBROUTINE KNDX(K, N, N0, N1, NMAX, DX, X)
      DIMENSION X(NMAX)
      COMMON /COM4/X0, XE
      N=2**(K-1)*N0
      DX=(XE-X0)/N
      N=N+1
      N1=N+K
      DO 10 I=1, N
10    X(I)=X0+(I-1)*DX
      RETURN
      END
      SUBROUTINE TRANS(N1, N2, P, H, RO, EPS, EDA, R)
      DIMENSION P(N2), H(N2), RO(N2), EPS(N2), EDA(N2), R(N2)
      DO 10 I=1, N1
      II=2*I-1
      P(I)=P(II)
      H(I)=H(II)
      R(I)=R(II)
      RO(I)=RO(II)
      EPS(I)=EPS(II)
10    EDA(I)=EDA(II)
      RETURN
      END
      SUBROUTINE OUPT(NX, NZ, X, T, T2)
      DIMENSION X(NX), T(NX, NZ)
      DO I=1, NX
      DO K=1, NZ
      T(I, K)=303.*(T(I, K)-1.0)
```

```
      IF(T(I, K).GT.T2)T2=T(I, K)
      END DO
      END DO
      WRITE(8, 20)NX, NX, NZ
20    FORMAT(15X, ' T(1, 1)-T(1, ', I4, ')-T(', I4, ', ', I2, ') ')
      DO I=1, NX
      WRITE(8, 30)X(I), (T(I, K), K=1, NZ)
      ENDDO
30    FORMAT(6(1X, E12.6))
      RETURN
      END
      SUBROUTINE ERROM(NX, NZ, T1, T, KT)
      DIMENSION T(NX, NZ), T1(NX)
      KT=2
      ERM=0.
      C1=1./FLOAT(NZ)
      DO 20 I=1, NX
      TT=0.
      DO 10 K=1, NZ
10    TT=TT+T(I, K)
      TT=C1*TT
      ER=ABS((TT-T1(I))/TT)
      IF(ER.GT.ERM)ERM=ER
20    T1(I)=TT
      IF(ERM.LT.0.003)KT=1
      RETURN
      END
      SUBROUTINE THERM(NX, NZ, DX, T, P, H)
      DIMENSION T(NX, NZ), P(NX), H(NX), T1(21), TI(21), TF(21), U(21),
      DU(21), W(21), EDA(21), RO(21), EDA1(21), EDA2(21), ROR(21), UU(21)
      DATA KK/0/
      IF(KK.NE.0)GOTO 4
      DO 2 K=1, NZ
2     T(1, K)=1.0
4     DO 30 I=2, NX
      KG=0
      DO 8 K=1, NZ
      TF(K)=T(I-1, K)
      IF(KK.NE.0)GOTO 6
      T1(K)=T(I-1, K)
      GOTO 8
6     T1(K)=T(I, K)
8     TI(K)=T1(K)
      P1=P(I)
      H1=H(I)
      DP=(P(I)-P(I-1))/DX
      CALL TBOUD(NX, NZ, I, CC1, CC2, T)
10    CALL EROEQ(NZ, T1, P1, EDA, RO, EDA1, EDA2, KG)
      CALL UCAL(NZ, DX, H1, EDA, RO, ROR, EDA1, EDA2, U, UU, DU, W, DP)
      CALL TCAL(NZ, DX, CC1, CC2, T1, TF, U, W, DU, H1, DP, EDA, RO)
      CALL ERRO(NZ, TI, T1, ETS)
```

```
      KG=KG+3
      IF(ETS.GT.1.E-4.AND.KG.LE.50)GOTO 10
      DO 20 K=1, NZ
      ROR(K)=RO(K)
      UU(K)=U(K)
 20   T(I, K)=T1(K)
 30   CONTINUE
      KK=1
      RETURN
      END
      SUBROUTINE TBOUD(NX, NZ, I, CC1, CC2, T)
      DIMENSION T(NX, NZ)
      CC1=0.
      CC2=0.
      DO 10 L=1, I-1
      DS=1./SQRT(FLOAT(I-L))
      IF(L.EQ.I-1)DS=1.1666667
      CC1=CC1+DS*(T(L, 2)-T(L, 1))
 10   CC2=CC2+DS*(T(L, NZ)-T(L, NZ-1))
      RETURN
      END
      SUBROUTINE ERRO(NZ, T0, T, ETS)
      DIMENSION T0(NZ), T(NZ)
      ETS=0.0
      DO 10 K=1, NZ
      IF(T(K).LT.1.E-5)ETS0=1.
      IF(T(K).GE.1.E-5)ETS0=ABS((T(K)-T0(K))/T(K))
      IF(ETS0.GT.ETS)ETS=ETS0
 10   T0(K)=T(K)
      RETURN
      END
      SUBROUTINE EROEQ(NZ, T, P, EDA, RO, EDA1, EDA2, KG)
      DIMENSION T(NZ), EDA(NZ), RO(NZ), EDA1(NZ), EDA2(NZ)
      COMMON /COM1/ENDA, A1, A2, A3, Z, O1, O2, O3
      COMMON /COM2/T0, EDA0, AK, AK1, AK2, CV, CV1, CV2, RO0, RO1, RO2, S0, D0
      COMMON /COM3/E1, PH, B, U1, U2, R, CC/COM6/FF
      DATA A4, A5/0.455445545, 0.544554455/
      IF(KG.NE.0)GOTO 20
      B1=(1.+A2*P)**Z
      B2=(A3+1.34*P)/(A3+P)
 20   DO 30 K=1, NZ
      EDA(K)=EXP(A1*(-1.+B1*((T(K)-A4)/A5)**S0))
 30   RO(K)=1+D0*T0*(T(K)-1.)
      CC1=0.5/(NZ-1.)
      CC2=1./(NZ-1.)
      C1=0.
      C2=0.
      DO 40 K=1, NZ
      IF(K.EQ.1)GOTO 32
      C1=C1+0.5/EDA(K)+0.5/EDA(K-1)
      C2=C2+CC1*((K-1.)/EDA(K)+(K-2.)/EDA(K-1))
 32   EDA1(K)=C1*CC2
```

```
40  EDA2(K)=C2*CC2
    IF(KG.NE.2)RETURN
    C1=0.
    C2=0.
    C3=0.
    DO 50 K=1, NZ
    IF(K.EQ.1)GOTO 50
    C1=C1+0.5*(RO(K)+RO(K-1))
    C2=C2+0.5*(RO(K)*EDA1(K)+RO(K)*EDA1(K-1))
    C3=C3+0.5*(RO(K)*EDA2(K)+RO(K)*EDA2(K-1))
50  CONTINUE
    B1=12.*CC2*(C1*EDA2(NZ)/EDA1(NZ)-C2)
60  B2=2.*CC2/(U1+U2)*(C1*(U1-U2)/EDA1(NZ)+C3*U1)
    RETURN
    END
    SUBROUTINE UCAL(NZ, DX, H, EDA, RO, ROR, EDA1, EDA2, U, UU, DU, W, DP)
    DIMENSION U(NZ), DU(NZ), W(NZ), ROR(NZ), UU(NZ), EDA(NZ), RO(NZ), EDA1
     (NZ), EDA2(NZ)
    COMMON /COM2/T0, EDA0, AK, AK1, AK2, CV, CV1, CV2, RO0, RO1, RO2, S0, D0
    COMMON /COM3/E1, PH, B, U1, U2, R, CC
    DATA KK/0/
    IF(KK.NE.0)GOTO 20
    A1=U1
    A2=PH*(B/R)**3/E1
    A3=U2-U1
20  CC1=A2*DP*H
    CC2=CC1*H
    CC3=A3/H
    CC4=1./EDA1(NZ)
    DO 30 K=1, NZ
    U(K)=A1+CC2*(EDA2(K)-CC4*EDA2(NZ)*EDA1(K))+ A3*CC4*EDA1(K)
    IF(U(K).LT.0.0)U(K)=0.
30  DU(K)=CC1/EDA(K)*((K-1.)/(NZ-1.)-CC4*EDA2(NZ))+CC3*CC4/EDA(K)
    A4=B/((NZ-1)*R*DX)
    C1=A4*H
    IF(KK.EQ.0)GOTO 50
    DO 40 K=2, NZ-1
    W(K)=(RO(K-1)*W(K-1)+C1*(RO(K)*U(K)-ROR(K)*UU(K)))/RO(K)
40  CONTINUE
50  KK=1
    RETURN
    END
    SUBROUTINE TCAL(NZ, DX, CC1, CC2, T, TF, U, W, DU, H, DP, EDA, RO)
    DIMENSION T(NZ), TF(NZ), U(NZ), DU(NZ), W(NZ), EDA(NZ), RO(NZ), A(4,
21), D(21), AA(2, 21)
    COMMON /COM2/T0, EDA0, AK, AK1, AK2, CV, CV1, CV2, RO0, RO1, RO2, S0, D0
    COMMON /COM3/E1, PH, B, U1, U2, R, CC
    DATA KK, CC5, PAI, TAU0/0, 0.6666667, 3.14159265, 4.E7/
    IF(KK.NE.0)GOTO 5
    KK=1
    TAU=TAU0*B*B/(E1*R*R)
    A2=-CV*RO0*E1*B**3/(EDA0*AK*R)
```

```
      A3=-E1*PH*B**3*D0/(AK*EDA0*T0*R)
      A4=-(E1*R)**2/(AK*EDA0*T0)
      A5=0.5*R/B*A2
      A6=AK*SQRT(EDA0*R/(PAI*RO1*CV1*U1*E1*AK1*B**3))
      A7=AK*SQRT(EDA0*R/(PAI*RO2*CV2*U2*E1*AK2*B**3))
5     CC3=A6*SQRT(DX)
      CC4=A7*SQRT(DX)
      DZ=H/(NZ-1.)
      DZ1=1./DZ
      DZ2=DZ1*DZ1
      CC6=A3*DP
      DO 10 K=2, NZ-1
      A(1, K)=DZ2+DZ1*A5*RO(K)*W(K)
      A(2, K)=-2.*DZ2+A2*RO(K)*U(K)/DX+CC6*U(K)/RO(K)
      A(3, K)=DZ2-DZ1*A5*RO(K)*W(K)
      AE=ABS(EDA(K)*DU(K))
10    A(4, K)=A4*ABS(DU(K))*AE+A2*RO(K)*U(K)*TF(K)/DX
      A(1, 1)=0.
      A(2, 1)=1.+2.*DZ1*CC3*CC5
      A(3, 1)=-2.*DZ1*CC3*CC5
      A(1, NZ)=-2.*DZ1*CC4*CC5
      A(2, NZ)=1.+2.*DZ1*CC4*CC5
      A(3, NZ)=0.
      A(4, 1)=1.+CC1*CC3*DZ1
      A(4, NZ)=1.-CC2*CC4*DZ1
      CALL TRA3(NZ, D, A, AA)
      DO 20 K=1, NZ
20    T(K)=(1.-CC)*T(K)+CC*D(K)
30    CONTINUE
      RETURN
      END
      SUBROUTINE TRA3(N, D, A, B)
      DIMENSION D(N), A(4, N), B(2, N)
      C=1./A(2, N)
      B(1, N)=-A(1, N)*C
      B(2, N)=A(4, N)*C
      DO 10 I=1, N-1
      IN=N-I
      IN1=IN+1
      C=1./(A(2, IN)+A(3, IN)*B(1, IN1))
      B(1, IN)=-A(1, IN)*C
10    B(2, IN)=(A(4, IN)-A(3, IN)*B(2, IN1))*C
      D(1)=B(2, 1)
      DO 20 I=2, N
20    D(I)=B(1, I)*D(I-1)+B(2, I)
      RETURN
      END
      BLOCK DATA
      COMMON /COM2/T0, EDA0, AK, AK1, AK2, CV, CV1, CV2, RO0, RO1, RO2, S0, D0
     DATA T0, EDA0, AK, AK1, AK2, CV, CV1, CV2, RO0, RO1, RO2, S0, D0/303., 0.08,
   0.14, 46., 46., 2000., 470., 470., 890., 7850., 7850., -1.1, -0.00065/
      END
```

Results

The characteristics of the thermal GEHL in the line contact can be analyzed based on the calculation results. Figure 11.8 shows the influence of the rheology parameter n on the film thickness and the pressure distribution of thermal GEHL in the line contact. Although heat produces certain influence on the pressure and film thickness, the thermal EHL and isothermal EHL have no significant difference in the basic characteristics.

Figure 11.9 shows that the temperature rises across the film under the given parameters. Curves A and E are the temperature rises of the two surfaces; the other three

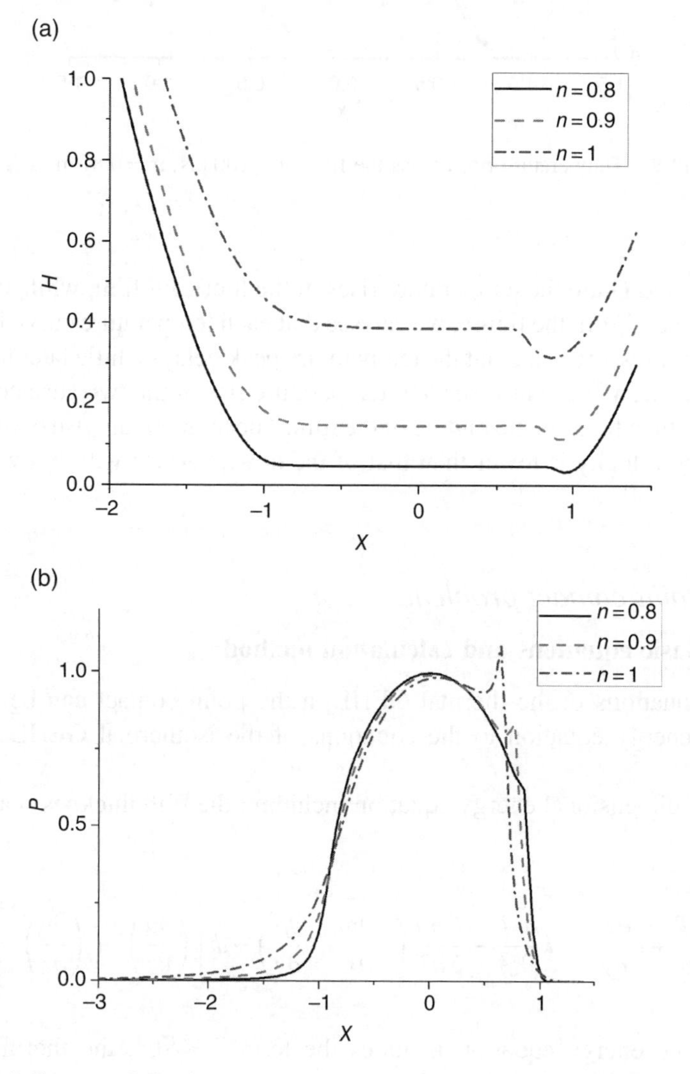

Figure 11.8 Influence of rheology parameter n on lubrication ($w = 100$ kN, $u_s = 0.87$ m·s^{-1}). (a) Film thickness and (b) pressure

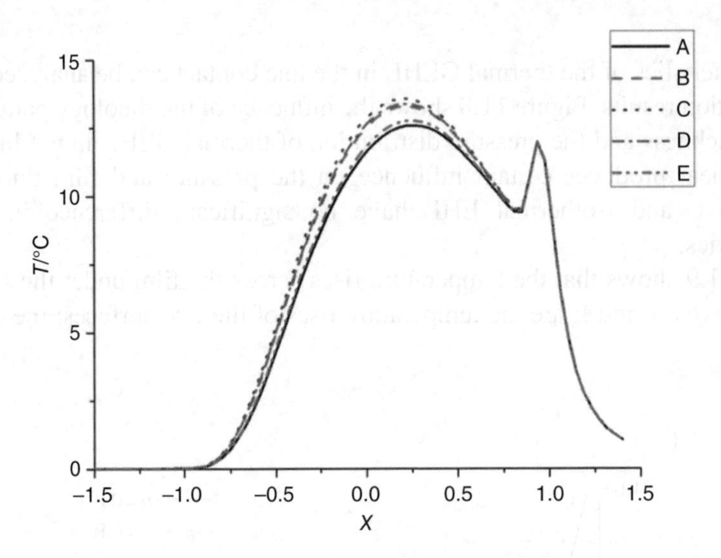

Figure 11.9 Temperature rise across the film ($w = 100$ kN, $u_s = 0.87$ m · s^{-1}, $n = 0.8$)

curves B, C, and D are the temperature rises in the lubricant film, while curve C is of the middle layer. From the figure, we can see that each temperature curve has a peak at the secondary pressure peak, but the temperature peak delays a little later than the pressure peak occurs. We can also see that temperature rise of the two surfaces are significantly lower than those in the middle of the film. The temperature rise of the up surface with a higher velocity is lower than that of the down surface with a lower velocity.

11.3.2 Point contact problem

11.3.2.1 Basic equations and calculation method

The basic equations of the thermal GEHL in the point contact can be obtained by adding the energy equation to the equations of the isothermal GEHL in the point contact.

The three-dimensional energy equation including the film thickness can be written as follows;

$$\rho c_p\left(u\frac{\partial T}{\partial x}+v\frac{\partial T}{\partial y}\right)=k\frac{\partial^2 T}{\partial z^2}-\frac{T}{\rho}\frac{\partial \rho}{\partial T}\left(u\frac{\partial p}{\partial x}+v\frac{\partial p}{\partial y}\right)+\phi\left[\left(\frac{\partial u}{\partial z}\right)^2+\left(\frac{\partial v}{\partial z}\right)^2\right] \tag{11.38}$$

Because the energy equation includes the term $\partial^2 T/\partial z^2$, the thermal boundary conditions of the down and up surfaces should be given. They are as follows:

$$T(x,y,0) = \frac{k}{\sqrt{\pi\rho_1 c_1 k_1 u_1}} \int_{-\infty}^{x} \left.\frac{\partial T}{\partial z}\right|_{x,y,0} \frac{ds}{\sqrt{x-s}} + T_0$$
$$T(x,y,h) = \frac{k}{\sqrt{\pi\rho_2 c_2 k_2 u_2}} \int_{-\infty}^{x} \left.\frac{\partial T}{\partial z}\right|_{x,y,h} \frac{ds}{\sqrt{x-s}} + T_0 \tag{11.39}$$

If the slip–roll ratio s is known, we can calculate the both surface velocities:

$$u_1 = 0.5(2+s)u_s$$
$$u_2 = 0.5(2-s)u_s \tag{11.40}$$

The nondimensional energy equation is as follows:

$$A\rho^* u^* \left(\frac{\partial T}{\partial X} + \alpha \frac{\partial T}{\partial Y}\right) + B\frac{T^*}{\rho^*}\frac{\partial \rho^*}{\partial T^*}\left(u^* \frac{\partial P}{\partial X} + \alpha v^* \frac{\partial P}{\partial Y}\right) + C\frac{\partial^2 T}{\partial Z^2} + D\phi^* \left(\frac{\partial u^*}{\partial Z}\right)^2 = 0 \tag{11.41}$$

where, $A = \rho_0 c_p u_s T_0/a$, $B = u_s p_H/a$, $C = -kT_0 R/a^2$, $D = -\phi_0 (u_s R/a^2)^2$.

The discrete nondimensional energy equation is as follows:

$$A\rho^* u^* \left(\frac{T_{i,j,k} - T_{i-1,j,k}}{\Delta X} + \alpha \frac{T_{i,j,k} - T_{i,j-1,k}}{\Delta Y}\right) + B\frac{T^*}{\rho^*}\frac{\partial \rho^*}{\partial T^*}\left(u^* \frac{P_{i,j} - P_{i-1,j}}{\Delta X} + \alpha v^* \frac{P_{i,j} - P_{i,j-1}}{\Delta Y}\right)$$
$$+ C\frac{T_{i,j,k+1} - 2T_{i,j,k} + T_{i,j,k-1}}{\Delta Z^2} + D\phi^* \left(\frac{u^*_{i,j,k+1} - u^*_{i,j,k}}{\Delta Z}\right)^2 = 0 \tag{11.42}$$

where, k is the node number in the film thickness direction, and ρ^*, ϕ^*, and $\partial\rho^*/\partial T^*$ can be calculated analytically by expressions, and hence their differences are not necessary.

The viscosity–pressure–temperature equation, density–pressure–temperature, and load equation are the same as explained earlier.

Since the temperature in the energy equation, the viscosity in the viscosity–temperature–pressure equations, and the elastic deformation in the film thickness equation are all dependent on pressure, the general numerical method is as follows:

1. Give an initial pressure (the Hertz contact pressure) and a temperature distribution (the uniform temperature field);
2. Calculate the film thickness and viscosity, and then substitute them into the Reynolds equation to solve the new pressure distribution;
3. Revise repeatedly the pressure distribution obtained by the previous iteration, and then substitute the pressure into the energy equation to get the temperature distribution;

4. Update the viscosity by the new temperature, solve the pressure iteratively, repeat the process until the difference between the pressure distributions obtained from two successive iterations is very small, and then stop the iteration.

And we will obtain the final film thickness, pressure, and temperature distribution.

11.3.2.2 Calculation program and results

Calculation diagram (Figure 11.10)

Calculation program

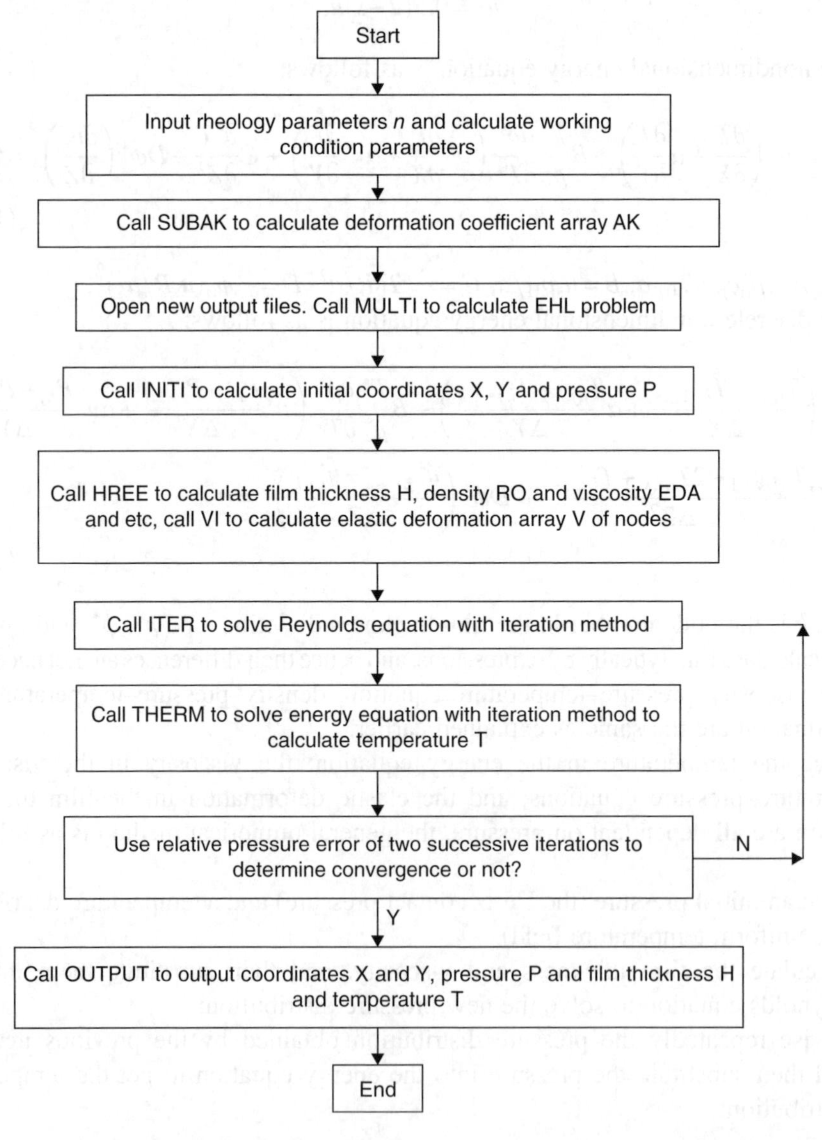

Figure 11.10 Diagram to calculate thermal GEHL in point contact

The parameters that are preassigned in this program are as follows:

Viscosity–pressure coefficient	$Z = 0.68$
Number of nodes	$N = 65 \times 65$
Load	W0 = 39.24 N
Equivalent modulus of elasticity	E1 = 2.21E11 Pa
Initial plastic viscosity	EDA0 = 0.058 Pa · s
Equivalent curvature radius	RX = RY = 0.05 m
Average velocity	US = 1.5 m · s^{-1}
Nondimensional starting coordinate in x direction	X0 = −2.5
Nondimensional ending coordinate in x direction	XE = 1.5
Nondimensional starting coordinate in y direction	Y0 = −2.0
Nondimensional ending coordinate in y direction	YE = 2.0
Number of layers cross the film thickness direction	NZ = 5
Iteration coefficient of temperature	CT = 0.31
Sliding–rolling ratio	AKC = 1.0.

In the BLOCK DATA, the following parameters are given:

Initial temperature	T0 = 303 K
Thermal conductivity of grease	AK0 = 0.14 W · m^{-1}·K
Thermal conductivity of down surface	AK1 = 46 W · m^{-1}·K
Thermal conductivity of up surface	AK2 = 46 W · m^{-1}·K;
Specific heat of grease	CV = 2000 J · kg^{-1}·K^{-1}
Specific heat of down surface	CV1 = 470 J · kg^{-1}·K^{-1}
Specific heat of up surface	CV2 = 470 J · kg^{-1}·K^{-1}
Initial density of grease	RO0 = 890 kg · m^{-3}
Density of down surface	RO1 = 7850 kg · m^{-3}
Density of up surface	RO2 = 7850 kg · m^{-3}
Viscosity–temperature coefficient	S0 = −1.1
Density–temperature coefficient	D0 = −0.00065.

The output parameters are as follows:

The pressure array P(I,J) will be saved in the file PRESS.DAT
The film thickness array H(I,J) will be saved in the file FILM.DAT
The temperature array T(I,J) will be saved in the file TEM.DAT.

Furthermore, after executing the program, one needs to input the rheology parameter n ($0.8 < n < 1$). If the user needs to change the above parameters, the program must be recompiled and relinked before executing.

```
      PROGRAM GREASEPOINTEHLT
      DIMENSION THETA(15), EALFA(15), EBETA(15)
     COMMON /COM1/Z, ENDA, AKC, HM0, HMC, EK, EAL, EBE, AD, AD1, KK1, KK2, KK3,
KK4, FN, FN1, FF
      COMMON /COM2/W0, E1, RX, B, PH, US, U1, U2, T1, T2, CT
      COMMON /COM3/T0, EDA0, AK, AK1, AK2, CV, CV1, CV2, RO0, RO1, RO2, S0, D0
```

```
      COMMON /COM4/A1, A2, A3 LMIN
      DATA PAI, Z, AKC, AD, AD1/3.14159265, 0.68, 1.0, 0.0, 0.0/
      DATA T0, EDA0, AK, AK1, AK2, CV, CV1, CV2, RO0, RO1, RO2, S0, D0/303.,
      0.058, 0.14, 46., 46., 2000., 470., 470., 890., 7850., 7850., -1.1,
      -0.00065/
      DATA N, NZ, RX, RY, X0, XE, W0, E1, US, CT/65, 5, 0.05, 0.05, -2.5, 1.5,
      39.24, 2.21E11, 1.5, 0.31/
      DATA THETA/10., 20., 30., 35., 40., 45., 50., 55., 60., 65., 70., 75., 80.,
      85., 90./
      DATA EALFA/6.612, 3.778, 2.731, 2.397, 2.136, 1.926, 1.754, 1.611,
      1.486, 1.378, 1.284, 1.202, 1.128, 1.061, 1.0/
      DATA EBETA/0.319, 0.408, 0.493, 0.53, 0.567, 0.604, 0.641, 0.678, 0.717,
      0.759, 0.802, 0.846, 0.893, 0.944, 1.0/
      DATA KK1, KK2, KK3, KK4, EAL, EBE/0, 0, 0, 0, 1.0, 1.0/
      WRITE(*, *)'n<=1 INPUT n=?'
      READ(*, *)FN
      FN1=1.0/FN
      FF=1.0/FN-1.0
      WRITE(*, *)"FF=", FF
      EK=RX/RY
      AA=0.5*(1./RX+1./RY)
      BB=0.5*ABS(1./RX-1./RY)
      CC=ACOS(BB/AA)*180.0/PAI
      DO I=1, 15
      IF(CC.LT.THETA(I))THEN
      WRITE(*, *)I
      EAL=EALFA(I-1)+(CC-THETA(I))*(EALFA(I)-EALFA(I-1))/(THETA(I)-THETA
      (I-1))
      EBE=EBETA(I-1)+(CC-THETA(I))*(EBETA(I)-EBETA(I-1))/(THETA(I)-THETA
      (I-1))

      GOTO 10
      ENDIF
      ENDDO
10    EA=EAL*(1.5*W0/AA/E1)**(1./3.0)
      EB=EBE*(1.5*W0/AA/E1)**(1./3.0)
      PH=1.5*W0/(EA*EB*PAI)
      OPEN(8, FILE='FILM.DAT', STATUS='UNKNOWN')
      OPEN(9, FILE='PRESS.DAT', STATUS='UNKNOWN')
      OPEN(10, FILE='TEM.DAT', STATUS='UNKNOWN')
      WRITE(*, *)"N, X0, XE, PH, E1, EDA0, RX, US"
      WRITE(*, *)N, X0, XE, PH, E1, EDA0, RX, US
      H00=0.0
      MM=N-1
      LMIN=ALOG(N-1.)/ALOG(2.)-1.99
      U=EDA0*(US/2.)**FN/(E1*RX**FN)
      U1=0.5*(2.+AKC)*U
      U2=0.5*(2.-AKC)*U
      A1=ALOG(EDA0)+9.67
      A2=5.1E-9*PH
      A3=0.59/(PH*1.E-9)
```

```
      B=PAI*PH*RX/E1
      W=2.*PAI*PH/(3.*E1)*(B/RX)**2
      ALFA=Z*5.1E-9*A1
      G=ALFA*E1
      AHM=1.0-EXP(-0.68*1.03)
      AHC=1.0-0.61*EXP(-0.73*1.03)
      HM0=3.63*(RX/B)**2*G**0.49*U**0.68*W**(-0.073)*AHM
      HMC=2.69*(RX/B)**2*G**0.53*U**0.67*W**(-0.067)*AHC
      ENDA=2.*U*(3.+FF)*2.0**(1.0+FF)*(E1/PH)**(1.0+FF)*(RX/B)**(3.0+FF)
      WRITE(*,*)"ENDA=",ENDA
      UTL=EDA0*US*RX/(B*B*2.E7)
      W0=2.0*PAI*EA*EB*PH/3.0
      T1=PH*B/RX
      T2=EDA0*US*RX/(B*B)
      WRITE(*,*)'          Wait please'
      CALL SUBAK(MM)
      CALL MULTI(N,NZ,X0,XE,H00)
      STOP
      END

      SUBROUTINE MULTI(N,NZ,X0,XE,H00)
      DIMENSION X(65),Y(65),H(4500),RO(4500),EPS(4500),EDA(4500),P
      (4500),POLD(4500),T(65,65,5)
      COMMON /COM1/Z,ENDA,AKC,HM0,HMC,EK,EAL,EBE,AD,AD1,KK1,KK2,KK3,
      KK4,FN,FN1,FF
      COMMON /COMT/TE(65,65)
      DATA MK,KTK,G00/200,1,2.0943951/
      G0=G00*EAL*EBE
      NX=N
      NY=N
      NN=(N+1)/2
      DO I=1,N
      DO J=1,N
      TE(I,J)=1.0
      DO K=1,5
      T(I,J,K)=1.0
      ENDDO
      ENDDO
      ENDDO
      CALL INITI(N,DX,X0,XE,X,Y,P,POLD)
      CALL HREE(N,DX,H00,G0,X,Y,H,RO,EPS,EDA,P)
      M=0
14    KK=15
      CALL ITER(N,KK,DX,H00,G0,X,Y,H,RO,EPS,EDA,P)
      CALL ERP(N,ER,P,POLD)
      ER=ER/KK
      WRITE(*,*)'ER=',ER
      CALL THERM(NX,NY,NZ,DX,P,H,T)
      CALL ERROM(NX,NY,NZ,T,ERM)
      M=M+1
      IF(M.LT.MK.AND.ER.GT.1.E-5)GOTO 14
```

```
      CALL OUPT(N, DX, X, Y, H, P, EDA, TMAX)
      RETURN
      END

      SUBROUTINE INITI(N, DX, X0, XE, X, Y, P, POLD)
      DIMENSION X(N), Y(N), P(N, N), POLD(N, N)
      NN=(N+1)/2
      DX=(XE-X0)/(N-1.)
      Y0=-0.5*(XE-X0)
      DO 5 I=1, N
      X(I)=X0+(I-1)*DX
      Y(I)=Y0+(I-1)*DX
5     CONTINUE
      DO 10 I=1, N
      D=1.-X(I)*X(I)
      DO 10 J=1, NN
      C=D-Y(J)*Y(J)
      IF(C.LE.0.0)P(I, J)=0.0
10    IF(C.GT.0.0)P(I, J)=SQRT(C)
      DO 20 I=1, N
      DO 20 J=NN+1, N
      JJ=N-J+1
20    P(I, J)=P(I, JJ)
      DO I=1, N
      DO J=1, N
      POLD(I, J)=P(I, J)
      ENDDO
      ENDDO
      RETURN
      END

      SUBROUTINE HREE(N, DX, H00, G0, X, Y, H, RO, EPS, EDA, P)
      DIMENSION X(N), Y(N), P(N, N), H(N, N), RO(N, N), EPS(N, N), EDA(N, N)
      DIMENSION W(150, 150), P0(150, 150), ROU(65, 65)
      COMMON /COM1/Z, ENDA, AKC, HM0, HMC, EK, EAL, EBE, AD, AD1, KK1, KK2, KK3,
      KK4, FN, FN1, FF
      COMMON /COM2/W0, E1, RX, B, PH, US, U1, U2, T1, T2, CT
      COMMON /COM3/T0, EDA0, AK, AK1, AK2, CV, CV1, CV2, RO0, RO1, RO2, S0, D0
      COMMON /COM4/A1, A2, A3, LMIN
      COMMON /COMT/TE(65, 65)
      DATA KR, NW, PAI, PAI1, DELTA/0, 150, 3.14159265, 0.2026423, 0.0/
      NN=(N+1)/2
      CALL VI(NW, N, DX, P, W)
      HMIN=1.E3
      DO 30 I=1, N
      DO 30 J=1, NN
      RAD=X(I)*X(I)+EK*Y(J)*Y(J)
      W1=0.5*RAD+DELTA
      ZZ=0.5*AD1*AD1+X(I)*ATAN(AD*PAI/180.0)
      IF(W1.LE.ZZ)W1=ZZ
```

```
      H0=W1+W(I, J)
      IF(H0.LT.HMIN)HMIN=H0
30    H(I, J)=H0
      IF(KK.EQ.0)THEN
      KG1=0
      H01=-HMIN+HM0
      DH=0.005*HM0
      H02=-HMIN
      H00=0.5*(H01+H02)
      ENDIF
      W1=0.0
      DO 32 I=1, N
      DO 32 J=1, N
32    W1=W1+P(I, J)
      W1=DX*DX*W1/G0
      DW=1.-W1
      IF(KK.EQ.0)THEN
      KK=1
      GOTO 50
      ENDIF
      IF(DW.LT.0.0)THEN
      KG1=1
      H00=AMIN1(H01, H00+DH)
      ENDIF
      IF(DW.GT.0.0)THEN
      KG2=2
      H00=AMAX1(H02, H00-DH)
      ENDIF
50    DO 60 I=1, N
      DO 60 J=1, NN
      H(I, J)=H00+H(I, J)
      CT1=((TE(I, J)-0.455445545)/0.544554455)**S0
      CT2=D0*T0*(TE(I, J)-1.)
      IF(P(I, J).LT.0.0)P(I, J)=0.0
      EDA1=EXP(A1*(-1.+(1.+A2*P(I, J))**Z*CT1))
      EDA(I, J)=EDA1
      RO(I, J)=1.+CT2
      EPS(I, J)=ENDA*RO(I, J)*H(I, J)**(2.+FN1)/(EDA(I, J)**FN1)
60    CONTINUE
      DO 70 J=NN+1, N
      JJ=N-J+1
      DO 70 I=1, N
      H(I, J)=H(I, JJ)
      RO(I, J)=RO(I, JJ)
      EDA(I, J)=EDA(I, JJ)
70    EPS(I, J)=EPS(I, JJ)
      RETURN
      END

      SUBROUTINE ITER(N, KK, DX, H00, G0, X, Y, H, RO, EPS, EDA, P)
      DIMENSION X(N), Y(N), P(N, N), H(N, N), RO(N, N), EPS(N, N), EDA(N, N)
```

```
      DIMENSION D(70), A(350), B(210), ID(70)
      COMMON /COM1/Z, ENDA, AKC, HM0, HMC, EK, EAL, EBE, AD, AD1, KK1, KK2, KK3,
      KK4, FN, FN1, FF
      COMMON /COMAK/AK(0:65, 0:65)
      DATA KG1, PAI1, C1, C2/0, 0.2026423, 0.27, 0.27/
      IF(KG1.NE.0)GOTO 2
      KG1=1
      AK00=AK(0, 0)
      AK10=AK(1, 0)
      AK20=AK(2, 0)
      BK00=AK00-AK10
      BK10=AK10-0.25*(AK00+2.*AK(1, 1)+AK(2, 0))
      BK20=AK20-0.25*(AK10+2.*AK(2, 1)+AK(3, 0))
2     NN=(N+1)/2
      MM=N-1
      DX1=1./DX
      DX2=DX*DX
      DX3=1./DX2
      DO 100 K=1, KK
      PMAX=0.0
      DO 70 J=2, NN
      J0=J-1
      J1=J+1
      IA=1
8     MM=N-IA
      IF(P(MM, J0).GT.1.E-6)GOTO 20
      IF(P(MM, J).GT.1.E-6)GOTO 20
      IF(P(MM, J1).GT.1.E-6)GOTO 20
      IA=IA+1
      IF(IA.LT.N)GOTO 8
      GOTO 70
20    IF(MM.LT.N-1)MM=MM+1
      DPDX1=ABS((P(2, J)-P(1, J))*DX1)**(FF)
      D2=0.5*(EPS(1, J)+EPS(2, J))*DPDX1
      DO 50 I=2, MM
      I0=I-1
      I1=I+1
      II=5*I0
      DPDX2=ABS((P(I1, J)-P(I, J))*DX1)**(FF)
      DPDY1=ABS((P(I, J)-P(I, J0))*DX1)**(FF)
      DPDY2=ABS((P(I, J1)-P(I, J))*DX1)**(FF)
      D1=D2
      D2=0.5*(EPS(I1, J)+EPS(I, J))*DPDX2
      D4=0.5*(EPS(I, J0)+EPS(I, J))*DPDY1
      D5=0.5*(EPS(I, J1)+EPS(I, J))*DPDY2
      P1=P(I0, J)
      P2=P(I1, J)
      P3=P(I, J)
      P4=P(I, J0)
      P5=P(I, J1)
      D3=D1+D2+D4+D5
```

```
    IF(H(I, J).LE.0.0)THEN
    ID(I)=0
    A(II+1)=0.0
    A(II+2)=0.0
    A(II+3)=1.0
    A(II+4)=0.0
    A(II+5)=1.0
    A(II-4)=0.0
    GOTO 50
    ENDIF
    ID(I)=1
    IF(J.EQ.NN)P5=P4
    A(II+1)=PAI1*(RO(I0, J)*AK10-RO(I, J)*AK20)
    A(II+2)=DX3*D1+PAI1*(RO(I0, J)*AK00-RO(I, J)*AK10)
    A(II+3)=-DX3*D3+PAI1*(RO(I0, J)*AK10-RO(I, J)*AK00)
    A(II+4)=DX3*D2+PAI1*(RO(I0, J)*AK20-RO(I, J)*AK10)
    A(II+5)=-DX3*(D1*P1+D2*P2+D4*P4+D5*P5-D3*P3)+DX1*(RO(I, J)*H(I, J)-
    RO(I0, J)*H(I0, J))
50  CONTINUE
    CALL TRA4(MM, D, A, B)
    DO 60 I=2, MM
    IF(ID(I).EQ.1)P(I, J)=P(I, J)+C1*D(I)
    IF(P(I, J).LT.0.0)P(I, J)=0.0
    IF(PMAX.LT.P(I, J))PMAX=P(I, J)
60  CONTINUE
70  CONTINUE
    DO 80 J=1, NN
    JJ=N+1-J
    DO 80 I=1, N
80  P(I, JJ)=P(I, J)
    CALL HREE(N, DX, H00, G0, X, Y, H, RO, EPS, EDA, P)
100 CONTINUE
    RETURN
    END

    SUBROUTINE TRA4(N, D, A, B)
    DIMENSION D(N), A(5, N), B(3, N)
    C=1./A(3, N)
    B(1, N)=-A(1, N)*C
    B(2, N)=-A(2, N)*C
    B(3, N)=A(5, N)*C
    DO 10 I=1, N-2
    IN=N-I
    IN1=IN+1
    C=1./(A(3, IN)+A(4, IN)*B(2, IN1))
    B(1, IN)=-A(1, IN)*C
    B(2, IN)=-(A(2, IN)+A(4, IN)*B(1, IN1))*C
10  B(3, IN)=(A(5, IN)-A(4, IN)*B(3, IN1))*C
    D(1)=0.0
    D(2)=B(3, 2)
    DO 20 I=3, N
```

```
20  D(I)=B(1, I)*D(I-2)+B(2, I)*D(I-1)+B(3, I)
    RETURN
    END

    SUBROUTINE VI(NW, N, DX, P, V)
    DIMENSION P(N, N), V(NW, NW)
    COMMON /COMAK/AK(0:65, 0:65)
    PAI1=0.2026423
    DO 40 I=1, N
    DO 40 J=1, N
    H0=0.0
    DO 30 K=1, N
    IK=IABS(I-K)
    DO 30 L=1, N
    JL=IABS(J-L)
30  H0=H0+AK(IK, JL)*P(K, L)
40  V(I, J)=H0*DX*PAI1
    RETURN
    END

    SUBROUTINE SUBAK(MM)
    COMMON /COMAK/AK(0:65, 0:65)
    S(X, Y)=X+SQRT(X**2+Y**2)
    DO 10 I=0, MM
    XP=I+0.5
    XM=I-0.5
    DO 10 J=0, I
    YP=J+0.5
    YM=J-0.5
    A1=S(YP, XP)/S(YM, XP)
    A2=S(XM, YM)/S(XP, YM)
    A3=S(YM, XM)/S(YP, XM)
    A4=S(XP, YP)/S(XM, YP)
    AK(I, J)=XP*ALOG(A1)+YM*ALOG(A2)+XM*ALOG(A3)+YP*ALOG(A4)
10  AK(J, I)=AK(I, J)
    RETURN
    END

    SUBROUTINE ERP(N, ER, P, POLD)
    DIMENSION P(N, N), POLD(N, N)
    ER=0.0
    SUM=0.0
    NN=(N+1)/2
    DO 10 I=1, N
    DO 10 J=1, NN
    ER=ER+ABS(P(I, J)-POLD(I, J))
    SUM=SUM+P(I, J)
10  CONTINUE
    ER=ER/SUM
    DO I=1, N
    DO J=1, N
```

```
      POLD(I, J)=P(I, J)
      ENDDO
      ENDDO
      RETURN
      END

      SUBROUTINE ERROM(NX, NY, NZ, T, ERM)
      DIMENSION T(NX, NY, NZ)
      COMMON /COMT/TE(65, 65)
      ERM=0.
      C1=1./FLOAT(NZ)
      DO 20 I=2, NX
      DO 20 J=2, NY
      TT=0.
      DO 10 K=1, NZ
10    TT=TT+T(I, J, K)
      TT=C1*TT
      ER=ABS((TT-TE(I, J))/TT)
      IF(ER.GT.ERM)ERM=ER
20    TE(I, J)=TT
      RETURN
      END

      SUBROUTINE OUPT(N, DX, X, Y, H, P, EDA, TMAX)
      DIMENSION X(N), Y(N), H(N, N), P(N, N), EDA(N, N)
      COMMON /COM1/Z, ENDA, AKC, HM0, HMC, EK, EAL, EBE, AD, AD1, KK1, KK2, KK3,
      KK4, FN, FN1, FF
      COMMON /COM2/W0, E1, RX, B, PH, US, U1, U2, T1, T2, CT
      COMMON /COMT/TE(65, 65)
      A=0.0
      WRITE(8, 40)A, (Y(I), I=1, N)
      DO I=1, N
      WRITE(8, 40)X(I), (H(I, J), J=1, N)
      ENDDO
      WRITE(9, 40)A, (Y(I), I=1, N)
      DO I=1, N
      WRITE(9, 40)X(I), (P(I, J), J=1, N)
      ENDDO
40    FORMAT(66(E12.6, 1X))
      WRITE(10, 60)A, (Y(I), I=1, N)
      TMAX=0.0
      DO I=1, N
      WRITE(10, 60)X(I), (273.0*(TE(I, JJ)-1.), JJ=1, N)
      DO J=1, N
      IF(TMAX.LT.273.0*(TE(I, J)-1.))TMAX=273.*(TE(I, J)-1.)
      ENDDO
      ENDDO
60    FORMAT(66(E12.6, 1X))
      HMIN=1.E3
      PMAX=0.0
      DO J=1, N
```

```
    DO I=2, N
    IF(H(I, J).LT.HMIN)HMIN=H(I, J)
    IF(P(I, J).GT.PMAX)PMAX=P(I, J)
    ENDDO
    ENDDO
    HMIN=HMIN*B*B/RX
    PMAX=PMAX*PH
    RETURN
    END

    SUBROUTINE THERM(NX, NY, NZ, DX, P, H, T)
    DIMENSION T(NX, NY, NZ), T1(21), TI(21), U(21), DU(21), UU(21), V(21),
    DV(21), VV(21), W(21), EDA(21), RO(21), EDA1(21), EDA2(21), ROR(21),
    P(NX, NX), H(NX, NX), TFX(21), TFY(21)
    COMMON /COM1/Z, ENDA, AKC, HM0, HMC, EK, EAL, EBE, AD, AD1, KK1, KK2, KK3,
    KK4, FN, FN1, FF
    IF(KK.NE.0)GOTO 4
    DO 2 K=1, NZ
    DO 1 J=1, NY
1   T(1, J, K)=1.0
    DO 2 I=1, NX
2   T(I, 1, K)=1.0
4   DO 30 I=2, NX
    DO 30 J=2, NY
    KG=0
    DO 6 K=1, NZ
    TFX(K)=T(I-1, J, K)
    TFY(K)=T(I, J-1, K)
    IF(KK.NE.0)GOTO 5
    T1(K)=T(I-1, J, K)
    GOTO 6
5   T1(K)=T(I, J, K)
6   TI(K)=T1(K)
    P1=P(I, J)
    H1=H(I, J)
    DPX=(P(I, J)-P(I-1, J))/DX
    DPY=(P(I, J)-P(I, J-1))/DX
    CALL TBOUD(NX, NY, NZ, I, J, CC1, CC2, T)
10  CALL EROEQ(NZ, T1, P1, H1, DPX, DPY, EDA, RO, EDA1, EDA2, KG)
    CALL UCAL(NZ, DX, H1, EDA, RO, ROR, EDA1, EDA2, U, UU, DU, V, VV, DV, W,
    DPX, DPY)
    CALL TCAL(NZ, DX, CC1, CC2, T1, TFX, TFY, U, V, W, DU, DV, H1, DPX, DPY,
    EDA, RO)
    CALL ERRO(NZ, TI, T1, ETS)
    KG=KG+3
    IF(ETS.GT.1.E-4.AND.KG.LE.50)GOTO 10
    DO 20 K=1, NZ
    ROR(K)=RO(K)
    UU(K)=U(K)
    VV(K)=V(K)
20  T(I, J, K)=T1(K)
```

```
30  CONTINUE
    KK=1
    RETURN
    END

    SUBROUTINE TBOUD(NX, NY, NZ, I, J, CC1, CC2, T)
    DIMENSION T(NX, NY, NZ)
    CC1=0.
    CC2=0.
    DO 10 L=1, I-1
    DS=1./SQRT(FLOAT(I-L))
    IF(L.EQ.I-1)DS=1.1666667
    CC1=CC1+DS*(T(L, J, 2)-T(L, J, 1))
10  CC2=CC2+DS*(T(L, J, NZ)-T(L, J, NZ-1))
    RETURN
    END

    SUBROUTINE ERRO(NZ, T0, T, ETS)
    DIMENSION T0(NZ), T(NZ)
    ETS=0.0
    DO 10 K=1, NZ
    IF(T(K).LT.1.E-5)ETS0=1.
    IF(T(K).GE.1.E-5)ETS0=ABS((T(K)-T0(K))/T(K))
    IF(ETS0.GT.ETS)ETS=ETS0
10  T0(K)=T(K)
    RETURN
    END

    SUBROUTINE EROEQ(NZ, T, P, H, DPX, DPY, EDA, RO, EDA1, EDA2, KG)
    DIMENSION T(NZ), EDA(NZ), RO(NZ), EDA1(NZ), EDA2(NZ)
    COMMON /COM1/Z, ENDA, AKC, HM0, HMC, EK, EAL, EBE, AD, AD1, KK1, KK2, KK3,
    KK4, FN, FN1, FF
    COMMON /COM2/W0, E1, RX, B, PH, US, U1, U2, T1, T2, CT
    COMMON /COM3/T0, EDA0, AK, AK1, AK2, CV, CV1, CV2, RO0, RO1, RO2, S0, D0
    COMMON /COM4/A1, A2, A3, LMIN
    DATA A4, A5/0.455445545, 0.544554455/
    IF(KG.NE.0)GOTO 20
    B1=(1.+A2*P)**Z
    B2=(A3+1.34*P)/(A3+P)
20  DO 30 K=1, NZ
    EDA3=EXP(A1*(-1.+B1*((T(K)-A4)/A5)**S0))
    EDA(K)=EDA3
30  RO(K)=B2+D0*T0*(T(K)-1.)
    CC1=0.5/(NZ-1.)
    CC2=1./(NZ-1.)
    C1=0.
    C2=0.
    DO 40 K=1, NZ
    IF(K.EQ.1)GOTO 32
    C1=C1+0.5/EDA(K)+0.5/EDA(K-1)
    C2=C2+CC1*((K-1.)/EDA(K)+(K-2.)/EDA(K-1))
```

```
32  EDA1(K)=C1*CC2
40  EDA2(K)=C2*CC2
    RETURN
    END

    SUBROUTINE UCAL(NZ, DX, H, EDA, RO, ROR, EDA1, EDA2, U, UU, DU, V, VV, DV,
    W, DPX, DPY)
    DIMENSION U(NZ), UU(NZ), DU(NZ), V(NZ), VV(NZ), DV(NZ), W(NZ), ROR(NZ),
    EDA(NZ), RO(NZ), EDA1(NZ), EDA2(NZ)
    COMMON /COM1/Z, ENDA, AKC, HM0, HMC, EK, EAL, EBE, AD, AD1, KK1, KK2, KK3,
    KK4, FN, FN1, FF
    COMMON /COM2/W0, E1, R, B, PH, US, U1, U2, T1, T2, CC
    COMMON /COM3/T0, EDA0, AK, AK1, AK2, CV, CV1, CV2, RO0, RO1, RO2, S0, D0
    IF(KK.NE.0)GOTO 20
    A1=U1
    A2=PH*(B/R)**3/E1
    A3=U2-U1
20  CUA=A2*DPX*H
    CUB=CUA*H
    CVA=A2*DPY*H
    CVB=CVA*H
    CC3=A3/H
    CC4=1./EDA1(NZ)
    DO 30 K=1, NZ
    U(K)=A1+CUB*(EDA2(K)-CC4*EDA2(NZ)*EDA1(K))+A3*CC4*EDA1(K)
    V(K)=CVB*(EDA2(K)-CC4*EDA2(NZ)*EDA1(K))
    DU(K)=CUA/EDA(K)*((K-1.)/(NZ-1.)-CC4*EDA2(NZ))+CC3*CC4/EDA(K)
30  DV(K)=CVA/EDA(K)*((K-1.)/(NZ-1.)-CC4*EDA2(NZ))
    A4=B/((NZ-1)*R*DX)
    C1=A4*H
    IF(KK.EQ.0)GOTO 50
    DO 40 K=2, NZ-1
    W(K)=(RO(K-1)*W(K-1)+C1*(RO(K)*(U(K)+V(K))-ROR(K)*(UU(K)+VV
    (K))))/RO(K)
40  CONTINUE
50  KK=1
    RETURN
    END

    SUBROUTINE TCAL(NZ, DX, CC1, CC2, T, TFX, TFY, U, V, W, DU, DV, H, DPX, DPY,
    EDA, RO)
    DIMENSION T(NZ), U(NZ), DU(NZ), V(NZ), DV(NZ), W(NZ), EDA(NZ), RO(NZ),
    A(4, 21), D(21), AA(2, 21), TFX(NZ), TFY(NZ)
    COMMON /COM1/Z, ENDA, AKC, HM0, HMC, EK, EAL, EBE, AD, AD1, KK1, KK2, KK3,
    KK4, FN, FN1, FF
    COMMON /COM2/W0, E1, R, B, PH, US, U1, U2, T1, T2, CC
    COMMON /COM3/T0, EDA0, AK, AK1, AK2, CV, CV1, CV2, RO0, RO1, RO2, S0, D0
    DATA CC5, PAI/0.6666667, 3.14159265/
    IF(KK.NE.0)GOTO 5
    KK=1
    A2=-CV*RO0*E1*B**3/(EDA0*AK*R)
```

```
      A3=-E1*PH*B**3*D0/(AK*EDA0*T0*R)
      A4=-(E1*R)**2/(AK*EDA0*T0)
      A5=0.5*R/B*A2
      A6=AK*SQRT(EDA0*R/(PAI*RO1*CV1*U1*E1*AK1*B**3))
      A7=AK*SQRT(EDA0*R/(PAI*RO2*CV2*U2*E1*AK2*B**3))
5     CC3=A6*SQRT(DX)
      CC4=A7*SQRT(DX)
      DZ=H/(NZ-1.)
      DZ1=1./DZ
      DZ2=DZ1*DZ1
      CC6=A3*DPX
      CC7=A3*DPY
      DO 10 K=2, NZ-1
      A(1, K)=DZ2+DZ1*A5*RO(K)*W(K)
      A(2, K)=-2.*DZ2+A2*RO(K)*(U(K)+V(K))/DX+(CC6*U(K)+CC7*V(K))/RO(K)
      A(3, K)=DZ2-DZ1*A5*RO(K)*W(K)
10    A(4, K)=A4*EDA(K)*(DU(K)**2+DV(K)**2)+A2*RO(K)*(U(K)*TFX(K)+V(K)
      *TFY(K))/DX
      A(1, 1)=0.
      A(2, 1)=1.+2.*DZ1*CC3*CC5
      A(3, 1)=-2.*DZ1*CC3*CC5
      A(1, NZ)=-2.*DZ1*CC4*CC5
      A(2, NZ)=1.+2.*DZ1*CC4*CC5
      A(3, NZ)=0.
      A(4, 1)=1.+CC1*CC3*DZ1
      A(4, NZ)=1.-CC2*CC4*DZ1
      CALL TRA3(NZ, D, A, AA)
      DO 20 K=1, NZ
      T(K)=(1.-CC)*T(K)+CC*D(K)
20    IF(T(K).LT.1.)T(K)=1.
30    CONTINUE
      RETURN
      END

      SUBROUTINE TRA3(N, D, A, B)
      DIMENSION D(N), A(4, N), B(2, N)
      C=1./A(2, N)
      B(1, N)=-A(1, N)*C
      B(2, N)=A(4, N)*C
      DO 10 I=1, N-1
      IN=N-I
      IN1=IN+1
      C=1./(A(2, IN)+A(3, IN)*B(1, IN1))
      B(1, IN)=-A(1, IN)*C
10    B(2, IN)=(A(4, IN)-A(3, IN)*B(2, IN1))*C
      D(1)=B(2, 1)
      DO 20 I=2, N
20    D(I)=B(1, I)*D(I-1)+B(2, I)
      RETURN
      END
```

Calculation results

Based on the program, the results obtained for the film thickness, pressure distribution, and temperature distribution are shown in Figure 11.11.

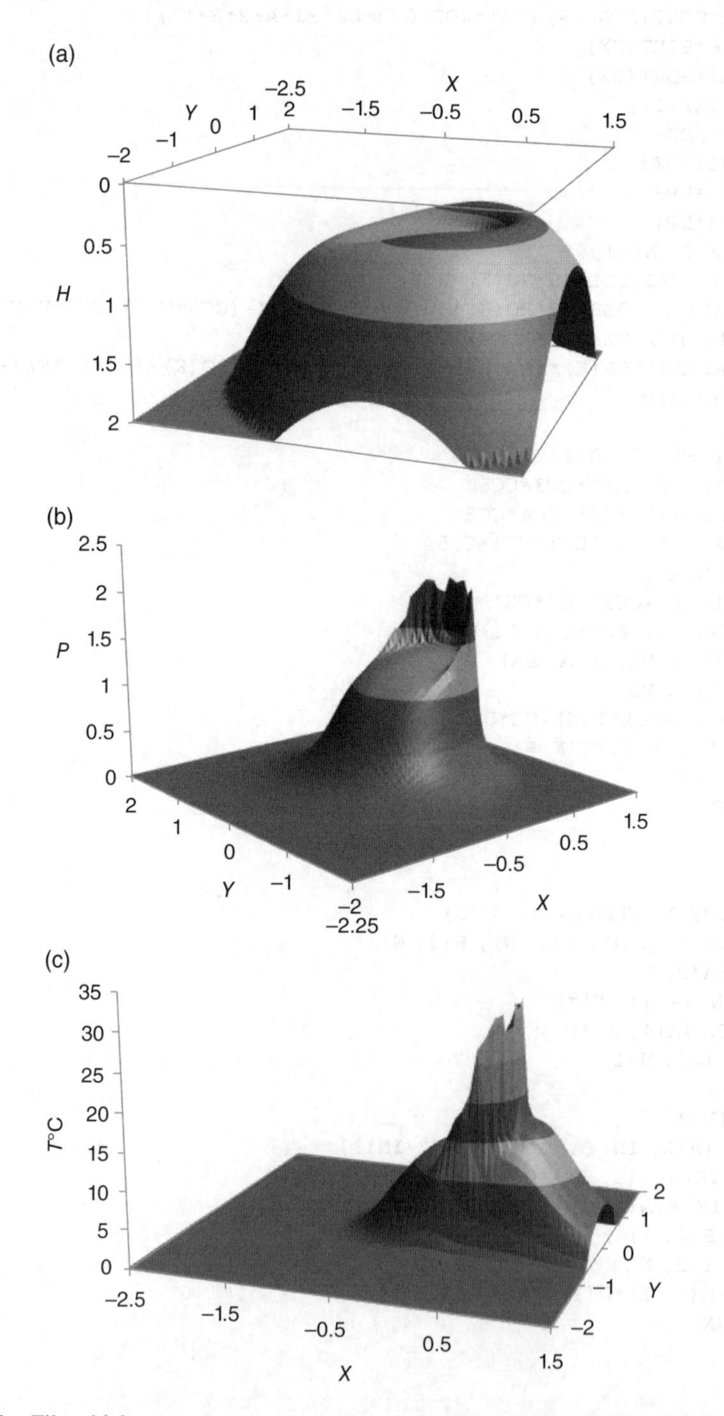

Figure 11.11 Film thickness, pressure, and temperature distribution of thermal GEHL in point contact. (a) Film thickness, (b) pressure distribution, and (c) temperature distribution

12

Numerical calculation method and program for EHL considering effect of electric double layer

12.1 Structure of electric double layer

At the interface between a solid and liquid, charge migration within the electric double layer (EDL) causes an electroviscous force to develop which manifests itself as an enhanced fluid viscosity. As shown in Figure 12.1, the EDL consists of a Stern layer and a diffuse layer. The Stern layer is a layer of ions that are strongly attracted to the solid surface and are immobile. The layer presents a potential ψ. The macro flow is observed at the boundary comprising of one or several molecules between the Stern layer and the diffuse layer, which is usually referred as the slipping plane. The electrical potential at the slipping plane, called the Zeta potential ζ, can be achieved theoretically according to the properties of water and ceramic.

12.2 Reynolds equation considering EDL effect

12.2.1 Modified Reynolds equation

Figure 12.2 shows the EDL hydrodynamic lubrication model. Let us assume that there is one EDL at each interface of solid and lubricant in lubrication regime. The bottom surface at $z=0$ moves with a velocity u_s in the direction x, and the upper surface is static. Therefore, the EDL potential in the direction z is as follows:

Numerical Calculation of Elastohydrodynamic Lubrication: Methods and Programs, First Edition.
Ping Huang.

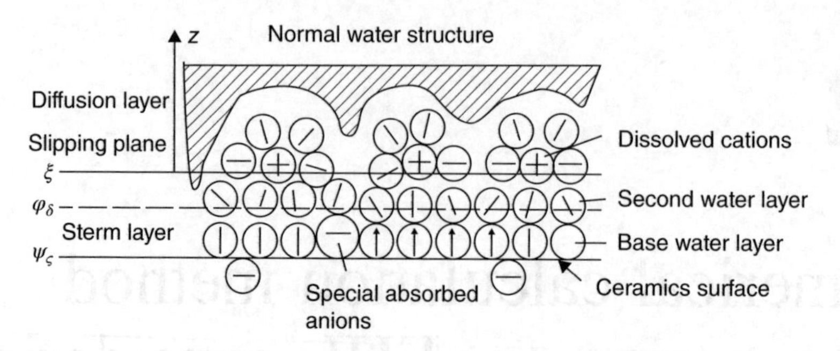

Figure 12.1 Structure of electric double layer

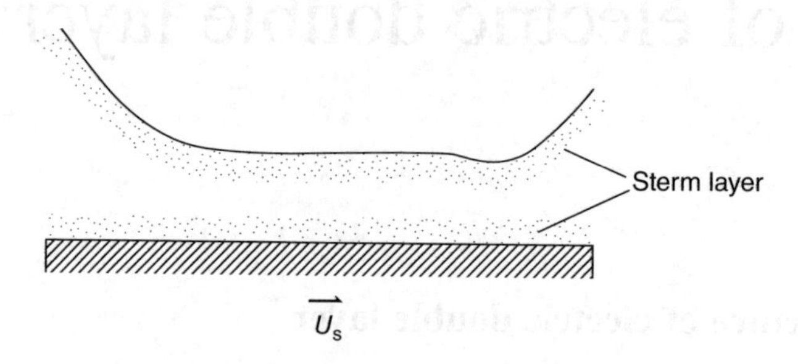

Figure 12.2 EHL model with electrical double layer

$$\psi=\begin{cases}\zeta\ \exp(-\chi z), & 0<z\le h/2\\ \zeta\ \exp[-\chi(h-z)], & h/2<z<h\end{cases} \tag{12.1}$$

where, h is the film thickness, $\chi-1$ is the Debye length, and ζ is the electrical potential on the slipping plane.

In the conventional lubrication theory, only the viscous force and the pressure of fluid are considered. However, the electroviscous force needs to be taken into account in the EDL lubrication. The viscous forces on the micro unit in the directions x and y can be calculated by the following equation:

$$\begin{cases}dF_x=\dfrac{\partial\tau_{zx}}{\partial z}dxdydz\\ dF_y=\dfrac{\partial\tau_{zy}}{\partial z}dxdydz\end{cases} \tag{12.2}$$

From the Newton's law of viscosity, the following equation can be obtained.

$$\tau_{zx}=\eta\frac{\partial u_x}{\partial z}$$

$$\tau_{zy}=\eta\frac{\partial u_y}{\partial z}$$

where, u_x and u_y are the fluid velocities, respectively, in the direction x and y and η is the fluid viscosity.

By substituting the above equation into Equation 12.2, we have

$$\begin{cases} dF_x = \eta \dfrac{\partial^2 u_x}{\partial z^2} dxdydz \\ dF_y = \eta \dfrac{\partial^2 u_y}{\partial z^2} dxdydz \end{cases} \tag{12.3}$$

The following equations present the expression of the pressure difference dP_x and dP_y at the distance dx and dy, respectively, in the direction x and y:

$$\begin{cases} dP_x = -\dfrac{\partial p}{\partial x} dxdydz \\ dP_y = -\dfrac{\partial p}{\partial y} dxdydz \end{cases} \tag{12.4}$$

And the electric field force differences dR_x and dR_y, respectively, in the direction x and y can be expressed as:

$$\begin{cases} dR_x = E_x \rho dxdydz \\ dR_y = E_y \rho dxdydz \end{cases} \tag{12.5}$$

where, E_x and E_y are the streaming potential gradient induced by the EDL inner flowing fluid, respectively, in the direction x and y, and ρ is the bulk charge density. The Helmholtz-Smoluchowski formula gives the relationship between the streaming potentials E_x and E_y and the fluid pressure gradients as follows:

$$\begin{cases} E_x = -\dfrac{\zeta \varepsilon}{4\pi \eta_a \lambda} \dfrac{\partial p}{\partial x} \\ E_y = -\dfrac{\zeta \varepsilon}{4\pi \eta_a \lambda} \dfrac{\partial p}{\partial y} \end{cases} \tag{12.6}$$

where, λ is the bulk electrical conductivity and ε is the absolute dielectric constant of fluid.

Equation 12.6 is reduced based on a capillary. Since the capillary diameter is far greater than the EDL thickness, the bulk viscosity η in the original equation is replaced by the apparent viscosity η_a considering EDL.

Then, the force-balance equation for the micro unit is obtained thus:

$$\begin{cases} dF_x + dP_x + dR_x = 0 \\ dF_y + dP_y + dR_y = 0 \end{cases} \tag{12.7}$$

Substituting Equations 12.3–12.5 into Equation 12.7 gives

$$\begin{cases} \eta \dfrac{\partial^2 u_x}{\partial z^2} - \dfrac{\partial p}{\partial x} + E_x \rho = 0 \\ \eta \dfrac{\partial^2 u_y}{\partial z^2} - \dfrac{\partial p}{\partial y} + E_y \rho = 0 \end{cases} \tag{12.8}$$

The EDL electrical potential ψ can be expressed as

$$\nabla^2 \psi = -\frac{4\pi\rho}{\varepsilon} \tag{12.9}$$

Considering that the EDL dimension is far larger in the directions x and y than in the direction z, which means $\dfrac{\partial^2}{\partial x^2} \ll \dfrac{\partial^2}{\partial z^2}$ and $\dfrac{\partial^2}{\partial y^2} \ll \dfrac{\partial^2}{\partial z^2}$, substitution of Equation 12.9 into Equation 12.8 and further simplification give the following equation:

$$\begin{cases} \eta \dfrac{\partial^2 u_x}{\partial z^2} - \dfrac{\partial p}{\partial x} - \dfrac{E_x \varepsilon}{4\pi} \dfrac{\partial^2 \psi}{\partial z^2} = 0 \\ \eta \dfrac{\partial^2 u_y}{\partial z^2} - \dfrac{\partial p}{\partial y} - \dfrac{E_y \varepsilon}{4\pi} \dfrac{\partial^2 \psi}{\partial z^2} = 0 \end{cases} \tag{12.10}$$

Let us assume that the pressure remains constant in the z direction, and hence the following expression is obtained after two integrations of the first formula of Equation 12.10.

$$\eta u_x - \frac{E_x \varepsilon}{4\pi} \psi = \frac{1}{2} \frac{\partial p}{\partial x} z^2 + Az + B \tag{12.11}$$

where, A and B are the integral constants that are determined by the following boundary conditions:

$$\begin{cases} u_x|_{z=0} = -u_s, \quad \psi|_{z=0} = \zeta \\ u_x|_{z=h} = 0, \quad \psi|_{z=h} = \zeta \end{cases}$$

On substituting the above boundary condition into Equation 12.11, we get A and B and are as follows:

$$A = -\frac{h}{2}\frac{\partial p}{\partial x} + \frac{\eta u_s}{h}$$

$$B = -\frac{E_x \varepsilon}{4\pi}\zeta - \eta u_s$$

Then, by substituting the above expressions of A and B into Equation 12.11, we get

$$\eta u_x = \frac{z^2}{2}\frac{\partial p}{\partial x} - \frac{hz}{2}\frac{\partial p}{\partial x} + \frac{E_x \varepsilon}{4\pi}(\psi - \zeta) - \eta\left(1 - \frac{z}{h}\right)u_s \tag{12.12}$$

Similarly, for the y direction, we have:

$$\eta u_y = \frac{z^2}{2}\frac{\partial p}{\partial y} - \frac{hz}{2}\frac{\partial p}{\partial y} + \frac{E_y \varepsilon}{4\pi}(\psi - \zeta) \tag{12.13}$$

In the reduction of Equation 12.13, the following boundary conditions are applied.

$$u_y\big|_{z=0} = 0; \ u_y\big|_{z=h} = 0$$

The flow rate in the x direction can be expressed as follows:

$$Q_x = \int_0^h u_x dz$$

By substituting Equation 12.12 into the above equation, we get

$$Q_x = \frac{1}{\eta}\left\{-\frac{h^3}{12}\frac{\partial p}{\partial x} - \frac{E_x \varepsilon \zeta}{4\pi}\left[h - \frac{2}{\chi}\left(1 - e^{-\chi h/2}\right)\right]\right\} - \frac{h u_s}{2} \tag{12.14}$$

In a similar way, the expression of flow rate in the y direction is obtained:

$$Q_y = \frac{1}{\eta}\left\{-\frac{h^3}{12}\frac{\partial p}{\partial y} - \frac{E_y \varepsilon \zeta}{4\pi}\left[h - \frac{2}{\chi}\left(1 - e^{-\chi h/2}\right)\right]\right\} \tag{12.15}$$

The law of conservation of mass for non-compressible fluid can be expressed as follows:

$$\frac{\partial Q_x}{\partial x} + \frac{\partial Q_y}{\partial y} = 0$$

Substituting Equations 12.14 and 12.15 into the above equation, we get

$$\frac{\partial}{\partial x}\left(\frac{h^3}{12\eta}\frac{\partial p}{\partial x}\right)+\frac{\partial}{\partial y}\left(\frac{h^3}{12\eta}\frac{\partial p}{\partial y}\right)=-\frac{u_s}{2}\frac{\partial h}{\partial x}-\frac{\partial}{\partial x}\left\{\frac{E_x\varepsilon\zeta}{4\pi\eta}\left[h-\frac{2}{\chi}\left(1-e^{-\chi h/2}\right)\right]\right\}$$
$$-\frac{\partial}{\partial y}\left\{\frac{E_y\varepsilon\zeta}{4\pi\eta}\left[h-\frac{2}{\chi}\left(1-e^{-\chi h/2}\right)\right]\right\} \tag{12.16}$$

Therefore, the modified Reynolds equation with considering the EDL effect is obtained after substituting Equation 12.6 into Equation 12.16 and is as follows:

$$\frac{\partial}{\partial x}\left(\frac{h^3}{12\eta_a}\frac{\partial p}{\partial x}\right)+\frac{\partial}{\partial y}\left(\frac{h^3}{12\eta_a}\frac{\partial p}{\partial y}\right)=-\frac{u_s}{2}\frac{\partial h}{\partial x} \tag{12.17}$$

where, the apparent viscosity η_a is expressed as

$$\eta_a=\eta+\frac{3\varepsilon^2\zeta^2\left\{h-\frac{2}{\chi}\left(1-e^{-\chi h/2}\right)\right\}}{4\pi^2\lambda h^3} \tag{12.18}$$

12.2.2 *Expression of electroviscosity*

Equation 12.18 indicates that the increase in apparent viscosity is proportional to the square of EDL ζ potential and reciprocal to the third power of h. This means that the EDL significantly affects lubrication with a very thin film. It is also presented that the ζ potential may enhance the carrying capacity of the lubrication film by inducing an additional electroviscosity on the fluid viscosity. Here, the electroviscosity is defined as:

$$\eta_e=\frac{3\varepsilon^2\zeta^2\left\{h-\frac{2}{\chi}\left(1-e^{-\chi h/2}\right)\right\}}{4\pi^2\lambda h^3} \tag{12.19}$$

Substituting Equations 12.19 into Equation 12.18, we get

$$\eta_a=\eta+\eta_e \tag{12.20}$$

Since the film thickness h is far larger than the EDL Debye thickness, which means $\chi h \gg 1$, Equation 12.18 can be simplified as:

$$\eta_a=\eta+\frac{3\varepsilon^2\zeta^2}{4\pi^2\lambda h^3} \tag{12.21}$$

Correspondingly, the electroviscosity can be expressed as:

$$\eta_e = \frac{3\varepsilon^2\zeta^2}{4\pi^2\lambda h^3} \tag{12.22}$$

12.3 Calculation program and example

12.3.1 Calculation program

Equation 12.22 is slightly modified to analyze EDL effect. A few modifications are made to the subprogram HREE, as they are the only those related to fluid viscosity computation. The modified parts are represented in bold for better understanding.

12.3.1.1 Calculation program in line contact

```
PROGRAM LINEEHLDEL
COMMON /COM1/ENDA,A1,A2,A3,Z,HM0,DH/COM2/EDA0/COM4/X0,XE/COM3/E1,
PH,B,R
DATA PAI,Z,P0/3.14159265,0.68,1.96E8/
DATA N,X0,XE,W,E1,EDA0,R,Us/130,
-4.0,1.5,1.0E5,2.2E11,0.001,0.05,1.5/
OPEN(8,FILE='OUT.DAT',STATUS='UNKNOWN')
W1=W/(E1*R)
PH=E1*SQRT(0.5*W1/PAI)
A1=(ALOG(EDA0)+9.67)
A2=PH/P0
A3=0.59/(PH*1.E-9)
B=4.*R*PH/E1
ALFA=Z*A1/P0
G=ALFA*E1
U=EDA0*US/(2.*E1*R)
CC1=SQRT(2.*U)
AM=2.*PAI*(PH/E1)**2/CC1
ENDA=3.*(PAI/AM)**2/8.
HM0=1.6*(R/B)**2*G**0.6*U**0.7*W1**(-0.13)
WRITE(*,*)N,X0,XE,W,E1,EDA0,R,US
CALL SUBAK(N)
CALL EHL(N)
STOP
END
SUBROUTINE EHL(N)
DIMENSION X(1100),P(1100),H(1100),RO(1100),POLD(1100),EPS(1100),
EDA(1100),V(1100)
COMMON /COM1/ENDA,A1,A2,A3,Z,HM0,DH/COM4/X0,XE
COMMON /COM3/E1,PH,B,RR
MK=1
```

```
      DX=(XE-X0)/(N-1.0)
      DO 10 I=1,N
      X(I)=X0+(I-1)*DX
      IF(ABS(X(I)).GE.1.0)P(I)=0.0
      IF(ABS(X(I)).LT.1.0)P(I)=SQRT(1.-X(I)*X(I))
10    CONTINUE
      CALL HREE(N,DX,X,P,H,RO,EPS,EDA,V)
      CALL FZ(N,P,POLD)
14    KK=19
      CALL ITER(N,KK,DX,X,P,H,RO,EPS,EDA,V)
      MK=MK+1
      CALL ERROP(N,P,POLD,ERP)
      WRITE(*,*)'ERP=',ERP
      IF(ERP.GT.1.E-5.AND.DH.GT.1.E-7)THEN
      IF(MK.GE.50)THEN
      MK=1
      DH=0.5*DH
      ENDIF
      GOTO 14
      ENDIF
      IF(DH.LE.1.E-7)WRITE(*,*)'Pressures are not convergent!!!'
      H2=1.E3
      P2=0.0
      DO 106 I=1,N
      IF(H(I).LT.H2)H2=H(I)
      IF(P(I).GT.P2)P2=P(I)
106   CONTINUE
      H3=H2*B*B/RR
      P3=P2*PH
110   FORMAT(6(1X,E12.6))
120   CONTINUE
      WRITE(*,*)'P2,H2,P3,H3=',P2,H2,P3,H3
      CALL OUTHP(N,X,P,H)
      RETURN
      END
      SUBROUTINE OUTHP(N,X,P,H)
      DIMENSION X(N),P(N),H(N)
      DO 10 I=1,N
      WRITE(8,20)X(I),P(I),H(I)
10    CONTINUE
20    FORMAT(1X,6(E12.6,1X))
      RETURN
      END
      SUBROUTINE HREE(N,DX,X,P,H,RO,EPS,EDA,V)
      DIMENSION X(N),P(N),H(N),RO(N),EPS(N),EDA(N),V(N)
      REAL*4 LAMBDA
      COMMON /COM1/ENDA,A1,A2,A3,Z,HM0,DH/COM2/EDA0/COMAK/AK(0:1100)/
      COM3/E1,PH,B,R
      DATA KK,PAI1,G0,PAI/0,0.318309886,1.570796325,3.14159265/
      DATA EPSILON,ZETA,LAMBDA/7.08E-10,0.01,1.9E-4/
      IF(KK.NE.0)GOTO 3
```

```
      H00=0.0
3     W1=0.0
      DO 4 I=1,N
4     W1=W1+P(I)
      C3=(DX*W1)/G0
      DW=1.-C3
      CALL VI(N,DX,P,V)
      HMIN=1.E3
      DO 30 I=1,N
      H0=0.5*X(I)*X(I)+V(I)
      IF(H0.LT.HMIN)HMIN=H0
      H(I)=H0
30    CONTINUE
      IF(KK.NE.0)GOTO 32
      KK=1
      DH=0.05*HM0
      H00=-HMIN+HM0
32    IF(DW.LT.0.0)H00=H00+DH
      IF(DW.GT.0.0)H00=H00-DH
      DO 60 I=1,N
      H(I)=H00+H(I)
      EDA(I)=EXP(A1*(-1.+(1.+A2*P(I))**Z))
      H1=H(I)*B*B/R
      EEDA=0.75*(EPSILON*ZETA/PAI/H1)**2/(H1*LAMBDA*EDA0)
      EDA(I)=EDA(I)+EEDA
      RO(I)=(A3+1.35*P(I))/(A3+P(I))
      EPS(I)=RO(I)*H(I)**3/(ENDA*EDA(I))
60    CONTINUE
      RETURN
      END
      SUBROUTINE ITER(N,KK,DX,X,P,H,RO,EPS,EDA,V)
      DIMENSION X(N),P(N),H(N),RO(N),EPS(N),EDA(N),V(N)
      COMMON /COMAK/AK(0:1100)
      DATA PAI1/0.318309886/
      DO 100 K=1,KK
      D2=0.5*(EPS(1)+EPS(2))
      D3=0.5*(EPS(2)+EPS(3))
      DO 70 I=2,N-1
      D1=D2
      D2=D3
      IF(I.NE.N-1)D3=0.5*(EPS(I+1)+EPS(I+2))
      D8=RO(I)*AK(0)*PAI1
      D9=RO(I-1)*AK(1)*PAI1
      D10=1.0/(D1+D2+(D9-D8)*DX)
      D11=D1*P(I-1)+D2*P(I+1)
      D12=(RO(I)*H(I)-RO(I-1)*H(I-1)+(D8-D9)*P(I))*DX
      P(I)=(D11-D12)*D10
      IF(P(I).LT.0.0)P(I)=0.0
70    CONTINUE
      CALL HREE(N,DX,X,P,H,RO,EPS,EDA,V)
100   CONTINUE
```

```
      RETURN
      END
      SUBROUTINE VI(N,DX,P,V)
      DIMENSION P(N),V(N)
      COMMON /COMAK/AK(0:1100)
      PAI1=0.318309886
      C=ALOG(DX)
      DO 10 I=1,N
      V(I)=0.0
      DO 10 J=1,N
      IJ=IABS(I-J)
10    V(I)=V(I)+(AK(IJ)+C)*DX*P(J)
      DO I=1,N
      V(I)=-PAI1*V(I)
      ENDDO
      RETURN
      END
      SUBROUTINE SUBAK(MM)
      COMMON /COMAK/AK(0:1100)
      DO 10 I=0,MM
10    AK(I)=(I+0.5)*(ALOG(ABS(I+0.5))-1.)-(I-0.5)*(ALOG(ABS
     (I-0.5))-1.)
      RETURN
      END
      SUBROUTINE FZ(N,P,POLD)
      DIMENSION P(N),POLD(N)
      DO 10 I=1,N
10    POLD(I)=P(I)
      RETURN
      END
      SUBROUTINE ERROP(N,P,POLD,ERP)
      DIMENSION P(N),POLD(N)
      SD=0.0
      SUM=0.0
      DO 10 I=1,N
      SD=SD+ABS(P(I)-POLD(I))
      POLD(I)=P(I)
10    SUM=SUM+P(I)
      ERP=SD/SUM
      RETURN
      END
```

12.3.1.2 Calculation program in point contact

The program of EHL in the point contact, which is based on the influence of EDL, is similar to that in the line contact. Therefore, only the modified statements are presented in bold for better understanding and comparison.

```
      PROGRAM POINTEHLDEL
      COMMON /COM1/ENDA,A1,A2,A3,Z,HM0,DH/COM3/E1,PH,B,RX,EDA0
      DATA PAI,Z/3.14159265,0.68/
      DATA N,PH,E1,EDA0,RX,US,X0,XE/33,0.5E9,2.21E11,0.001,0.02,1.5,
      -2.5,1.5/
      OPEN(4,FILE='OUT.DAT',STATUS='UNKNOWN')
      OPEN(8,FILE='FILM.DAT',STATUS='UNKNOWN')
      OPEN(10,FILE='PRESSURE.DAT',STATUS='UNKNOWN')
      MM=N-1
      A1=ALOG(EDA0)+9.67
      A2=5.1E-9*PH
      A3=0.59/(PH*1.E-9)
      U=EDA0*US/(2.*E1*RX)
      B=PAI*PH*RX/E1
      W0=2.*PAI*PH/(3.*E1)*(B/RX)**2
      ALFA=Z*5.1E-9*A1
      G=ALFA*E1
      HM0=3.63*(RX/B)**2*G**0.49*U**0.68*W0**(-0.073)
      ENDA=12.*U*(E1/PH)*(RX/B)**3
      WRITE(*,*)N,X0,XE,W0,PH,E1,EDA0,RX,US
      WRITE(4,*)N,X0,XE,W0,PH,E1,EDA0,RX,US
      WRITE(*,*)'          Wait please'
      CALL SUBAK(MM)
      CALL EHL(N,X0,XE)
      STOP
      END
      SUBROUTINE EHL(N,X0,XE)
     DIMENSION X(65),Y(65),H(4500),RO(4500),EPS(4500),EDA(4500),P(4500),
     POLD(4500),V(4500)
      COMMON /COM1/ENDA,A1,A2,A3,Z,HM0,DH
      DATA MK,G0/1,2.0943951/
      CALL INITI(N,DX,X0,XE,X,Y,P,POLD)
      KK=0
      CALL HREE(N,DX,KK,H00,G0,X,Y,H,RO,EPS,EDA,P,V)
 14   KK=15
      CALL ITER(N,KK,DX,H00,G0,X,Y,H,RO,EPS,EDA,P,V)
      MK=MK+1
      CALL ERP(N,ER,P,POLD)
      WRITE(*,*)'ER=',ER
      IF(ER.GT.1.E-5.AND.DH.GT.1.E-7)THEN
      IF(MK.GE.20)THEN
      MK=1
      DH=0.5*DH
      ENDIF
      GOTO 14
      ENDIF
      IF(DH.LE.1.E-7)WRITE(*,*)'Pressures are not convergent!!!'
      CALL OUTPUT(N,DX,X,Y,H,P)
      RETURN
      END
```

```
      SUBROUTINE ERP(N,ER,P,POLD)
      DIMENSION P(N,N),POLD(N,N)
      ER=0.0
      SUM=0.0
      DO 10 I=1,N
      DO 10 J=1,N
      ER=ER+ABS(P(I,J)-POLD(I,J))
      POLD(I,J)=P(I,J)
      SUM=SUM+P(I,J)
 10   CONTINUE
      ER=ER/SUM
      RETURN
      END
      SUBROUTINE INITI(N,DX,X0,XE,X,Y,P,POLD)
      DIMENSION X(N),Y(N),P(N,N),POLD(N,N)
      NN=(N+1)/2
      DX=(XE-X0)/(N-1.)
      Y0=-0.5*(XE-X0)
      DO 5 I=1,N
      X(I)=X0+(I-1)*DX
      Y(I)=Y0+(I-1)*DX
 5    CONTINUE
      DO I=1,N
      D=1.-X(I)*X(I)
      DO J=1,NN
      C=D-Y(J)*Y(J)
      IF(C.LE.0.0)P(I,J)=0.0
      IF(C.GT.0.0)P(I,J)=SQRT(C)
      POLD(I,J)=P(I,J)
      ENDDO
      ENDDO
      RETURN
      END
      SUBROUTINE HREE(N,DX,KK,H00,G0,X,Y,H,RO,EPS,EDA,P,V)
      DIMENSION X(N),Y(N),P(N,N),H(N,N),RO(N,N),EPS(N,N),EDA(N,N),V(N,N)
      REAL*4 LAMBDA
      COMMON /COM1/ENDA,A1,A2,A3,Z,HM0,DH/COMAK/AK(0:65,0:65)/COM3/E1,PH,
      B,R,EDA0
      DATA PAI,PAI1/3.14159265,0.2026423/
      DATA EPSILON,ZETA,LAMBDA/7.08E-10,0.01,1.9E-4/
      NN=(N+1)/2
      CALL VI(N,DX,P,V)
      HMIN=1.E3
      DO 30 I=1,N
      DO 30 J=1,NN
      RAD=X(I)*X(I)+Y(J)*Y(J)
      W1=0.5*RAD
      H0=W1+V(I,J)
      IF(H0.LT.HMIN)HMIN=H0
 30   H(I,J)=H0
      IF(KK.EQ.0)THEN
```

```
      KK=1
      DH=0.01*HM0
      H00=-HMIN+HM0
      ENDIF
      W1=0.0
      DO 32 I=1,N
      DO 32 J=1,N
32    W1=W1+P(I,J)
      W1=DX*DX*W1/G0
      DW=1.-W1
      IF(DW.LT.0.0)H00=H00+DH
      IF(DW.GT.0.0)H00=H00-DH
      DO 60 I=1,N
      DO 60 J=1,NN
      H(I,J)=H00+H(I,J)
      EDA1=EXP(A1*(-1.+(1.+A2*P(I,J))**Z))
      EDA(I,J)=EDA1
      H1=H(I,J)*B*B/R
      EEDA=0.75*(EPSILON*ZETA/PAI/H1)**2/(H1*LAMBDA*EDA0)
      EDA(I,J)=EDA(I,J)+EEDA
      RO(I,J)=(A3+1.34*P(I,J))/(A3+P(I,J))
60    EPS(I,J)=RO(I,J)*H(I,J)**3/(ENDA*EDA1)
      DO 70 J=NN+1,N
      JJ=N-J+1
      DO 70 I=1,N
      H(I,J)=H(I,JJ)
      RO(I,J)=RO(I,JJ)
      EDA(I,J)=EDA(I,JJ)
70    EPS(I,J)=EPS(I,JJ)
      RETURN
      END
      SUBROUTINE ITER(N,KK,DX,H00,G0,X,Y,H,RO,EPS,EDA,P,V)
      DIMENSION X(N),Y(N),P(N,N),H(N,N),RO(N,N),EPS(N,N),EDA(N,N),V(N,N)
      COMMON /COMAK/AK(0:65,0:65)
      DATA KG1,PAI/0,3.14159265/
      IF(KG1.NE.0)GOTO 2
      KG1=1
      AK00=AK(0,0)
      AK10=AK(1,0)
2     NN=(N+1)/2
      DO 100 K=1,KK
      DO 70 J=2,NN
      J0=J-1
      J1=J+1
      D2=0.5*(EPS(1,J)+EPS(2,J))
      DO 70 I=2,N-1
      I0=I-1
      I1=I+1
      D1=D2
      D2=0.5*(EPS(I1,J)+EPS(I,J))
      D4=0.5*(EPS(I,J0)+EPS(I,J))
```

```
      D5=0.5*(EPS(I,J1)+EPS(I,J))
      D8=2.0*RO(I,J)*AK00/PAI**2
      D9=2.0*RO(I0,J)*AK10/PAI**2
      D10=D1+D2+D4+D5+D8*DX-D9*DX
      D11=D1*P(I0,J)+D2*P(I1,J)+D4*P(I,J0)+D5*P(I,J1)
      D12=(RO(I,J)*H(I,J)-D8*P(I,J)-RO(I0,J)*H(I0,J)+D9*P(I,J))*DX
      P(I,J)=(D11-D12)/D10
      IF(P(I,J).LT.0.0)P(I,J)=0.0
70    CONTINUE
      DO 80 J=1,NN
      JJ=N+1-J
      DO 80 I=1,N
80    P(I,JJ)=P(I,J)
      CALL HREE(N,DX,KK,H00,G0,X,Y,H,RO,EPS,EDA,P,V)
100   CONTINUE
      RETURN
      END
      SUBROUTINE VI(N,DX,P,V)
      DIMENSION P(N,N),V(N,N)
      COMMON /COMAK/AK(0:65,0:65)
      PAI1=0.2026423
      DO 40 I=1,N
      DO 40 J=1,N
      H0=0.0
      DO 30 K=1,N
      IK=IABS(I-K)
      DO 30 L=1,N
      JL=IABS(J-L)
30    H0=H0+AK(IK,JL)*P(K,L)
40    V(I,J)=H0*DX*PAI1
      RETURN
      END
      SUBROUTINE SUBAK(MM)
      COMMON /COMAK/AK(0:65,0:65)
      S(X,Y)=X+SQRT(X**2+Y**2)
      DO 10 I=0,MM
      XP=I+0.5
      XM=I-0.5
      DO 10 J=0,I
      YP=J+0.5
      YM=J-0.5
      A1=S(YP,XP)/S(YM,XP)
      A2=S(XM,YM)/S(XP,YM)
      A3=S(YM,XM)/S(YP,XM)
      A4=S(XP,YP)/S(XM,YP)
      AK(I,J)=XP*ALOG(A1)+YM*ALOG(A2)+XM*ALOG(A3)+YP*ALOG(A4)
10    AK(J,I)=AK(I,J)
      RETURN
      END
      SUBROUTINE OUTPUT(N,DX,X,Y,H,P)
      DIMENSION X(N),Y(N),H(N,N),P(N,N)
```

```
      A=0.0
      WRITE(8,110)A,(Y(I),I=1,N)
      DO I=1,N
      WRITE(8,110)X(I),(H(I,J),J=1,N)
      ENDDO
      WRITE(10,110)A,(Y(I),I=1,N)
      DO I=1,N
      WRITE(10,110)X(I),(P(I,J),J=1,N)
      ENDDO
110 FORMAT(66(E12.6,1X))
      RETURN
      END
```

12.3.2 Example

Taking the line contact calculation as an example, if we neglect the temperature influence on lubrication and the temperature is maintained at the room temperature (25°C), the initial viscosity of water is $\eta_0 = 0.001$ Pa·s, the dielectric constant $\varepsilon = 7.08 \times 10^{-10}$ F·m^{-1}, the conductivity $\lambda = 1.9 \times 10^{-4}$ S·m^{-1} and the EDL potential $\zeta_e = 0.01$ V, the results are given in Figure 12.3.

As shown in Figure 12.3, while the influence of EDL is considered, the pressure distribution does not change significantly but the film thickness changes. Because of the low viscosity, the calculation of the film thickness is much thin so that the EDL effect is evident. As shown in the figure, when the influence of EDL is considered, because the apparent viscosity η_a significantly increased, the film thickness increased significantly, although pressure changes a little. It tends to the pressure of hydrodynamic lubrication.

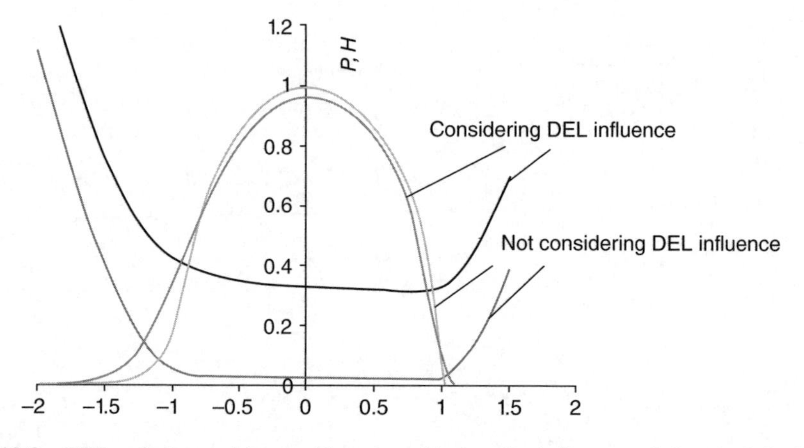

Figure 12.3 EHL solutions with or without considering the influence of electrical double layer in line contact

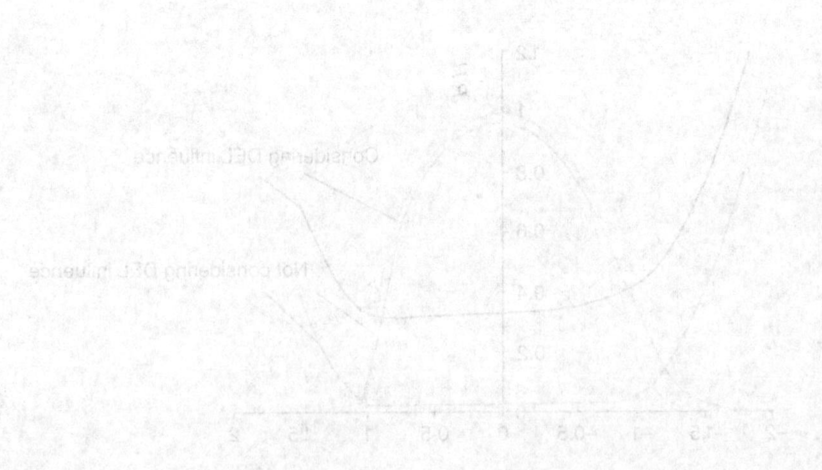

13

Numerical calculation method and program for time-dependent EHL in line contact

13.1 Time-dependent EHL Reynolds equation

13.1.1 Nondimensional Reynolds equation

In Chapter 1, the time-dependent elastohydrodynamic lubrication (EHL) Reynolds equation is given as follows:

$$\frac{\partial}{\partial x}\left(\frac{\rho h^3}{\eta}\frac{\partial p}{\partial x}\right) = 12u_s\frac{\partial(\rho h)}{\partial x} + 12\rho\frac{\partial h}{\partial t} \tag{13.1}$$

As the squeezing effect caused by rolling is derived by EHL Reynolds equation, it is not further included in the second term of Equation 13.1 on the right-hand side while deriving the time-dependent effect. In Equation 13.1, only the time-dependent effect caused by the center moving up and down, the effect of the whole film thickness, is considered.

The nondimensional Reynolds equation with the consideration of the time-dependent effect is as follows:

$$\frac{d}{dX}\left(\varepsilon\frac{dP}{dX}\right) = \frac{d(\rho^* H)}{dX} + \frac{d(\rho^* H)}{dT} \tag{13.2}$$

Numerical Calculation of Elastohydrodynamic Lubrication: Methods and Programs, First Edition. Ping Huang.

13.1.2 Discrete Reynolds equation

The discrete equation of Equation 13.2 is as follows:

$$\frac{\varepsilon_{i-1/2}P_{i-1}^{k+1}-\left(\varepsilon_{i-1/2}+\varepsilon_{i+1/2}\right)P_{i}^{k+1}+\varepsilon_{i+1/2}P_{i-1}^{k+1}}{\Delta X^{2}}=\frac{\rho_{i}^{*}H_{i}^{k+1}-\rho_{i-1}^{*}H_{i-1}^{k+1}}{\Delta X}$$
$$+\frac{\rho_{i}^{*}H_{i}^{k+1}-\rho_{i}^{k*}H_{i}^{k}}{\Delta T} \tag{13.3}$$

The last term in Equation 13.3 shows that the calculations of the pressure and film thickness at the current step are based on the density and film thickness at the previous step. Therefore, an EHL solution must be obtained in advance.

13.2 Numerical calculation method and program

13.2.1 Iteration method

Firstly, Equation 13.3 is written in the residual form as follows:

$$\left(\varepsilon_{i-1/2}+\varepsilon_{i+1/2}\right)P_{i}^{k+1}=\varepsilon_{i-1/2}P_{i-1}^{k+1}+\varepsilon_{i+1/2}P_{i-1}^{k+1}-\Delta X\left(\rho_{i}^{*}H_{i}^{k+1}-\rho_{i-1}^{*}H_{i-1}^{k+1}\right)$$
$$-(\Delta X)^{2}\left(\rho_{i}^{*}H_{i}^{k+1}-\rho_{i}^{k*}H_{i}^{k}\right)/\Delta T \tag{13.4}$$

If

$$r_{i}=\varepsilon_{i-1/2}P_{i-1}^{k+1}+\varepsilon_{i+1/2}P_{i-1}^{k+1}-\Delta X\left(\rho_{i}^{*}H_{i}^{k+1}-\rho_{i-1}^{*}H_{i-1}^{k+1}\right)-(\Delta X)^{2}\left(\rho_{i}^{*}H_{i}^{k+1}-\rho_{i}^{k*}H_{i}^{k}\right)/\Delta T \tag{13.5}$$

Equation 13.4 becomes

$$\left(\varepsilon_{i-1/2}+\varepsilon_{i+1/2}\right)P_{i}^{k+1}=r_{i} \tag{13.6}$$

It should be noted that the term P_{i}^{k+1} is present in all the terms of the film thickness as $H_{i}^{k+1}=H_{1}^{k+1}+D_{i}^{j}P_{j}^{k+1}$ and $H_{i-1}^{k+1}=H_{2}^{k+1}+D_{i-1}^{j}P_{j}^{k+1}$. By extracting the same term P_{i}^{k+1} from the film thickness, we can obtain Equation 13.7.

$$\begin{cases} H_{i}^{k+1}=H_{1}^{k+1}+D_{i}^{j}P_{j}^{k+1}-a_{1}P_{i}^{k+1}+a_{1}P_{i}^{k+1}=H_{i}^{k+1}-a_{1}P_{i}^{k+1}+a_{1}P_{i}^{k+1} \\ H_{i-1}^{k+1}=H_{1}^{k+1}+D_{i-1}^{j}P_{j}^{k+1}-a_{2}P_{i}^{k+1}+a_{2}P_{i}^{k+1}=H_{i-1}^{k+1}-a_{2}P_{i}^{k+1}+a_{2}P_{i}^{k+1} \end{cases} \tag{13.7}$$

As partial extraction shows a better result than that of all being extracted, it is used in the following program.

By subtracting the term P_i^{k+1} on both the ends of Equation 13.7, we get the following:

$$\left\{ \left(\varepsilon_{i-1/2} + \varepsilon_{i+1/2}\right) - \Delta X \rho_i^* a_1 + \Delta X a_2 \rho_{i-1}^* - (\Delta X)^2 a_1 \rho_i^* / \Delta T \right\} P_i^{k+1}$$
$$= r_i - \Delta X \rho_i^* a_1 P_i^{k+1} + \Delta X \rho_{i-1}^* a_2 P_i^{k+1} - (\Delta X)^2 a_1 \rho_i^* P_i^{k+1} / \Delta T \tag{13.8}$$

Finally, the following iteration equation is obtained:

$$P_i^{k+1} = \frac{r_i - \Delta X \rho_i^* a_1 P_i^{k+1} + \Delta X \rho_{i-1}^* a_2 P_i^{k+1} - (\Delta X)^2 a_1 \rho_i^* P_i^{k+1} / \Delta T}{\left(\varepsilon_{i-1/2} + \varepsilon_{i+1/2}\right) - \Delta X \rho_i^* a_1 + \Delta X a_2 \rho_{i-1}^* - (\Delta X)^2 a_1 \rho_i^* / \Delta T} \tag{13.9}$$

Because the term P_i^{k+1} is subtracted from the first equation given in Equation 13.7 but not from the second one, the iterative effect of Equation 13.9 is much better than that obtained from Equation 13.7.

Furthermore, because the time-dependent load in an EHL problem varies with time, the size of the contact region also varies at different time intervals. Therefore, the dimensionless results must be in the same scale so that they could be compared with each other.

13.2.2 Calculation diagram

The diagram to calculate the time-dependent EHL is shown in Figure 13.1.

13.2.3 Calculation program

In the main subroutine EHL for time-dependent EHL, the calculation program for the steady-state EHL is located from the very beginning to label 30, which is similar to the normal EHL. The available steady-state EHL result is same as the initial result of the time-dependent EHL calculation.

In Subroutine EHL, W0 is the dimensional load and DT0 is the time increment. DT is the nondimensional time increment. In addition, KH is added as a switching variable. When KH = 0, it is transmitted into Subroutine HREE to reset the increment DH of the film thickness and calculate H0 again.

The pressure distribution and film thickness calculated at each step not only serve as the results of that particular step but also serve as the initial values for the subsequent steps to calculate the pressure and film thickness in order to be convergent and save time.

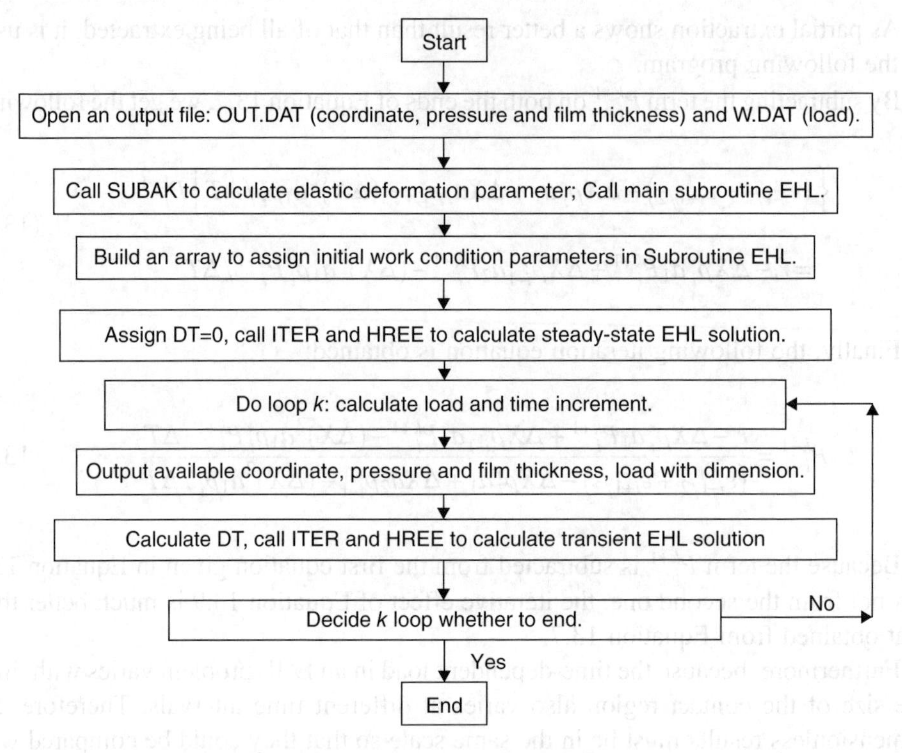

Figure 13.1 Diagram to calculate time-dependent EHL

In addition, it should be noticed that the load in the time-dependent EHL is variable. Some invariable parameters in a steady state, such as the Hertz contact stress PH and the half width of the contact area B, also vary with the load. Therefore, all the parameters and their related variables must be calculated again at the beginning of a new iteration. All the work is given in the label 100.

Finally, it should be noted that according to Equation 13.9, the iteration calculation in Subroutine ITER is the same as the steady state when DT = 0, otherwise $DT \neq 0$.

```
PROGRAM LINEEHLD
OPEN(8,FILE='OUT.DAT',STATUS='UNKNOWN')
OPEN(9,FILE='W.DAT',STATUS='UNKNOWN')
N=129
CALL SUBAK(N)
CALL EHL(N)
STOP
END
SUBROUTINE EHL(N)
DIMENSION X(129),P(129),H(129),RO(129),POLD(129),EPS(129),EDA
(129),V(129)
```

```
      DIMENSION X0(129),P0(129),H0(129),RO0(129),EPS0(129)
      COMMON /COM1/ENDA,A1,A2,A3,Z,HM0,DH
      COMMON /COMHREE/KH
      DATA PAI,Z,P00/3.14159265,0.68,1.96E8/
      DATA XS,XE,W0,E1,EDA0,R,US/-4.0,1.4,2.0E5,2.21E11,0.05,0.05,1.5/
      DATA K,N1,N2,DT0,DT/0,12,129,0.02,0.0/
      A1=(ALOG(EDA0)+9.67)
      DX=(XE-XS)/(N-1.0)
      W1=W0/(E1*R)
      PH=E1*SQRT(0.5*W1/PAI)
      A2=PH/P00
      A3=0.59/(PH*1.E-9)
      B=4.*R*PH/E1
      ALFA=Z*A1/P00
      G=ALFA*E1
      U=EDA0*US/(2.*E1*R)
      CC1=SQRT(2.*U)
      AM=2.*PAI*(PH/E1)**2/CC1
      ENDA=3.*(PAI/AM)**2/8.
      HM0=1.6*(R/B)**2*G**0.6*U**0.7*W1**(-0.13)
      WRITE(*,*)N,XS,XE,E1,EDA0,R,US
      WRITE(*,*)'W0,PH,B=',W0,PH,B
      WRITE(9,*)K*PAI/6.0,DT,W0
      MK=1
      DO 10 I=1,N
      X(I)=XS+(I-1)*DX
      IF(ABS(X(I)).GE.1.0)P(I)=0.0
10    IF(ABS(X(I)).LT.1.0)P(I)=SQRT(1.-X(I)*X(I))
      KH=0
      CALL HREE(N,DX,X,P,H,RO,EPS,EDA,V)
      CALL FZ(N,P,POLD)
      DT=0.0
20    IK=19
      CALL ITER(N,IK,DX,DT,X,P,H,RO,EPS,EDA,P0,H0,RO0,V)
      MK=MK+1
      CALL ERROP(N,P,POLD,ERP)
      IF(ERP.GT.1.E-5.AND.DH.GT.1.E-7)THEN
      IF(MK.GE.50)THEN
      WRITE(*,*)'ERP',ERP
      MK=1
      DH=0.5*DH
      ENDIF
      GOTO 20
      ENDIF
      WRITE(*,*)'ERP,DH=',ERP,DH
      IF(DH.LE.1.E-7)WRITE(*,*)'Pressures are not convergent!!!'
      WRITE(*,*)'INITIAL PRESSURE OBTAINED'
      DO 30 I=1,N
30    WRITE(8,120)B*X(I),PH*P(I),B*B*H(I)/R
      DO 100 K=1,N1
```

```
      B0=B
      PH0=PH
      DO 50 I=1,N
      X0(I)=B0/B*X(I)
      P0(I)=PH0/PH*P(I)
50    H0(I)=B/B0*B/B0*H(I)
      W0=W0+1.0E5*(SIN(K*PAI/6.0)-SIN((K-1)*PAI/6.0))
      W1=W0/(E1*R)
      PH=E1*SQRT(0.5*W1/PAI)
      A2=PH/P00
      A3=0.59/(PH*1.E-9)
      B=4.*R*PH/E1
      AM=2.*PAI*(PH/E1)**2/CC1
      ENDA=3.*(PAI/AM)**2/8.
      HM0=1.6*(R/B)**2*G**0.6*U**0.7*W1**(-0.13)
      DT=DT0*US/B
      WRITE(*,*)'W0,PH,B=',W0,PH,B,DT
      WRITE(9,*)K*PAI/6.0,DT,W0
      P(1)=0.0
      DO 70 I=2,N-1
      P(I)=0.0
      DO 60 J=1,N-1
      IF(X0(J).LE.X(I).AND.X0(J+1).GE.X(I))THEN
      P(I)=P0(J)+(P0(J+1)-P0(J))*(X(I)-X0(J))/(X0(J+1)-X0(J))
      GOTO 60
      ENDIF
60    CONTINUE
70    CONTINUE
      P(N)=0.0
      KH=0
      CALL HREE(N,DX,X,P,H,RO,EPS,EDA,V)
      CALL FZ(N,P,P0)
      CALL FZ(N,RO,RO0)
      CALL FZ(N,P,POLD)
      MK=1
80    IK=19
      CALL ITER(N,IK,DX,DT,X,P,H,RO,EPS,EDA,P0,H0,RO0,V)
      MK=MK+1
      CALL ERROP(N,P,POLD,ERP)
      IF(ERP.GT.1.E-5.AND.DH.GT.1.E-7)THEN
      IF(MK.GE.50)THEN
      WRITE(*,*)'ERP=',ERP
      MK=1
      DH=0.5*DH
      ENDIF
      GOTO 80
      ENDIF
      WRITE(*,*)'ERP,DH=',ERP,DH
      IF(DH.LE.1.E-7)WRITE(*,*)'Pressures are not convergent!!!'
      DO 90 I=1,N
```

```
90      WRITE(8,120)B*X(I),PH*P(I),B*B*H(I)/R
100     CONTINUE
120     FORMAT(1X,3(E12.6,1X))
        RETURN
        END
        SUBROUTINE HREE(N,DX,X,P,H,RO,EPS,EDA,V)
        DIMENSION X(N),P(N),H(N),RO(N),EPS(N),EDA(N),V(N)
        COMMON /COM1/ENDA,A1,A2,A3,Z,HM0,DH/COMAK/AK(0:129)
        COMMON /COMHREE/KH
        DATA PAI1,G0/0.318309886,1.570796325/
        W1=0.0
        DO 4 I=1,N
4       W1=W1+P(I)
        C3=(DX*W1)/G0
        DW=1.-C3
        CALL VI(N,DX,P,V)
        HMIN=1.E3
        DO 30 I=1,N
        H0=0.5*X(I)*X(I)+V(I)
        IF(H0.LT.HMIN)HMIN=H0
        H(I)=H0
30      CONTINUE
        IF(KH.NE.0)GOTO 32
        KH=1
        DH=0.005*HM0
        H00=-HMIN+HM0
32      IF(DW.LT.0.0)H00=H00+DH
        IF(DW.GT.0.0)H00=H00-DH
        DO 100 I=1,N
        H(I)=H00+H(I)
        EDA(I)=EXP(A1*(-1.+(1.+A2*P(I))**Z))
        RO(I)=(A3+1.34*P(I))/(A3+P(I))
        EPS(I)=RO(I)*H(I)**3/(ENDA*EDA(I))
100     CONTINUE
        RETURN
        END
        SUBROUTINE ITER(N,IK,DX,DT,X,P,H,RO,EPS,EDA,P0,H0,RO0,V)
        DIMENSION X(N),P(N),H(N),RO(N),EPS(N),EDA(N),P0(N),H0(N),RO0(N),
        V(N)
        COMMON /COMAK/AK(0:129)
        DATA PAI1/0.318309886/
        AT=0.0
        IF(DT.GT.0.1)AT=DX*DX/DT
        DO 100 K=1,IK
        D2=0.5*(EPS(1)+EPS(2))
        D3=0.5*(EPS(2)+EPS(3))
        DO 70 I=2,N-1
        D1=D2
        D2=D3
        IF(I.NE.N-1)D3=0.5*(EPS(I+1)+EPS(I+2))
```

```
      D8=(1.-AT)*RO(I)*AK(0)*PAI1
      D9=RO(I-1)*AK(1)*PAI1
      D10=1.0/(D1+D2+(D9-D8)*DX)
      D11=D1*P(I-1)+D2*P(I+1)
      D12=((1.-AT)*RO(I)*H(I)-RO(I-1)*H(I-1)+(D8-D9)*P(I))*DX
      D13=AT*RO0(I)*H0(I)
      P(I)=(D11-D12+D13)*D10
      IF(P(I).LT.0.0)P(I)=0.0
70    CONTINUE
      CALL HREE(N,DX,X,P,H,RO,EPS,EDA,V)
100   CONTINUE
      RETURN
      END
      SUBROUTINE VI(N,DX,P,V)
      DIMENSION P(N),V(N)
      COMMON /COMAK/AK(0:129)
      DATA PAI1/0.318309886/
      C=ALOG(DX)
      DO 10 I=1,N
      V(I)=0.0
      DO 10 J=1,N
      IJ=IABS(I-J)
10    V(I)=V(I)+(AK(IJ)+C)*DX*P(J)
      DO 20 I=1,N
20    V(I)=-PAI1*V(I)
      RETURN
      END
      SUBROUTINE SUBAK(MM)
      COMMON /COMAK/AK(0:129)
      DO 10 I=0,MM
10    AK(I)=(I+0.5)*(ALOG(ABS(I+0.5))-1.)-(I-0.5)*(ALOG
      (ABS(I-0.5))-1.)
      RETURN
      END
      SUBROUTINE FZ(N,P,POLD)
      DIMENSION P(N),POLD(N)
      DO 10 I=1,N
10    POLD(I)=P(I)
      RETURN
      END
      SUBROUTINE ERROP(N,P,POLD,ERP)
      DIMENSION P(N),POLD(N)
      ERP=0.0
      SUM=0.0
      DO 10 I=1,N
      ERP=ERP+ABS(P(I)-POLD(I))
      POLD(I)=P(I)
10    SUM=SUM+P(I)
      ERP=ERP/SUM
      RETURN
      END
```

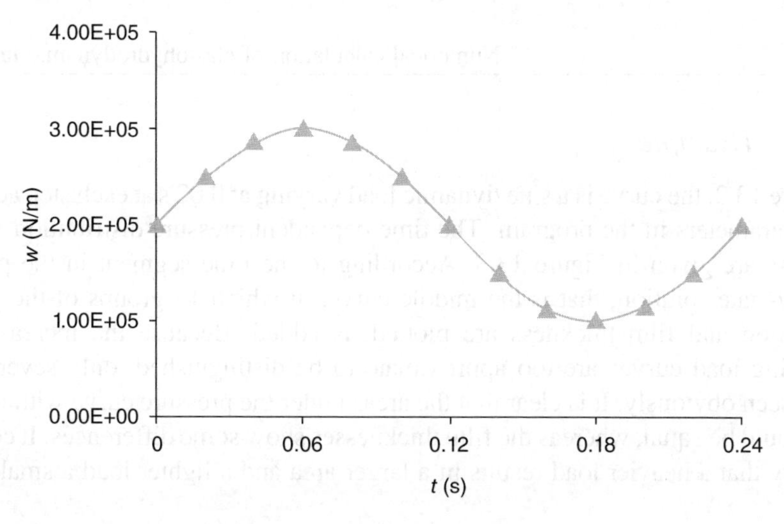

Figure 13.2 Curve of load with time

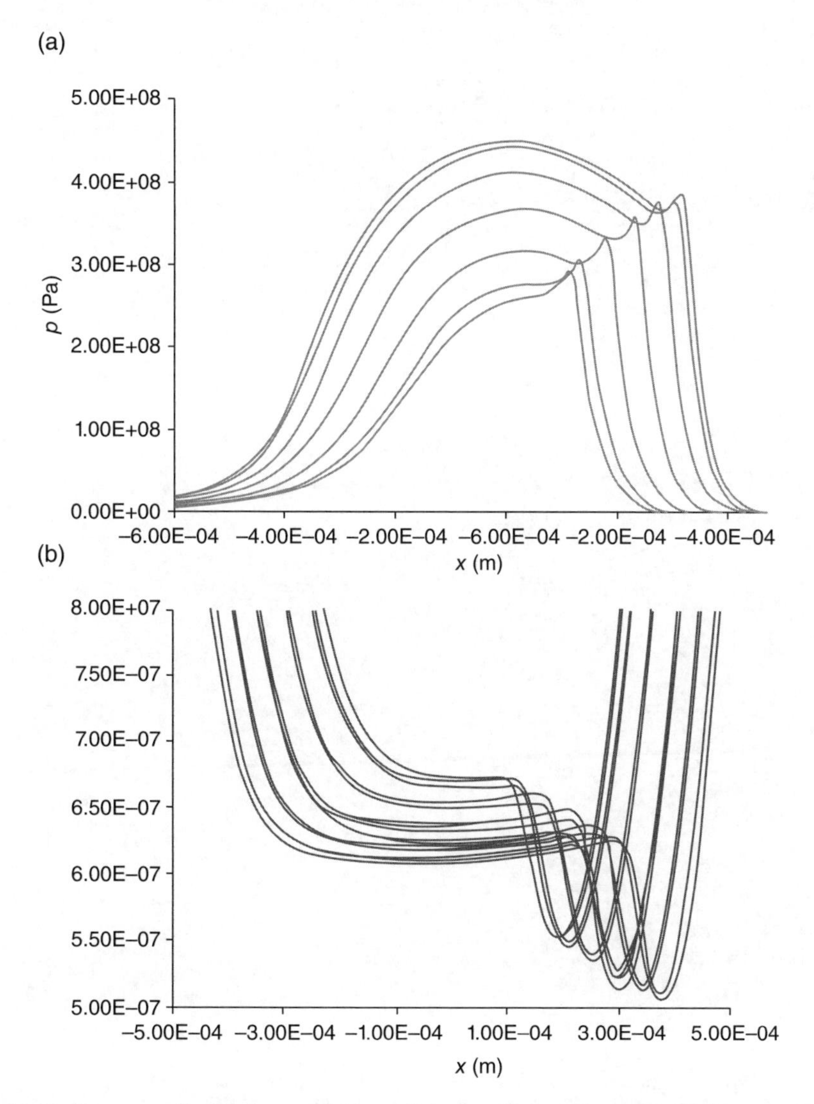

Figure 13.3 Pressure distribution and film thickness of time-dependent EHL. (a) Curves of pressure distribution with load (b) Curves of film thickness with load

13.2.4 Example

In Figure 13.2, the curve is a sine dynamic load varying at 0.02 s at each step according to the parameters in the program. The time-dependent pressure distribution and film thickness are given in Figure 13.3. According to the time segment in the program, a steady-state solution, that is the middle curve, in which 13 groups of the pressure distribution and film thickness are plotted, is added. Because the increasing and decreasing load curves are too approximate to be distinguished, only seven curves can be seen obviously. It is clear that the areas under the pressure curve with the same load should be equal, whereas the film thicknesses show some differences. It conforms to reality that a heavier load results in a larger area and a lighter load a smaller area.

14

Numerical calculation method and program for isothermal EHL with rough surface

14.1 Film thickness equation with surface roughness

In Chapter 1, we have given the formula for the surface film thickness with the elastic deformation to calculate the smooth surface elastohydrodynamic lubrication (EHL) problems. By adding the roughness on a smooth surface, we can use this formula to calculate rough surface EHL problems. The film thickness equation of rough surfaces in the line contact is given as follows:

$$h = h_0 + \frac{x^2}{2R} + v(x) + r(x) \tag{14.1}$$

The film thickness equation of rough surfaces in the point contact is as follows:

$$h = h_0 + \frac{x^2 + y^2}{2R} + v(x, y) + r(x, y) \tag{14.2}$$

In Equations 14.1 and 14.2, $r(x)$ and $r(x, y)$ are the surface roughness functions, which may be analytical or numerical. In order to change the roughness size, generally they are written as follows:

$$\delta(x) = \frac{r(x)}{a} \tag{14.3}$$

Numerical Calculation of Elastohydrodynamic Lubrication: Methods and Programs, First Edition. Ping Huang.

and

$$\delta(x,y) = \frac{r(x,y)}{a} \tag{14.4}$$

where, a is the maximum roughness, which can be obtained while $r(x)$ or $r(x, y)$ has been known; $\delta(x)$ and $\delta(x, y)$ are the roughness functions with the maximum to be equal to 1.

Note, in the present program, the film thickness is made nondimensional by using (b^2/R).

14.2 Calculation for rough surface EHL problem with Newton–Raphson method

14.2.1 Sine roughness

14.2.1.1 Program modification

The Newton–Raphson method used to solve EHL problem in the line contact has been described in detail in Chapter 5. As the Newton–Raphson method needs a relatively accurate initial start solution, but the Hertz pressure solution is far away from the true solution so as easily to cause the iteration divergent, we must increase the roughness amplitude gradually in order to obtain a relatively larger roughness EHL solution. Therefore, in each calculation, a small roughness amplitude is added to the former roughness EHL solution as the beginning of the latter solution so as to avoid divergence. The reader can solve more difficult EHL problems in a similar way.

Therefore, we made some changes to modify the smooth surface EHL program and those modified statements are represented in bold letters in the following source program.

1. In the main program, in Label 36, we added a statement to set the roughness amplitude DA = 0 first. Moreover, we added the next sentence with Label 10 as the starting to change DA.
2. In the center of the program, a sine roughness function with DA × sin(2πX) is added on the smooth surface film thickness. On the interval between X = ±1, there are two peaks and two valleys.
3. At the end of the program, we added a loop of statements to change DA so as to realize the calculation loop to continue or stop. When DA is equal to or greater than 0.16, the calculation will stop.

14.2.1.2 Calculation program

```
   DIMENSION X(101),P(101),H(101),RO(101),EDA(101),POLD(101),V(101)
   DIMENSION D(0:101),C(102,102),B(102),IP(102),DP(102)
   COMMON /COMAK/AK(0:1100)
   DATA X0,XE,E,RO0,EDA0,U0,W0,Z,R/-2.0,1.5,2.21E11,1.0,0.02,1.5,
   0.5E5,0.68,0.02/
   DATA N,KG,I1/101,0,87/
   DATA PAI,G,PAI1/3.14159265,5000.0,0.318309886/
   OPEN(8,FILE='OUT.DAT',STATUS='UNKNOWN')
   N1=N-1
   N2=N+1
   W=W0/E/R
   U=EDA0*U0/E/R
   AKK=0.75*PAI*PAI*U/(W*W)
   PH=E*SQRT(W/2.0/PAI)
   B0=R*SQRT(8.0*W/PAI)
   A1=0.6*PH*1.E-9
   A2=1.7*PH*1.E-9
   A3=5.1*PH*1.E-9
   A4=ALOG(EDA0)+9.67
   A5=-0.5/PAI
   A6=A3*Z*A4
   A7=2.3E-9*PH
   HMIN=2.65*G**0.54*U**0.7/W**0.13*(R/B0)**2
   DX=(XE-X0)/(N1)
   CX=ALOG(DX)
   II=87
   DO I=1,N
   X(I)=X0+(I-1)*DX
   P(I)=0.0
   IF(ABS(X(I)).LE.1.0)P(I)=SQRT(1.0-X(I)*X(I))
   IF(X(I).LE.1.0)IE=I
   ENDDO
   CALL SUBAK(N2)
   DO I=0,N
   D(I)=-(AK(I)+CX)*DX*PAI1
   ENDDO
   DA=0.0
10 C(N2,1)=0.0
   C(N2,N2)=0.0
   DO I=2,N
   C(N2,I)=DX
   ENDDO
   DO I=1,N1
   POLD(I)=P(I)
   DP(I)=(P(I+1)-P(I))/DX
   ENDDO
   POLD(N)=0.0
   DP(N)=(P(N)-P(N1))/DX
```

```
CALL VI(N,DX,P,V)
IF(KG.EQ.0)H0=100.0
DO I=1,N
H(I)=0.5*X(I)*X(I)+V(I)+DA*SIN(X(I)*2.0*PAI)
IF(KG.EQ.0.AND.H(I).LT.H0)H0=H(I)
RO(I)=1.0+A1*P(I)/(1.0+A2*P(I))
EDA(I)=EXP(A4*(-1.0+(1.0+A3*P(I))**Z))
ENDDO
IF(KG.EQ.0)THEN
H0=-H0+HMIN
ENDIF
DO I=1,N
H(I)=H0+H(I)
ENDDO
IF(KG.EQ.0)THEN
ROEHE=RO(IE)*H(IE)
ENDIF
DO I=1,N
C(I,1)=AKK*EDA(I)/RO(I)
DO J=2,N
IJ=IABS(I-J)
D1=3.0*H(I)**2*DP(I)*D(IJ)
D2=H(I)**3*(DELTA(I+1,J)-DELTA(I,J))/DX
D3=-AKK*A6*(1.0+A3*P(I))**(Z-1.0)*EDA(I)*DELTA(I,J)*(H(I)-ROEHE/
RO(I))
D4=-AKK*EDA(I)*(D(IJ)+ROEHE*A1*DELTA(I,J)/(1.0+A7*P(I))**2)
C(I,J)=D1+D2+D3+D4
ENDDO
C(I,N2)=3.0*H(I)*H(I)*DP(I)-AKK*EDA(I)
B(I)=-H(I)**3*DP(I)+AKK*EDA(I)*(H(I)-ROEHE/RO(I))
ENDDO
B(N2)=0.5*PAI
DO I=1,N
B(N2)=B(N2)-P(I)*C(N2,I)
ENDDO
DO I=I1,N
IF(P(I).LE.0.0)THEN
B(I)=0.0
DO J=1,N2
C(I,J)=0.0
C(J,I)=0.0
ENDDO
C(I,I)=1.0
ENDIF
ENDDO
CALL INV(N2,C,IP,IDET)
IF(IDET.EQ.0)GOTO 20
DO J=1,N2
DP(J)=0.0
DO I=1,N2
DP(J)=DP(J)+C(J,I)*B(I)
```

```
      ENDDO
      ENDDO
      ROEHE=RO(II)*H(II)
      DO I=2,N1
      P(I)=P(I)+DP(I)
      IF(P(I).LT.0.0)P(I)=0.0
      ENDDO
      H0=H0-B(N2)*DP(N2)
      ER=0.0
      SUM=0.0
      DO I=1,N
      SUM=SUM+P(I)
      ER=ER+ABS(P(I)-POLD(I))
      ENDDO
      ER=ER/SUM
      WRITE(*,*)'ER,DA=',ER,DA
      KG=KG+1
      IF(ER.GT.1.E-5.AND.KG.LT.100)GOTO 10
      KG=1
      DA=DA+0.05
      I1=21
      DO I=1,N
      WRITE(8,100)X(I),P(I),H(I)
100   FORMAT(6(E12.6,1X))
      ENDDO
      IF(DA.LT.0.16)GOTO 10
20    STOP
      END
      FUNCTION DELTA(I,J)
      DELTA=0.0
      IF(I.EQ.J)DELTA=1.0
      RETURN
      END
      SUBROUTINE INV(N,A,IP,IDET)
      DIMENSION A(N,N), IP(N)
      IDET=1
      EPS=1.E-6
      DO K=1,N
      P=0
      I0=K
      IP(K)=K
      DO I=K,N
      IF(ABS(A(I,K)).GT.ABS(P))THEN
      P=A(I,K)
      I0=I
      IP(K)=I
      ENDIF
      ENDDO
      IF(ABS(P).LE.EPS)THEN
      IDET=0
      GOTO 10
```

```
      ENDIF
      IF(I0.NE.K)THEN
      DO J=1,N
      S=A(K,J)
      A(K,J)=A(I0,J)
      A(I0,J)=S
      ENDDO
      ENDIF
      A(K,K)=1./P
      DO I=1,N
      IF(I.NE.K)THEN
      A(I,K)=-A(I,K)*A(K,K)
      DO J=1,N
      IF(J.NE.K)THEN
      A(I,J)=A(I,J)+A(I,K)*A(K,J)
      ENDIF
      ENDDO
      ENDIF
      ENDDO
      DO J=1,N
      IF(J.NE.K)THEN
      A(K,J)=A(K,K)*A(K,J)
      ENDIF
      ENDDO
      ENDDO
      DO K=N-1,1,-1
      IR=IP(K)
      IF(IR.NE.K)THEN
      DO I=1,N
      S=A(I,IR)
      A(I,IR)=A(I,K)
      A(I,K)=S
      ENDDO
      ENDIF
      ENDDO
10    RETURN
      END
      SUBROUTINE VI(N,DX,P,V)
      DIMENSION P(N),V(N)
      COMMON /COMAK/AK(0:1100)
      DATA PAI1/0.318309886/
      C=ALOG(DX)
      DO 10 I=1,N
      V(I)=0.0
      DO 10 J=1,N
      IJ=IABS(I-J)
10    V(I)=V(I)+(AK(IJ)+C)*DX*P(J)
      DO I=1,N
      V(I)=-PAI1*V(I)
      ENDDO
      RETURN
```

```
      END
      SUBROUTINE SUBAK(MM)
      COMMON /COMAK/AK(0:1100)
      DO 10 I=0,MM
10    AK(I)=(I+0.5)*(ALOG(ABS(I+0.5))-1.)-(I-0.5)*(ALOG(ABS(I-0.5))-1.)
      RETURN
      END
```

14.2.1.3 Example

According to the parameters given in the above program, the results obtained are shown in Figure 14.1. In the figure, the curves correspond to the roughness amplitude DA = 0, 0.5, 0.10, and 0.15. As the load is fixed, we can see that with the increase in the roughness amplitude, the film thicknesses change a little. However, the pressures change significantly. When we continuously increase the roughness amplitude DA, the calculation will be divergent.

14.2.2 Single depression

The author had solved a deep rough surface EHL problem early with the Newton–Raphson method, which can be found in Ref. [8]. To solve a deep roughness problem, one needs to pay attention to the following two problems:

1. For rough peaks, pressure will increase. Although such a change will bring some difficulties in calculation, no special treatment is needed. However, for a deep

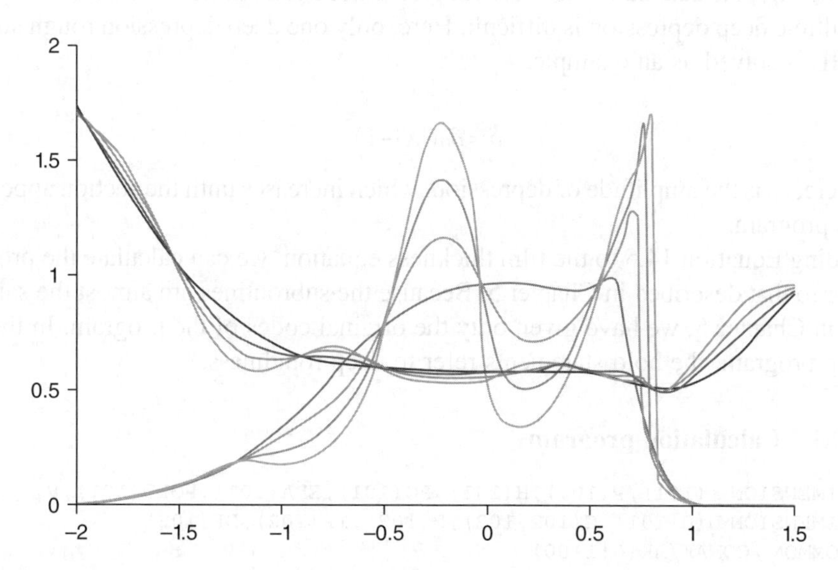

Figure 14.1 Pressure and film thickness of EHL solutions with sine roughness by Newton–Raphson method

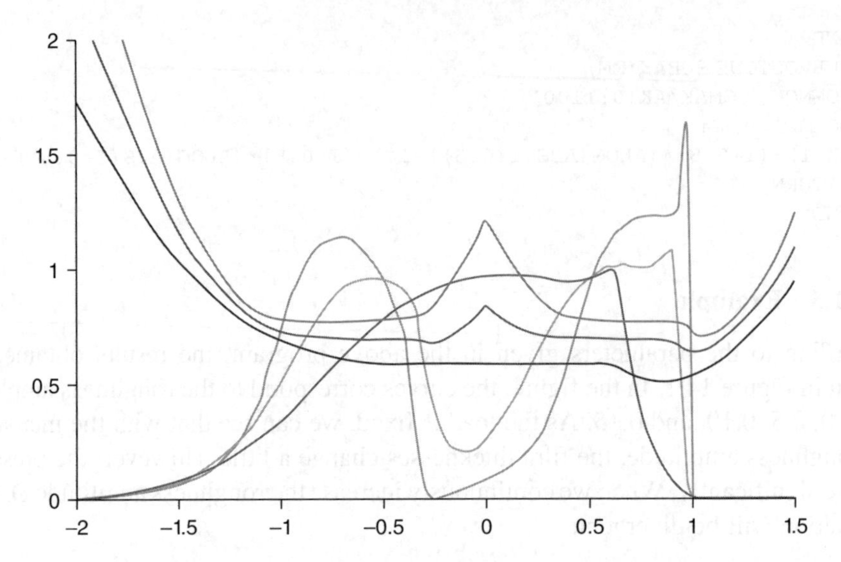

Figure 14.2 Pressure and film thickness with single deep depression

depression, the pressure may be reduced to zero or even negative. However, according to the common assumption, negative pressure must be set to zero. So, we have to modify the coefficient matrix to set the corresponding row and column coefficients as well as the right term to be zero and maintain the main element to be equal to one (see Section 5.2), and the modifications made are represented by the bold statements in the following program.

2. Generally, the calculation to solve rough surface EHL problems automatically with multiple deep depression is difficult. Here, only one deep depression rough surface EHL is solved as an example.

$$\delta = \delta_0(|X|-1)^2 \tag{14.5}$$

where, δ_0 is the amplitude of depression, which increases until the section appears in the program.

Adding Equation 14.5 to the film thickness equation, we can calculate the problem similar to that described in Chapter 5. Because the subroutines are almost the same as those in Chapter 5, we have given only the original codes of the program. In the following program, the bold statements refer to deep roughness.

14.2.2.1 Calculation program

```
DIMENSION X(101),P(101),H(101),RO(101),EDA(101),POLD(101),V(101)
DIMENSION D(0:101),C(102,102),B(102),IP(102),DP(102)
COMMON /COMAK/AK(0:1100)
DATA X0,XE,E,RO0,EDA0,U0,W0,Z,R/-2.0,1.5,2.21E11,1.0,0.02,1.5,
0.5E5,0.68,0.02/
```

```
      DATA N,KG,I1/101,0,87/
      DATA PAI,G,PAI1/3.14159265,5000.0,0.318309886/
      OPEN(8,FILE='OUT.DAT',STATUS='UNKNOWN')
      N1=N-1
      N2=N+1
      W=W0/E/R
      U=EDA0*U0/E/R
      AKK=0.75*PAI*PAI*U/(W*W)
      PH=E*SQRT(W/2.0/PAI)
      B0=R*SQRT(8.0*W/PAI)
      A1=0.6*PH*1.E-9
      A2=1.7*PH*1.E-9
      A3=5.1*PH*1.E-9
      A4=ALOG(EDA0)+9.67
      A5=-0.5/PAI
      A6=A3*Z*A4
      A7=2.3E-9*PH
      HMIN=2.65*G**0.54*U**0.7/W**0.13*(R/B0)**2
      DX=(XE-X0)/(N1)
      CX=ALOG(DX)
      II=87
      DO I=1,N
      X(I)=X0+(I-1)*DX
      P(I)=0.0
      IF(ABS(X(I)).LE.1.0)P(I)=SQRT(1.0-X(I)*X(I))
      IF(X(I).LE.1.0)IE=I
      ENDDO
      CALL SUBAK(N2)
      DO I=0,N
      D(I)=-(AK(I)+CX)*DX*PAI1
      ENDDO
      DA=0.0
10    C(N2,1)=0.0
      C(N2,N2)=0.0
      DO I=2,N
      C(N2,I)=DX
      ENDDO
      DO I=1,N1
      POLD(I)=P(I)
      DP(I)=(P(I+1)-P(I))/DX
      ENDDO
      POLD(N)=0.0
      DP(N)=(P(N)-P(N1))/DX
      CALL VI(N,DX,P,V)
      IF(KG.EQ.0)H0=100.0
      DO I=1,N
      H(I)=0.5*X(I)*X(I)+V(I)+DA*(ABS(X(I))-1.0)**2
      IF(KG.EQ.0.AND.H(I).LT.H0)H0=H(I)
      RO(I)=1.0+A1*P(I)/(1.0+A2*P(I))
      EDA(I)=EXP(A4*(-1.0+(1.0+A3*P(I))**Z))
      ENDDO
```

```
IF(KG.EQ.0)THEN
H0=-H0+HMIN
ENDIF
DO I=1,N
H(I)=H0+H(I)
ENDDO
IF(KG.EQ.0)THEN
ROEHE=RO(IE)*H(IE)
ENDIF
DO I=1,N
C(I,1)=AKK*EDA(I)/RO(I)
DO J=2,N
IJ=IABS(I-J)
D1=3.0*H(I)**2*DP(I)*D(IJ)
D2=H(I)**3*(DELTA(I+1,J)-DELTA(I,J))/DX
D3=-AKK*A6*(1.0+A3*P(I))**(Z-1.0)*EDA(I)*DELTA(I,J)*(H(I)-ROEHE/
RO(I))
D4=-AKK*EDA(I)*(D(IJ)+ROEHE*A1*DELTA(I,J)/(1.0+A7*P(I))**2)
C(I,J)=D1+D2+D3+D4
ENDDO
C(I,N2)=3.0*H(I)*H(I)*DP(I)-AKK*EDA(I)
B(I)=-H(I)**3*DP(I)+AKK*EDA(I)*(H(I)-ROEHE/RO(I))
ENDDO
B(N2)=0.5*PAI
DO I=1,N
B(N2)=B(N2)-P(I)*C(N2,I)
ENDDO
DO I=I1,N
IF(P(I).LE.0.0)THEN
B(I)=0.0
DO J=1,N2
C(I,J)=0.0
C(J,I)=0.0
ENDDO
C(I,I)=1.0
ENDIF
ENDDO
CALL INV(N2,C,IP,IDET)
IF(IDET.EQ.0)GOTO 20
DO J=1,N2
DP(J)=0.0
DO I=1,N2
DP(J)=DP(J)+C(J,I)*B(I)
ENDDO
ENDDO
ROEHE=RO(II)*H(II)
DO I=2,N1
P(I)=P(I)+DP(I)
IF(P(I).LT.0.0)P(I)=0.0
ENDDO
H0=H0-B(N2)*DP(N2)
```

```
      ER=0.0
      SUM=0.0
      DO I=1,N
      SUM=SUM+P(I)
      ER=ER+ABS(P(I)-POLD(I))
      ENDDO
      ER=ER/SUM
      WRITE(*,*)'ER,DA=',ER,DA
      KG=KG+1
      IF(ER.GT.1.E-5.AND.KG.LT.100)GOTO 10
      KG=1
      DA=DA+0.5
      I1=21
      DO I=1,N
      WRITE(8,100)X(I),P(I),H(I)
100   FORMAT(6(E12.6,1X))
      ENDDO
      IF(DA.LT.1.1)GOTO 10
20    STOP
      END
      FUNCTION DELTA(I,J)
      DELTA=0.0
      IF(I.EQ.J)DELTA=1.0
      RETURN
      END
      SUBROUTINE INV(N,A,IP,IDET)
      DIMENSION A(N,N), IP(N)
      IDET=1
      EPS=1.E-6
      DO K=1,N
      P=0
      I0=K
      IP(K)=K
      DO I=K,N
      IF(ABS(A(I,K)).GT.ABS(P))THEN
      P=A(I,K)
      I0=I
      IP(K)=I
      ENDIF
      ENDDO
      IF(ABS(P).LE.EPS)THEN
      IDET=0
      GOTO 10
      ENDIF
      IF(I0.NE.K)THEN
      DO J=1,N
      S=A(K,J)
      A(K,J)=A(I0,J)
      A(I0,J)=S
      ENDDO
      ENDIF
```

```
      A(K,K)=1./P
      DO I=1,N
      IF(I.NE.K)THEN
      A(I,K)=-A(I,K)*A(K,K)
      DO J=1,N
      IF(J.NE.K)THEN
      A(I,J)=A(I,J)+A(I,K)*A(K,J)
      ENDIF
      ENDDO
      ENDIF
      ENDDO
      DO J=1,N
      IF(J.NE.K)THEN
      A(K,J)=A(K,K)*A(K,J)
      ENDIF
      ENDDO
      ENDDO
      DO K=N-1,1,-1
      IR=IP(K)
      IF(IR.NE.K)THEN
      DO I=1,N
      S=A(I,IR)
      A(I,IR)=A(I,K)
      A(I,K)=S
      ENDDO
      ENDIF
      ENDDO
10    RETURN
      END
      SUBROUTINE VI(N,DX,P,V)
      DIMENSION P(N),V(N)
      COMMON /COMAK/AK(0:1100)
      DATA PAI1/0.318309886/
      C=ALOG(DX)
      DO 10 I=1,N
      V(I)=0.0
      DO 10 J=1,N
      IJ=IABS(I-J)
10    V(I)=V(I)+(AK(IJ)+C)*DX*P(J)
      DO I=1,N
      V(I)=-PAI1*V(I)
      ENDDO
      RETURN
      END
      SUBROUTINE SUBAK(MM)
      COMMON /COMAK/AK(0:1100)
      DO 10 I=0,MM
10    AK(I)=(I+0.5)*(ALOG(ABS(I+0.5))-1.)-(I-0.5)*(ALOG(ABS
      (I-0.5))-1.)
      RETURN
      END
```

14.2.2.2 Example

According to the given conditions, the calculation results are shown in Figure 14.2.

14.3 EHL solution with random roughness in line contact

14.3.1 Description of program

For the random rough surface EHL problem in the line contact, Subroutine HREE, in which the film thickness is calculated, is modified, and the other subroutines are also modified a little. Therefore, we have explained only these modifications here.

1. In Subroutine HREE, a vector ROU is provided as the dummy argument, whose number is equal to the number of nodes.
2. For convenience, in the running program subdirectory, roughness has been computed in advance and it has been saved in the file ROUGH2.DAT for reading into the subroutine.
3. An integer KR is used to guarantee that the roughness is read only once. That is, when $KR = 0$, read the roughness in the file ROUGH2.DAT, and when $KR \neq 0$, skip the reading statement.
4. Add the roughness to the smooth surface film thickness equation. Because the ready roughness is nondimensional, DA is used as the amplitude of the roughness in the calculation.

14.3.2 Calculation program

```
PROGRAM LINEEHLROUGH
     COMMON /COM1/ENDA,A1,A2,A3,Z,HM0,DH/COM2/EDA0/COM4/X0,XE/COM3/E1,
     PH,B,R
     DATA PAI,Z,P0/3.14159265,0.68,1.96E8/
     DATA N,X0,XE,W,E1,EDA0,R,Us/129,-
     4.0,1.4,1.0E5,2.21E11,0.05,0.05,1.5/
     OPEN(8,FILE='OUT.DAT',STATUS='UNKNOWN')
     W1=W/(E1*R)
     PH=E1*SQRT(0.5*W1/PAI)
     A1=(ALOG(EDA0)+9.67)
     A2=PH/P0
     A3=0.59/(PH*1.E-9)
     B=4.*R*PH/E1
     ALFA=Z*A1/P0
     G=ALFA*E1
     U=EDA0*US/(2.*E1*R)
     CC1=SQRT(2.*U)
     AM=2.*PAI*(PH/E1)**2/CC1
     ENDA=3.*(PAI/AM)**2/8.
     HM0=1.6*(R/B)**2*G**0.6*U**0.7*W1**(-0.13)
     WRITE(*,*)N,X0,XE,W,E1,EDA0,R,US
```

```
      CALL SUBAK(N)
      CALL EHL(N)
      STOP
      END
      SUBROUTINE EHL(N)
      DIMENSION X(1100),P(1100),H(1100),RO(1100),POLD(1100),EPS(1100),
      EDA(1100),V(1100),ROU(1100)
      COMMON /COM1/ENDA,A1,A2,A3,Z,HM0,DH/COM4/X0,XE
      COMMON /COM3/E1,PH,B,RR
      MK=1
      DX=(XE-X0)/(N-1.0)
      DO 10 I=1,N
      X(I)=X0+(I-1)*DX
      IF(ABS(X(I)).GE.1.0)P(I)=0.0
      IF(ABS(X(I)).LT.1.0)P(I)=SQRT(1.-X(I)*X(I))
10    CONTINUE
      CALL HREE(N,DX,X,P,H,RO,EPS,EDA,V,ROU)
      CALL FZ(N,P,POLD)
14    KK=19
      CALL ITER(N,KK,DX,X,P,H,RO,EPS,EDA,V,ROU)
      MK=MK+1
      CALL ERROP(N,P,POLD,ERP)
      WRITE(*,*)'ERP=',ERP
      IF(ERP.GT.1.E-5.AND.DH.GT.1.E-6)THEN
      IF(MK.GE.50)THEN
      MK=1
      DH=0.5*DH
      ENDIF
      GOTO 14
      ENDIF
      IF(DH.LE.1.E-6)WRITE(*,*)'Pressures are not convergent!!!'
      H2=1.E3
      P2=0.0
      DO 106 I=1,N
      IF(H(I).LT.H2)H2=H(I)
      IF(P(I).GT.P2)P2=P(I)
106   CONTINUE
      H3=H2*B*B/RR
      P3=P2*PH
      WRITE(*,*)'P2,H2,P3,H3=',P2,H2,P3,H3
      CALL OUTHP(N,X,P,H,ROU)
      RETURN
      END
      SUBROUTINE OUTHP(N,X,P,H,ROU)
      DIMENSION X(N),P(N),H(N),ROU(N)
      DO 10 I=1,N
      WRITE(8,20)X(I),P(I),H(I),ROU(I)
10    CONTINUE
20    FORMAT(1X,6(E12.6,1X))
      RETURN
      END
```

```
      SUBROUTINE HREE(N,DX,X,P,H,RO,EPS,EDA,V,ROU)
      DIMENSION X(N),P(N),H(N),RO(N),EPS(N),EDA(N),V(N),ROU(N)
      COMMON /COM1/ENDA,A1,A2,A3,Z,HM0,DH/COM2/EDA0/COMAK/AK(0:1100)
      DATA KK,PAI1,G0/0,0.318309886,1.570796325/
      DATA KR,DA/0,0.1/
      W1=0.0
      DO 4 I=1,N
4      W1=W1+P(I)
      C3=(DX*W1)/G0
      DW=1.-C3
      CALL VI(N,DX,P,V)
      HMIN=1.E3
      IF(KR.EQ.0)THEN
      OPEN(12,FILE='ROUGH2.DAT',STATUS='UNKNOWN')
      DO I=1,N
      READ(12,5)ROU(I)
      ENDDO
5     FORMAT(1X,F10.6)
      CLOSE(12)
      KR=1
      ENDIF
      DO 30 I=1,N
      H0=0.5*X(I)*X(I)+V(I)+DA*ROU(I)
      IF(H0.LT.HMIN)HMIN=H0
      H(I)=H0
30    CONTINUE
      IF(KK.EQ.0)THEN
      KK=1
      DH=0.005*HM0
      H00=-HMIN+HM0
      ENDIF
      IF(DW.LT.0.0)H00=H00+DH
      IF(DW.GT.0.0)H00=H00-DH
      DO 60 I=1,N
      H(I)=H00+H(I)
      EDA(I)=EXP(A1*(-1.+(1.+A2*P(I))**Z))
      RO(I)=(A3+1.34*P(I))/(A3+P(I))
      EPS(I)=RO(I)*H(I)**3/(ENDA*EDA(I))
60    CONTINUE
      RETURN
      END
      SUBROUTINE ITER(N,KK,DX,X,P,H,RO,EPS,EDA,V,ROU)
      DIMENSION X(N),P(N),H(N),RO(N),EPS(N),EDA(N),V(N),ROU(N)
      COMMON /COMAK/AK(0:1100)
      DATA PAI1/0.318309886/
      DO 100 K=1,KK
      D2=0.5*(EPS(1)+EPS(2))
      D3=0.5*(EPS(2)+EPS(3))
      DO 70 I=2,N-1
      D1=D2
      D2=D3
```

```
      IF(I.NE.N-1)D3=0.5*(EPS(I+1)+EPS(I+2))
      D8=RO(I)*AK(0)*PAI1
      D9=RO(I-1)*AK(1)*PAI1
      D10=1.0/(D1+D2+(D9-D8)*DX)
      D11=D1*P(I-1)+D2*P(I+1)
      D12=(RO(I)*H(I)-RO(I-1)*H(I-1)+(D8-D9)*P(I))*DX
      P(I)=(D11-D12)*D10
      IF(P(I).LT.0.0)P(I)=0.0
70    CONTINUE
      CALL HREE(N,DX,X,P,H,RO,EPS,EDA,V,ROU)
100   CONTINUE
      RETURN
      END
      SUBROUTINE VI(N,DX,P,V)
      DIMENSION P(N),V(N)
      COMMON /COMAK/AK(0:1100)
      DATA PAI1/0.318309886/
      C=ALOG(DX)
      DO 10 I=1,N
      V(I)=0.0
      DO 10 J=1,N
      IJ=IABS(I-J)
10    V(I)=V(I)+(AK(IJ)+C)*DX*P(J)
      DO I=1,N
      V(I)=-PAI1*V(I)
      ENDDO
      RETURN
      END
      SUBROUTINE SUBAK(MM)
      COMMON /COMAK/AK(0:1100)
      DO 10 I=0,MM
10    AK(I)=(I+0.5)*(ALOG(ABS(I+0.5))-1.)-(I-0.5)*(ALOG(ABS(I-0.5))-1.)
      RETURN
      END
      SUBROUTINE FZ(N,P,POLD)
      DIMENSION P(N),POLD(N)
      DO 10 I=1,N
10    POLD(I)=P(I)
      RETURN
      END
      SUBROUTINE ERROP(N,P,POLD,ERP)
      DIMENSION P(N),POLD(N)
      ERP=0.0
      SUM=0.0
      DO 10 I=1,N
      ERP=ERP+ABS(P(I)-POLD(I))
      POLD(I)=P(I)
10    SUM=SUM+P(I)
      ERP=ERP/SUM
      RETURN
      END
```

14.3.3 Example

According to the conditions in the program, the given roughness amplitude DA is equal to 0.1. The roughness, pressure distribution, and film thickness in the line contact are shown in Figure 14.3.

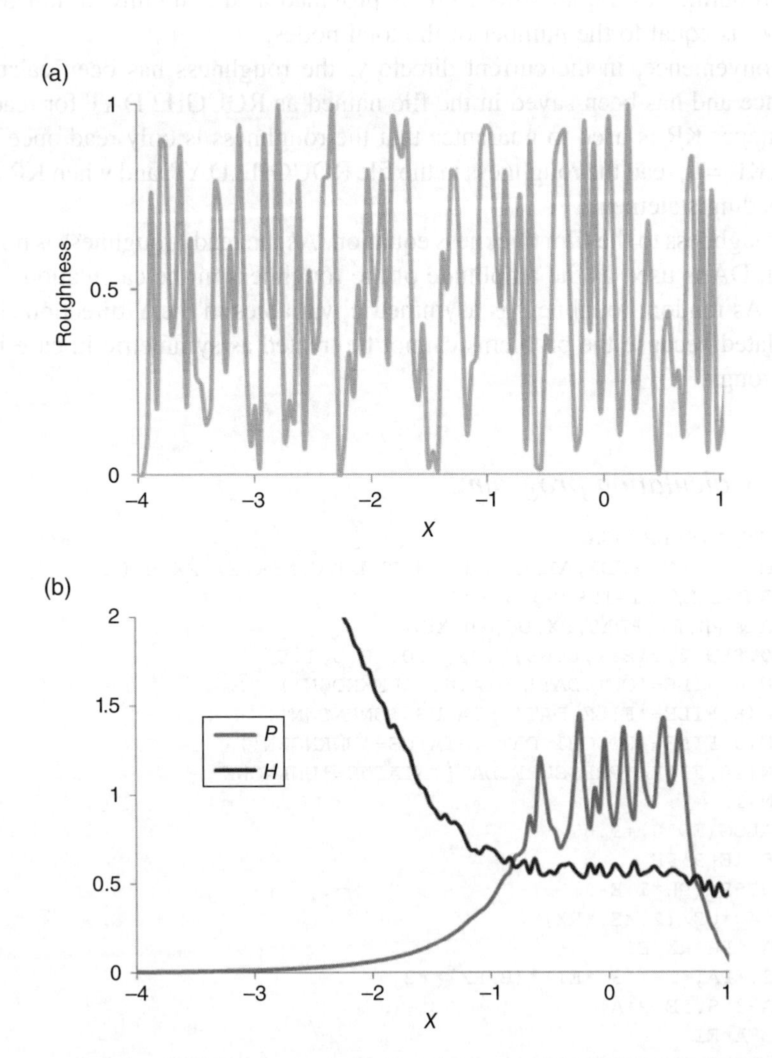

Figure 14.3 Random roughness, pressure distribution, and film thickness of rough surface EHL in line contact. (a) Random roughness and (b) pressure distribution and film thickness

14.4 EHL solution with random roughness in point contact

14.4.1 Description of program

For the random rough surface EHL problem in the point contact, Subroutine HREE, in which the film thickness is calculated, is modified, and the other subroutines are also modified a little. Therefore, we have explained only these subroutines and are as follows:

1. In Subroutine HREE, an array ROU is provided as the dummy argument, whose number is equal to the number of the total nodes.
2. For convenience, in the current directory, the roughness has been calculated in advance and has been saved in the file named as ROUGH2.DAT for reading.
3. An integer KR is used to guarantee that the roughness is only read once. That is, when KR = 0, read the roughness in the file ROUGH2.DAT, and when KR = 0, skip the reading statement.
4. Add roughness to the film thickness equation. As the ready roughness is nondimensional, DA is used as the amplitude of the roughness in the calculation.
5. Note: As random roughness is asymmetric, variables in the Y direction should be calculated because the problems cannot be treated as symmetric in case if results are wrong.

14.4.2 Calculation program

```
PROGRAM POINTEHLROUGH
    COMMON /COM1/ENDA,A1,A2,A3,Z,HM0,DH/COMER/E1,RX,B,PH
    DATA PAI,Z/3.14159265,0.68/
    DATA N,PH,E1,EDA0,RX,US,X0,XE/
    65,0.8E9,2.21E11,0.05,0.02,1.0,-2.5,1.5/
    OPEN(4,FILE='OUT.DAT',STATUS='UNKNOWN')
    OPEN(8,FILE='FILM.DAT',STATUS='UNKNOWN')
    OPEN(9,FILE='ROUGH1.DAT',STATUS='UNKNOWN')
    OPEN(10,FILE='PRESSURE.DAT',STATUS='UNKNOWN')
    MM=N-1
    A1=ALOG(EDA0)+9.67
    A2=5.1E-9*PH
    A3=0.59/(PH*1.E-9)
    U=EDA0*US/(2.*E1*RX)
    B=PAI*PH*RX/E1
    W0=2.*PAI*PH/(3.*E1)*(B/RX)**2
    ALFA=Z*5.1E-9*A1
    G=ALFA*E1
    HM0=3.63*(RX/B)**2*G**0.49*U**0.68*W0**(-0.073)
    ENDA=12.*U*(E1/PH)*(RX/B)**3
    WRITE(*,*)N,X0,XE,W0,PH,E1,EDA0,RX,US
    WRITE(4,*)N,X0,XE,W0,PH,E1,EDA0,RX,US
    WRITE(*,*)'         Wait please'
    CALL SUBAK(MM)
```

```
      CALL EHL(N,X0,XE)
      STOP
      END
      SUBROUTINE EHL(N,X0,XE)
      DIMENSION X(65),Y(65),H(4500),RO(4500),EPS(4500),EDA(4500),
      P(4500),POLD(4500),V(4500),ROU(4500)
      COMMON /COM1/ENDA,A1,A2,A3,Z,HM0,DH
      DATA MK,G0/1,2.0943951/
      CALL INITI(N,DX,X0,XE,X,Y,P,POLD)
      KK=0
      CALL HREE(N,DX,KK,H00,G0,X,Y,H,RO,EPS,EDA,P,V,ROU)
   14 KK=15
      CALL ITER(N,KK,DX,H00,G0,X,Y,H,RO,EPS,EDA,P,V,ROU)
      MK=MK+1
      CALL ERP(N,ER,P,POLD)
      WRITE(*,*)'ER=',ER
      IF(ER.GT.1.E-5.AND.DH.GT.1.E-7)THEN
      IF(MK.GE.10)THEN
      MK=1
      DH=0.5*DH
      ENDIF
      GOTO 14
      ENDIF
      IF(DH.LE.1.E-7)WRITE(*,*)'Pressures are not convergent!!!'
      CALL OUTPUT(N,DX,X,Y,H,P,ROU)
      RETURN
      END
      SUBROUTINE ERP(N,ER,P,POLD)
      DIMENSION P(N,N),POLD(N,N)
      ER=0.0
      SUM=0.0
      DO 10 I=1,N
      DO 10 J=1,N
      ER=ER+ABS(P(I,J)-POLD(I,J))
      POLD(I,J)=P(I,J)
      SUM=SUM+P(I,J)
   10 CONTINUE
      ER=ER/SUM
      RETURN
      END
      SUBROUTINE INITI(N,DX,X0,XE,X,Y,P,POLD)
      DIMENSION X(N),Y(N),P(N,N),POLD(N,N)
      DX=(XE-X0)/(N-1.)
      Y0=-0.5*(XE-X0)
      DO 5 I=1,N
      X(I)=X0+(I-1)*DX
      Y(I)=Y0+(I-1)*DX
    5 CONTINUE
      DO I=1,N
      D=1.-X(I)*X(I)
      DO J=1,N
```

```
      C=D-Y(J)*Y(J)
      IF(C.LE.0.0)P(I,J)=0.0
      IF(C.GT.0.0)P(I,J)=SQRT(C)
      POLD(I,J)=P(I,J)
      ENDDO
      ENDDO
      RETURN
      END
      SUBROUTINE HREE(N,DX,KK,H00,G0,X,Y,H,RO,EPS,EDA,P,V,ROU)
      DIMENSION X(N),Y(N),P(N,N),H(N,N),RO(N,N),EPS(N,N),EDA(N,N),
      V(N,N),ROU(N,N)
      COMMON /COM1/ENDA,A1,A2,A3,Z,HM0,DH/COMAK/AK(0:65,0:65)/COMER/E1,
      R,B,PH
      DATA PAI,PAI1/3.14159265,0.2026423/
      DATA KR,DA/0,0.1/
      CALL VI(N,DX,P,V)
      HMIN=1.E3
      IF(KR.EQ.0)THEN
      OPEN(12,FILE='ROUGH2.DAT',STATUS='UNKNOWN')
      DO I=1,N
      DO J=1,N
      READ(12,100)ROU(I,J)
      ENDDO
100   FORMAT(1X,F10.6)
      ENDDO
      CLOSE(12)
      KR=1
      ENDIF
      DO 30 I=1,N
      DO 30 J=1,N
      RAD=X(I)*X(I)+Y(J)*Y(J)+DA*ROU(I,J)
      W1=0.5*RAD
      H0=W1+V(I,J)
      IF(H0.LT.HMIN)HMIN=H0
30    H(I,J)=H0
      IF(KK.EQ.0)THEN
      KK=1
      DH=0.005*HM0
      H00=-HMIN+0.5*HM0
      ENDIF
      W1=0.0
      DO 32 I=1,N
      DO 32 J=1,N
32    W1=W1+P(I,J)
      W1=DX*DX*W1/G0
      DW=1.-W1
      IF(DW.LT.0.0)H00=H00+DH
      IF(DW.GT.0.0)H00=H00-DH
      DO 60 I=1,N
      DO 60 J=1,N
      H(I,J)=H00+H(I,J)
```

```
      IF(H(I,J).LT.0.0)THEN
      P(I,J)=P(I,J)+4.*SQRT(-H(I,J))**3*(B**3/R)*(E1/PH)/3.0
      H(I,J)=0.0
      ENDIF
      EDA1=EXP(A1*(-1.+(1.+A2*P(I,J))**Z))
      EDA(I,J)=EDA1
      RO(I,J)=(A3+1.34*P(I,J))/(A3+P(I,J))
60    EPS(I,J)=RO(I,J)*H(I,J)**3/(ENDA*EDA1)
      RETURN
      END
      SUBROUTINE ITER(N,KK,DX,H00,G0,X,Y,H,RO,EPS,EDA,P,V,ROU)
      DIMENSION X(N),Y(N),P(N,N),H(N,N),RO(N,N),EPS(N,N),EDA(N,N),
      V(N,N),ROU(N,N)
      COMMON /COMAK/AK(0:65,0:65)
      DATA KG1,PAI/0,3.14159265/
      IF(KG1.NE.0)GOTO 2
      KG1=1
      AK00=AK(0,0)
      AK10=AK(1,0)
2     DO 100 K=1,KK
      DO 70 J=2,N-1
      J0=J-1
      J1=J+1
      D2=0.5*(EPS(1,J)+EPS(2,J))
      DO 70 I=2,N-1
      IF(H(I,J).LE.0.0)GOTO 60
      I0=I-1
      I1=I+1
      D1=D2
      D2=0.5*(EPS(I1,J)+EPS(I,J))
      D4=0.5*(EPS(I,J0)+EPS(I,J))
      D5=0.5*(EPS(I,J1)+EPS(I,J))
      D8=2.0*RO(I,J)*AK00/PAI**2
      D9=2.0*RO(I0,J)*AK10/PAI**2
      D10=D1+D2+D4+D5+D8*DX-D9*DX
      D11=D1*P(I0,J)+D2*P(I1,J)+D4*P(I,J0)+D5*P(I,J1)
      D12=(RO(I,J)*H(I,J)-D8*P(I,J)-RO(I0,J)*H(I0,J)+D9*P(I,J))*DX
      P(I,J)=(D11-D12)/D10
      IF(P(I,J).LT.0.0)P(I,J)=0.0
60    CONTINUE
70    CONTINUE
      CALL HREE(N,DX,KK,H00,G0,X,Y,H,RO,EPS,EDA,P,V,ROU)
100   CONTINUE
      RETURN
      END
      SUBROUTINE VI(N,DX,P,V)
      DIMENSION P(N,N),V(N,N)
      COMMON /COMAK/AK(0:65,0:65)
      PAI1=0.2026423
      DO 40 I=1,N
      DO 40 J=1,N
```

```
      H0=0.0
      DO 30 K=1,N
      IK=IABS(I-K)
      DO 30 L=1,N
      JL=IABS(J-L)
30    H0=H0+AK(IK,JL)*P(K,L)
40    V(I,J)=H0*DX*PAI1
      RETURN
      END
      SUBROUTINE SUBAK(MM)
      COMMON /COMAK/AK(0:65,0:65)
      S(X,Y)=X+SQRT(X**2+Y**2)
      DO 10 I=0,MM
      XP=I+0.5
      XM=I-0.5
      DO 10 J=0,I
      YP=J+0.5
      YM=J-0.5
      A1=S(YP,XP)/S(YM,XP)
      A2=S(XM,YM)/S(XP,YM)
      A3=S(YM,XM)/S(YP,XM)
      A4=S(XP,YP)/S(XM,YP)
      AK(I,J)=XP*ALOG(A1)+YM*ALOG(A2)+XM*ALOG(A3)+YP*ALOG(A4)
10    AK(J,I)=AK(I,J)
      RETURN
      END
      SUBROUTINE OUTPUT(N,DX,X,Y,H,P,ROU)
      DIMENSION X(N),Y(N),H(N,N),P(N,N),ROU(N,N)
      A=0.0
      WRITE(8,110)A,(Y(I),I=1,N)
      DO I=1,N
      WRITE(8,110)X(I),(H(I,J),J=1,N)
      ENDDO
      WRITE(9,110)A,(Y(I),I=1,N)
      DO I=1,N
      WRITE(9,110)X(I),(ROU(I,J),J=1,N)
      ENDDO
      WRITE(10,110)A,(Y(I),I=1,N)
      DO I=1,N
      WRITE(10,110)X(I),(P(I,J),J=1,N)
      ENDDO
110   FORMAT(66(E12.6,1X))
      RETURN
      END
```

14.4.3 Example

According to the conditions of the program, roughness amplitude is equal to 0.1. The roughness, pressure distribution, and film thickness of EHL in the point contact are shown in Figure 14.4.

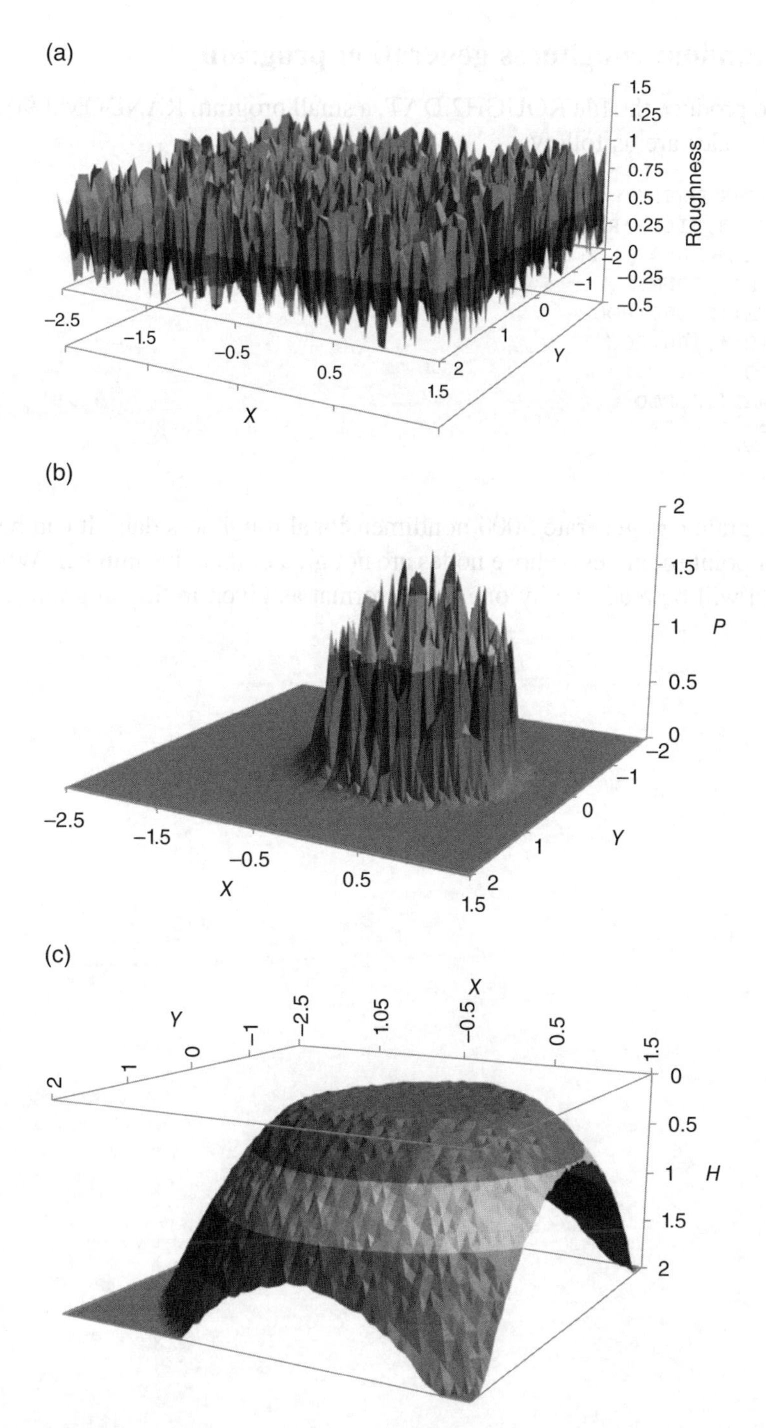

Figure 14.4 Roughness, pressure distribution, and film with random roughness of EHL in point contact. (a) Roughness, (b) pressure distribution, and (c) film thickness

14.5 Random roughness generation program

In order to produce the file ROUGH2.DAT, a small program RANDOM.F90 is given. Its source codes are as follows.

```
    INTEGER IDUM
    OPEN(12,FILE='ROUGH2.DAT',STATUS='UNKNOWN')
    DO I=1,5000
    ROU=RAN(IDUM)
    WRITE(12,100)ROU
    WRITE(*,100)ROU
    ENDDO
100 FORMAT(1X,F10.6)
    STOP
    END
```

The program can generate 5000 nondimensional roughness data. It can be used as the line or point roughness whose nodes are not greater than this number. When using it, the data will be read one by one in the format as given in the program.

15

Numerical calculation method and program for micropolar fluid EHL

15.1 Reynolds equation of micropolar fluid EHL

In lubrication theory, one of the most basic assumption is to neglect the internal microstructure of the lubricant. However, under certain conditions, some lubricants, such as with long-chain molecule polymer or with solid particles, do not satisfy the hypothesis. Here, we discuss the elastohydrodynamic lubrication (EHL) problem of micropolar fluid because its lubrication analysis is not entirely consistent with the classical EHL theory. A micropolar fluid has a population structure consisting of particles with an individual quality and a speed.

The micropolar fluid model ignores the deformation of the microelements but still keeps the micro motion of particles. Therefore, the continuous medium theory still can be used. However, due to the length of the particle, the motion analysis of the fluid should be used on the particle rotation. Thus, the micropolar fluid is a non-Newtonian fluid.

Singh and Sinha derived three-dimensional Reynolds equation for micropolar fluids problems [9]. The numerical method and programs explained in this chapter are based on the equations derived by them.

15.1.1 Basic equations

The basic equations of EHL of micropolar fluid are derived under the incompressibility condition. In the Cartesian coordinates, the expressions of three-dimensional steady-fluid mechanics for micropolar fluid are given by the following equations.

Numerical Calculation of Elastohydrodynamic Lubrication: Methods and Programs, First Edition.
Ping Huang.

Continuity equation:

$$\frac{\partial \rho u}{\partial x}+\frac{\partial \rho v}{\partial y}+\frac{\partial \rho w}{\partial z}=0 \tag{15.1}$$

Momentum equations:

$$\begin{aligned}
&\frac{1}{2}(2\mu+\chi)\left(\frac{\partial^2 u}{\partial x^2}+\frac{\partial^2 u}{\partial y^2}+\frac{\partial^2 u}{\partial z^2}\right)+\chi\left(\frac{\partial \omega_3}{\partial y}-\frac{\partial \omega_2}{\partial z}\right)-\frac{\partial p}{\partial x}=\rho\left(u\frac{\partial u}{\partial x}+u\frac{\partial u}{\partial y}+u\frac{\partial u}{\partial z}\right)\\
&\frac{1}{2}(2\mu+\chi)\left(\frac{\partial^2 v}{\partial x^2}+\frac{\partial^2 v}{\partial y^2}+\frac{\partial^2 v}{\partial z^2}\right)+\chi\left(\frac{\partial \omega_1}{\partial z}-\frac{\partial \omega_3}{\partial x}\right)-\frac{\partial p}{\partial y}=\rho\left(v\frac{\partial v}{\partial x}+v\frac{\partial v}{\partial y}+v\frac{\partial v}{\partial z}\right)\\
&\frac{1}{2}(2\mu+\chi)\left(\frac{\partial^2 w}{\partial x^2}+\frac{\partial^2 w}{\partial y^2}+\frac{\partial^2 w}{\partial z^2}\right)+\chi\left(\frac{\partial \omega_2}{\partial x}-\frac{\partial \omega_1}{\partial y}\right)-\frac{\partial p}{\partial z}=\rho\left(w\frac{\partial w}{\partial x}+w\frac{\partial w}{\partial y}+w\frac{\partial w}{\partial z}\right)\\
&\gamma\left(\frac{\partial^2 \omega_1}{\partial x^2}+\frac{\partial^2 \omega_1}{\partial y^2}+\frac{\partial^2 \omega_1}{\partial z^2}\right)+\chi\left(\frac{\partial w}{\partial y}-\frac{\partial v}{\partial z}\right)-2\chi\omega_1=\rho J\left(u\frac{\partial \omega_1}{\partial x}+v\frac{\partial \omega_1}{\partial y}+w\frac{\partial \omega_1}{\partial z}\right)\\
&\gamma\left(\frac{\partial^2 \omega_2}{\partial x^2}+\frac{\partial^2 \omega_2}{\partial y^2}+\frac{\partial^2 \omega_2}{\partial z^2}\right)+\chi\left(\frac{\partial u}{\partial z}-\frac{\partial w}{\partial x}\right)-2\chi\omega_2=\rho J\left(u\frac{\partial \omega_2}{\partial x}+v\frac{\partial \omega_2}{\partial y}+w\frac{\partial \omega_2}{\partial z}\right)\\
&\gamma\left(\frac{\partial^2 \omega_3}{\partial x^2}+\frac{\partial^2 \omega_3}{\partial y^2}+\frac{\partial^2 \omega_3}{\partial z^2}\right)+\chi\left(\frac{\partial v}{\partial x}-\frac{\partial u}{\partial y}\right)-2\chi\omega_3=\rho J\left(u\frac{\partial \omega_3}{\partial x}+v\frac{\partial \omega_3}{\partial y}+w\frac{\partial \omega_3}{\partial z}\right)
\end{aligned} \tag{15.2}$$

Because a micropolar fluid molecule has a characteristic length L, it not only has a translational motion but also has a rotational motion. Therefore, in Equation 15.2, there are three translational momentum equations and three angular momentum equations.

15.1.2 Reynolds equation

First, the variables should be nondimensionalized. Let:

$$X=\frac{x}{a};\ Y=\frac{y}{b};\ Z=\frac{z}{h}$$

$$\bar{u}=\frac{u}{u_s};\ \bar{v}=\frac{v}{u_s};\ \bar{w}=\frac{w}{u_s}$$

$$\bar{\omega}_i=\frac{\omega_i h}{u_s};\ P=\frac{p h_1^2}{(\mu+\chi/2)u_s a}$$

$$\delta_1=\frac{h}{a};\ \delta_2=\frac{h}{b};\ \delta_3=\frac{h_1}{a};\ \delta_4=\frac{b}{a}$$

$$\xi=\frac{h_1}{h};\ L=\frac{h_1}{l};\ l=\left(\frac{\gamma}{4\mu}\right)^{1/2};\ N=\left\{\frac{\chi}{(2\mu+\chi)}\right\}^{1/2}$$

In order to obtain the corresponding Reynolds equation of micropolar fluid, the assumptions while derivating the Reynolds equation are also used. They are: (1) laminar flow; (2) neglecting mass force; (3) that the film thickness is very thin, comparing with the length and width; (4) without interface sliding; and (5) smooth surface without pores. First, the dimensional analysis is carried out as follows.

$$\begin{aligned}
&\mathrm{Re}=2\rho hu_s/(2\mu+x)\ll 1;\ Re'=\rho ju_s h/4\mu l^2\ll 1\\
&\frac{\partial}{\partial x}\ll\frac{\partial}{\partial z};\ \frac{\partial}{\partial y}\ll\frac{\partial}{\partial z}\\
&w\ll u;\ w\ll v;\ \delta_1\ll 1;\ \delta_2\ll 1;\ \delta_3\ll 1\\
&\xi\approx O(1);\ \delta_4\approx O(1)
\end{aligned}$$

Compared with the variables in Equation 15.2 and ignore the higher order infinitesimals, the dimensional form of Equation 15.3 can be simplified as follows:

$$\begin{aligned}
&\frac{\partial^2\bar{u}}{\partial Z^2}-2N^2\frac{\partial\bar{\omega}_2}{\partial Z}-\frac{1}{\xi^2}\frac{\partial P}{\partial X}=0\\
&\frac{\partial^2\bar{v}}{\partial Z^2}+2N^2\frac{\partial\bar{\omega}_1}{\partial Z}-\frac{1}{\xi^2}\frac{\partial P}{\partial Y}=0\\
&\frac{\partial P}{\partial Z}=0\\
&\frac{\partial^2\bar{\omega}_1}{\partial Z^2}-\frac{N^2L^2}{2(1-N^2)\xi^2}\frac{\partial\bar{v}}{\partial Z}-\frac{N^2L^2}{(1-N^2)\xi^2}\bar{\omega}_1=0\\
&\frac{\partial^2\bar{\omega}_2}{\partial Z^2}+\frac{N^2L^2}{2(1-N^2)\xi^2}\frac{\partial\bar{u}}{\partial Z}-\frac{N^2L^2}{(1-N^2)\xi^2}\bar{\omega}_2=0\\
&\frac{\partial^2\bar{\omega}_3}{\partial Z^2}-\frac{N^2L^2}{(1-N^2)\xi^2}\bar{\omega}_3=0
\end{aligned}\tag{15.3}$$

From the third equation of Equation 15.3, we know $P=P(X, Y)$ independent of Z. With the sixth equation of Equation 15.3 and the boundary conditions, we can obtain

$\bar{\omega}_z = 0$. Equation 15.3 can be changed to the dimensional form and by substituting the third and the sixth equations in the other equations, we have:

$$
\begin{aligned}
&\left(\mu+\frac{1}{2}\chi\right)\frac{\partial^2 u}{\partial z^2}-\chi\frac{\partial \omega_2}{\partial z}-\frac{\partial p}{\partial x}=0\\
&\left(\mu+\frac{1}{2}\chi\right)\frac{\partial^2 v}{\partial z^2}+\chi\frac{\partial \omega_1}{\partial z}-\frac{\partial p}{\partial y}=0\\
&\frac{\partial^2 \omega_1}{\partial z^2}-2\chi\omega_1-\chi\frac{\partial v}{\partial z}=0\\
&\frac{\partial^2 \omega_2}{\partial z^2}-2\chi\omega_2+\chi\frac{\partial u}{\partial z}=0
\end{aligned}
\tag{15.4}
$$

When the boundary conditions are given, we can derive the velocity formulas. Usually, the velocity boundary conditions are as follows:

$$
\begin{aligned}
&\text{when } z=0 \quad u=u_{11}, v=u_{12}, \omega_1=0, \omega_2=0\\
&\text{when } z=h \quad u=u_{21}, v=u_{22}, \omega_1=0, \omega_2=0
\end{aligned}
$$

It can be verified that Equation 15.4 has the following velocities under the above velocity boundary conditions.

$$
\begin{aligned}
u&=\frac{1}{\eta}\left(\frac{\partial p}{\partial x}\frac{z^2}{2}+A_{11}z\right)-\frac{2N^2}{m}\{A_{21}\sinh(mz)+A_{31}\sinh(mz)\}+A_{41}\\
v&=\frac{1}{\eta}\left(\frac{\partial p}{\partial y}\frac{z^2}{2}+A_{12}z\right)-\frac{2N^2}{m}\{A_{22}\sinh(mz)+A_{32}\sinh(mz)\}+A_{42}\\
\omega_1&=\left\{-\frac{1}{2\eta}\left(\frac{\partial p}{\partial z}z+A_{12}\right)-A_{22}\cosh(mz)+A_{32}\sinh(mz)\right\}\\
\omega_2&=\left\{-\frac{1}{2\eta}\left(\frac{\partial p}{\partial z}z+A_{11}\right)-A_{21}\cosh(mz)+A_{31}\sinh(mz)\right\}
\end{aligned}
\tag{15.5}
$$

In Equation 15.5:

$$
A_{1j}=\frac{1}{A_5}\left\{(u_{2j}-u_{1j})\sinh(mh)-\frac{h}{2\eta}\frac{\partial p}{\partial x_j}\left[h\sinh(mh)+\frac{2N^2}{m}(1-\cosh(mh))\right]\right\}
$$

$$
A_{2j}=\frac{1}{\eta A_5}\left\{\frac{(u_{2j}-u_{1j})1-\cosh(mh)}{2}-\frac{h}{2\eta}\frac{\partial p}{\partial x_j}\left[\frac{h}{2}\{\cosh(mh)-1\}+\left\{h-\frac{2N^2}{m}\sinh(mh)\right\}\right]\right\}
$$

$$A_{3j} = \frac{A_{1j}}{2\eta}$$

$$A_{4j} = u_{1j} + \frac{2N^2}{m} A_{3j}$$

$$A_5 = \frac{h}{\eta}\left[\sinh(mh) - \frac{2N^2}{mh}\{\cosh(mh) - 1\}\right]$$

where, the subscript j is equal to 1 or 2, $x_1 = x$, $x_2 = y$, $m = N/l$, and $\eta = \mu + \chi/2$ is the viscosity of micropolar fluid.

By integrating the velocities of Equation 15.5, we obtain the following flow equations:

$$\begin{aligned} q_x &= \int_0^h \rho u dz = \frac{\rho h}{2}(u_{11} + u_{21}) - \frac{\rho f(N,l,h)}{12\eta}\frac{\partial p}{\partial x} \\ q_y &= \int_0^h \rho v dz = \frac{\rho h}{2}(u_{11} + u_{21}) - \frac{\rho f(N,l,h)}{12\eta}\frac{\partial p}{\partial y} \end{aligned} \tag{15.6}$$

where,

$$f(N,l,h) = h^3 + 12l^2h - 6Nlh^2 \coth\left(\frac{Nh}{2l}\right) \tag{15.7}$$

Finally, by substituting Equation 15.3 in the integrated continuity equation (Eq. 15.1), we get the modified general Reynolds equation for polar fluid and is as follows:

$$\begin{aligned} \frac{\partial}{\partial x}\left(\frac{\rho f(N,l,h)}{12\eta}\frac{\partial p}{\partial x}\right) + \frac{\partial}{\partial y}\left(\frac{\rho f(N,l,h)}{12\eta}\frac{\partial p}{\partial y}\right) &= \frac{1}{2}\frac{\partial}{\partial x}\{\rho(u_{11} + u_{21})h\} + \frac{1}{2}\frac{\partial}{\partial y}\{\rho(u_{12} + u_{22})h\} \\ &\quad - u_{21}\frac{\partial \rho h}{\partial x} - u_{22}\frac{\partial \rho h}{\partial y} + \rho w_{\mathrm{h}} - \rho w_0 \end{aligned} \tag{15.8}$$

It can be seen that if we consider a micropolar fluid, it is equivalent to changing h^3 in the classical Reynolds equation to $f(N, l, h)$. If the length l of a micropolar fluid tends to be zero (i.e., $L \to \infty$), $f(N, l, h) \to h^3$, and $\eta \to \mu$. Therefore, Equation 15.8 will become the classical Reynolds equation.

If we neglect the velocities in the y direction, that is, u_{21}, u_{12}, and u_{22}, and assume the velocity in the x direction to be a constant, the two-dimensional steady EHL Reynolds equation of micropolar fluid will be as follows:

$$\frac{\partial}{\partial x}\left(\frac{\rho f(N,l,h)}{\eta}\frac{\partial p}{\partial x}\right)+\frac{\partial}{\partial y}\left(\frac{\rho f(N,l,h)}{\eta}\frac{\partial p}{\partial y}\right)=12u_{\mathrm{s}}\frac{\partial(\rho h)}{\partial x} \tag{15.9}$$

Compared with the classical Reynolds equation, the Reynolds equation of micropolar fluid is obtained only to replace h^3 into $f(N,l,h)=h^3+12l^2h-6Nlh^2\coth(Nh/2l)$.

Furthermore, if the items in the y direction are not considered, Equation 15.7 can be converted into the one-dimensional EHL Reynolds equation of micropolar fluid:

$$\frac{d}{dx}\left(\frac{\rho f(N,l,h)}{\eta}\frac{dp}{dx}\right)=12u_{\mathrm{s}}\frac{d(\rho h)}{dx} \tag{15.10}$$

15.2 Calculation program for EHL of micropolar fluid in line contact

15.2.1 Description of program

As Subroutine HREE involves only the calculation of the film thickness, we only need to modify a few statements of HREE in the original EHL program.

1. In the DATA statement, add two new parameters AN and AL, respectively. They correspond to N and L in Equation 15.7.
2. In calculation of the coefficient EPS, $f(N,l,h)=h^3+12l^2h-6Nlh^2\coth(Nh/2l)$ is used to replace h^3. Furthermore, although it will overflow while $l=0$ in f calculation, the coth function is zero. Therefore, this calculation is divided into three sentences in the program.
3. Add the coth calculation function.

15.2.2 Calculation program

```
PROGRAM MICROPOLORLINEEHL
   COMMON /COM1/ENDA,A1,A2,A3,Z,HM0,DH/COM2/EDA0/COM4/X0,XE/COM3/E1,PH,
B,R
     DATA PAI,Z,P0/3.14159265,0.68,1.96E8/
   DATA N,X0,XE,W,E1,EDA0,R,Us/130,-4.0,1.5,1.0E5,2.2E11,0.05,0.05,1.5/
     OPEN(8,FILE='OUT.DAT',STATUS='UNKNOWN')
     W1=W/(E1*R)
     PH=E1*SQRT(0.5*W1/PAI)
     A1=(ALOG(EDA0)+9.67)
```

```
      A2=PH/P0
      A3=0.59/(PH*1.E-9)
      B=4.*R*PH/E1
      ALFA=Z*A1/P0
      G=ALFA*E1
      U=EDA0*US/(2.*E1*R)
      CC1=SQRT(2.*U)
      AM=2.*PAI*(PH/E1)**2/CC1
      ENDA=3.*(PAI/AM)**2/8.
      HM0=1.6*(R/B)**2*G**0.6*U**0.7*W1**(-0.13)
      WRITE(*,*)N,X0,XE,W,E1,EDA0,R,US
      CALL SUBAK(N)
      CALL EHL(N)
      STOP
      END
      SUBROUTINE EHL(N)
      DIMENSION X(1100),P(1100),H(1100),RO(1100),POLD(1100),EPS(1100),
      EDA(1100),V(1100)
      COMMON /COM1/ENDA,A1,A2,A3,Z,HM0,DH/COM4/X0,XE
      COMMON /COM3/E1,PH,B,RR
      MK=1
      DX=(XE-X0)/(N-1.0)
      DO 10 I=1,N
      X(I)=X0+(I-1)*DX
      IF(ABS(X(I)).GE.1.0)P(I)=0.0
      IF(ABS(X(I)).LT.1.0)P(I)=SQRT(1.-X(I)*X(I))
10    CONTINUE
      CALL HREE(N,DX,X,P,H,RO,EPS,EDA,V)
      CALL FZ(N,P,POLD)
14    KK=19
      CALL ITER(N,KK,DX,X,P,H,RO,EPS,EDA,V)
      MK=MK+1
      CALL ERROP(N,P,POLD,ERP)
      WRITE(*,*)'ERP=',ERP
      IF(ERP.GT.1.E-5.AND.DH.GT.1.E-6)THEN
      IF(MK.GE.50)THEN
      MK=1
      DH=0.5*DH
      ENDIF
      GOTO 14
      ENDIF
      IF(DH.LE.1.E-6)WRITE(*,*)'Pressures are not convergent!!!'
      H2=1.E3
      P2=0.0
      DO 106 I=1,N
      IF(H(I).LT.H2)H2=H(I)
      IF(P(I).GT.P2)P2=P(I)
106   CONTINUE
      H3=H2*B*B/RR
      P3=P2*PH
110   FORMAT(6(1X,E12.6))
```

```
120 CONTINUE
    WRITE(*,*)'P2,H2,P3,H3=',P2,H2,P3,H3
    CALL OUTHP(N,X,P,H)
    RETURN
    END
    SUBROUTINE OUTHP(N,X,P,H)
    DIMENSION X(N),P(N),H(N)
    DO 10 I=1,N
    WRITE(8,20)X(I),P(I),H(I)
10  CONTINUE
20  FORMAT(1X,6(E12.6,1X))
    RETURN
    END
    SUBROUTINE HREE(N,DX,X,P,H,RO,EPS,EDA,V)
    DIMENSION X(N),P(N),H(N),RO(N),EPS(N),EDA(N),V(N)
    COMMON /COM1/ENDA,A1,A2,A3,Z,HM0,DH/COM2/EDA0/COMAK/AK(0:1100)
    DATA KK,PAI1,G0/0,0.318309886,1.570796325/
    DATA AN,AL/0.56,0.25/
    IF(KK.NE.0)GOTO 3
    H00=0.0
3   W1=0.0
    DO 4 I=1,N
4   W1=W1+P(I)
    C3=(DX*W1)/G0
    DW=1.-C3
    CALL VI(N,DX,P,V)
    HMIN=1.E3
    DO 30 I=1,N
    H0=0.5*X(I)*X(I)+V(I)
    IF(H0.LT.HMIN)HMIN=H0
    H(I)=H0
30  CONTINUE
    IF(KK.NE.0)GOTO 32
    KK=1
    DH=0.005*HM0
    H00=-HMIN+HM0
32  IF(DW.LT.0.0)H00=H00+DH
    IF(DW.GT.0.0)H00=H00-DH
    DO 60 I=1,N
    H(I)=H00+H(I)
    EDA(I)=EXP(A1*(-1.+(1.+A2*P(I))**Z))
    RO(I)=(A3+1.35*P(I))/(A3+P(I))
    CA=1.0
    IF(AL.NE.0.0)CA=COTH(0.5*AN*H(I)/AL)
    EPS(I)=RO(I)*(H(I)**3+12.0*AL*AL*H(I)-6.0*AN*AL*H(I)*H(I)*CA)/
    (ENDA*EDA(I))
60  CONTINUE
    RETURN
    END
    SUBROUTINE ITER(N,KK,DX,X,P,H,RO,EPS,EDA,V)
    DIMENSION X(N),P(N),H(N),RO(N),EPS(N),EDA(N),V(N)
```

```
      COMMON /COMAK/AK(0:1100)
      DATA PAI1/0.318309886/
      DO 100 K=1,KK
      D2=0.5*(EPS(1)+EPS(2))
      D3=0.5*(EPS(2)+EPS(3))
      DO 70 I=2,N-1
      D1=D2
      D2=D3
      IF(I.NE.N-1)D3=0.5*(EPS(I+1)+EPS(I+2))
      D8=RO(I)*AK(0)*PAI1
      D9=RO(I-1)*AK(1)*PAI1
      D10=1.0/(D1+D2+(D9-D8)*DX)
      D11=D1*P(I-1)+D2*P(I+1)
      D12=(RO(I)*H(I)-RO(I-1)*H(I-1)+(D8-D9)*P(I))*DX
      P(I)=(D11-D12)*D10
      IF(P(I).LT.0.0)P(I)=0.0
70    CONTINUE
      CALL HREE(N,DX,X,P,H,RO,EPS,EDA,V)
100   CONTINUE
      RETURN
      END
      SUBROUTINE VI(N,DX,P,V)
      DIMENSION P(N),V(N)
      COMMON /COMAK/AK(0:1100)
      PAI1=0.318309886
      C=ALOG(DX)
      DO 10 I=1,N
      V(I)=0.0
      DO 10 J=1,N
      IJ=IABS(I-J)
10    V(I)=V(I)+(AK(IJ)+C)*DX*P(J)
      DO I=1,N
      V(I)=-PAI1*V(I)
      ENDDO
      RETURN
      END
      SUBROUTINE SUBAK(MM)
      COMMON /COMAK/AK(0:1100)
      DO 10 I=0,MM
10    AK(I)=(I+0.5)*(ALOG(ABS(I+0.5))-1.)-(I-0.5)*(ALOG(ABS(I-0.5))-1.)
      RETURN
      END
      SUBROUTINE FZ(N,P,POLD)
      DIMENSION P(N),POLD(N)
      DO 10 I=1,N
10    POLD(I)=P(I)
      RETURN
      END
      SUBROUTINE ERROP(N,P,POLD,ERP)
      DIMENSION P(N),POLD(N)
      SD=0.0
```

```
      SUM=0.0
      DO 10 I=1,N
      SD=SD+ABS(P(I)-POLD(I))
      POLD(I)=P(I)
10    SUM=SUM+P(I)
      ERP=SD/SUM
      RETURN
      END
      FUNCTION COTH(X)
      COTH=(EXP(X)+EXP(-X))/(EXP(X)-EXP(-X))
      RETURN
      END
```

15.2.3 Example

The pressure and film thickness are obtained according to the given parameters in the above program, that is, AN = 0.56, AL = 0.25. The results are shown in Figure 15.1. It can be seen that under the same conditions, both film thickness and pressure of EHL increased when the effect of micropolar fluid is considered compared with those when the effect is not considered.

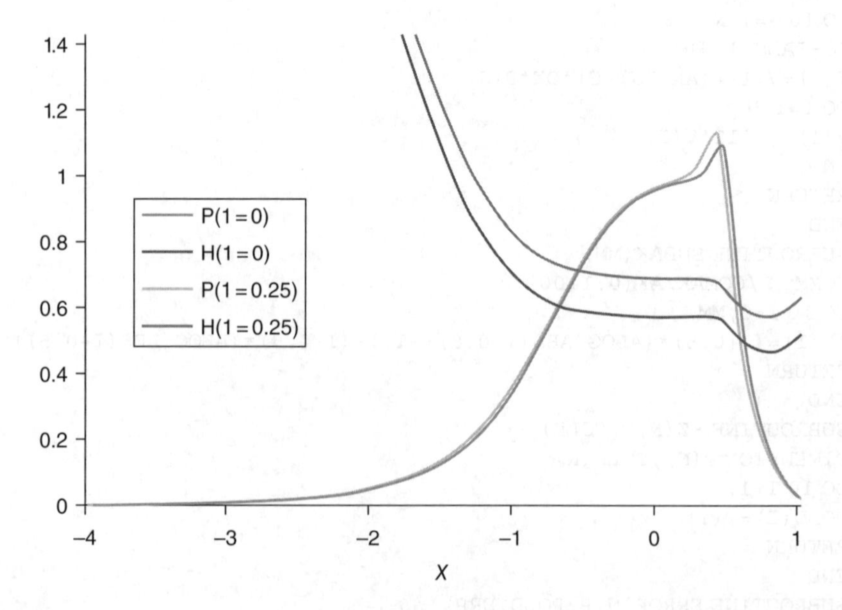

Figure 15.1 Pressure distribution and film thickness of EHL when micropolar effect is considered and not considered, respectively, in line contact

15.3 Calculation program for EHL with micropolar fluid in point contact

The calculation program for EHL of micropolar fluid in the point contact is similar to that in the line contact.

15.3.1 Calculation program

```
PROGRAM MICROPOLORPOINTEHL
    COMMON /COM1/ENDA,A1,A2,A3,Z,HM0,DH
    DATA PAI,Z/3.14159265,0.68/
    DATA N,PH,E1,EDA0,RX,US,X0,XE/
65,0.8E9,2.21E11,0.05,0.02,1.0,-2.5,1.5/
    OPEN(4,FILE='OUT.DAT',STATUS='UNKNOWN')
    OPEN(8,FILE='FILM.DAT',STATUS='UNKNOWN')
    OPEN(10,FILE='PRESSURE.DAT',STATUS='UNKNOWN')
    MM=N-1
    A1=ALOG(EDA0)+9.67
    A2=5.1E-9*PH
    A3=0.59/(PH*1.E-9)
    U=EDA0*US/(2.*E1*RX)
    B=PAI*PH*RX/E1
    W0=2.*PAI*PH/(3.*E1)*(B/RX)**2
    ALFA=Z*5.1E-9*A1
    G=ALFA*E1
    HM0=3.63*(RX/B)**2*G**0.49*U**0.68*W0**(-0.073)
    ENDA=12.*U*(E1/PH)*(RX/B)**3
    WRITE(*,*)N,X0,XE,W0,PH,E1,EDA0,RX,US
    WRITE(4,*)N,X0,XE,W0,PH,E1,EDA0,RX,US
    WRITE(*,*)'          Wait please'
    CALL SUBAK(MM)
    CALL EHL(N,X0,XE)
    STOP
    END

    SUBROUTINE EHL(N,X0,XE)
    DIMENSION X(65),Y(65),H(4500),RO(4500),EPS(4500),EDA(4500),P(4500),
POLD(4500),V(4500)
    COMMON /COM1/ENDA,A1,A2,A3,Z,HM0,DH
    DATA MK,G0/1,2.0943951/
    CALL INITI(N,DX,X0,XE,X,Y,P,POLD)
    KK=0
    CALL HREE(N,DX,KK,H00,G0,X,Y,H,RO,EPS,EDA,P,V)
14  KK=15
    CALL ITER(N,KK,DX,H00,G0,X,Y,H,RO,EPS,EDA,P,V)
    MK=MK+1
    CALL ERP(N,ER,P,POLD)
    WRITE(*,*)'ER=',ER
    IF(ER.GT.1.E-5.AND.DH.GT.1.E-6)THEN
```

```
      IF(MK.GE.10)THEN
      MK=1
      DH=0.5*DH
      ENDIF
      GOTO 14
      ENDIF
      IF(DH.LE.1.E-6)WRITE(*,*)'Pressures are not convergent!!!'
      CALL OUTPUT(N,DX,X,Y,H,P)
      RETURN
      END

      SUBROUTINE ERP(N,ER,P,POLD)
      DIMENSION P(N,N),POLD(N,N)
      ER=0.0
      SUM=0.0
      DO 10 I=1,N
      DO 10 J=1,N
      ER=ER+ABS(P(I,J)-POLD(I,J))
      POLD(I,J)=P(I,J)
      SUM=SUM+P(I,J)
10    CONTINUE
      ER=ER/SUM
      RETURN
      END

      SUBROUTINE INITI(N,DX,X0,XE,X,Y,P,POLD)
      DIMENSION X(N),Y(N),P(N,N),POLD(N,N)
      NN=(N+1)/2
      DX=(XE-X0)/(N-1.)
      Y0=-0.5*(XE-X0)
      DO 5 I=1,N
      X(I)=X0+(I-1)*DX
      Y(I)=Y0+(I-1)*DX
5     CONTINUE
      DO I=1,N
      D=1.-X(I)*X(I)
      DO J=1,NN
      C=D-Y(J)*Y(J)
      IF(C.LE.0.0)P(I,J)=0.0
      IF(C.GT.0.0)P(I,J)=SQRT(C)
      POLD(I,J)=P(I,J)
      ENDDO
      ENDDO
      RETURN
      END

      SUBROUTINE HREE(N,DX,KK,H00,G0,X,Y,H,RO,EPS,EDA,P,V)
      DIMENSION X(N),Y(N),P(N,N),H(N,N),RO(N,N),EPS(N,N),EDA(N,N),V(N,N)
      COMMON /COM1/ENDA,A1,A2,A3,Z,HM0,DH/COMAK/AK(0:65,0:65)
      DATA PAI,PAI1/3.14159265,0.2026423/
      DATA AN,AL/0.56,0.25/
      NN=(N+1)/2
```

```
      CALL VI(N,DX,P,V)
      HMIN=1.E3
      DO 30 I=1,N
      DO 30 J=1,NN
      RAD=X(I)*X(I)+Y(J)*Y(J)
      W1=0.5*RAD
      H0=W1+V(I,J)
      IF(H0.LT.HMIN)HMIN=H0
30    H(I,J)=H0
      IF(KK.EQ.0)THEN
      KK=1
      DH=0.05*HM0
      H00=-HMIN+HM0
      ENDIF
      W1=0.0
      DO 32 I=1,N
      DO 32 J=1,N
32    W1=W1+P(I,J)
      W1=DX*DX*W1/G0
      DW=1.-W1
      IF(DW.LT.0.0)H00=H00+DH
      IF(DW.GT.0.0)H00=H00-DH
      DO 60 I=1,N
      DO 60 J=1,NN
      H(I,J)=H00+H(I,J)
      IF(P(I,J).LT.0.0)P(I,J)=0.0
      EDA1=EXP(A1*(-1.+(1.+A2*P(I,J))**Z))
      EDA(I,J)=EDA1
      RO(I,J)=(A3+1.34*P(I,J))/(A3+P(I,J))
      CA=1.0
      IF(AL.NE.0.0)CA=COTH(0.5*AN*H(I,J)/AL)
60    EPS(I,J)=RO(I,J)*(H(I,J)**3+12.0*AL*AL*H(I,J)-6.0*AN*AL*H(I,J)*H
      (I,J)*CA)/(ENDA*EDA1)
      DO 70 J=NN+1,N
      JJ=N-J+1
      DO 70 I=1,N
      H(I,J)=H(I,JJ)
      RO(I,J)=RO(I,JJ)
      EDA(I,J)=EDA(I,JJ)
70    EPS(I,J)=EPS(I,JJ)
      RETURN
      END
      SUBROUTINE ITER(N,KK,DX,H00,G0,X,Y,H,RO,EPS,EDA,P,V)
      DIMENSION X(N),Y(N),P(N,N),H(N,N),RO(N,N),EPS(N,N),EDA(N,N),V(N,N)
      COMMON /COMAK/AK(0:65,0:65)
      DATA KG1,PAI/0,3.14159265/
      IF(KG1.NE.0)GOTO 2
      KG1=1
      AK00=AK(0,0)
      AK10=AK(1,0)
2     NN=(N+1)/2
      DO 100 K=1,KK
```

```
      DO 70 J=2,NN
      J0=J-1
      J1=J+1
      D2=0.5*(EPS(1,J)+EPS(2,J))
      DO 70 I=2,N-1
      I0=I-1
      I1=I+1
      D1=D2
      D2=0.5*(EPS(I1,J)+EPS(I,J))
      D4=0.5*(EPS(I,J0)+EPS(I,J))
      D5=0.5*(EPS(I,J1)+EPS(I,J))
      D8=2.0*RO(I,J)*AK00/PAI**2
      D9=2.0*RO(I0,J)*AK10/PAI**2
      D10=D1+D2+D4+D5+D8*DX-D9*DX
      D11=D1*P(I0,J)+D2*P(I1,J)+D4*P(I,J0)+D5*P(I,J1)
      D12=(RO(I,J)*H(I,J)-D8*P(I,J)-RO(I0,J)*H(I0,J)+D9*P(I,J))*DX
      P(I,J)=(D11-D12)/D10
      IF(P(I,J).LT.0.0)P(I,J)=0.0
70    CONTINUE
      DO 80 J=1,NN
      JJ=N+1-J
      DO 80 I=1,N
80    P(I,JJ)=P(I,J)
      CALL HREE(N,DX,KK,H00,G0,X,Y,H,RO,EPS,EDA,P,V)
100   CONTINUE
      RETURN
      END
      SUBROUTINE VI(N,DX,P,V)
      DIMENSION P(N,N),V(N,N)
      COMMON /COMAK/AK(0:65,0:65)
      PAI1=0.2026423
      DO 40 I=1,N
      DO 40 J=1,N
      H0=0.0
      DO 30 K=1,N
      IK=IABS(I-K)
      DO 30 L=1,N
      JL=IABS(J-L)
30    H0=H0+AK(IK,JL)*P(K,L)
40    V(I,J)=H0*DX*PAI1
      RETURN
      END
      SUBROUTINE SUBAK(MM)
      COMMON /COMAK/AK(0:65,0:65)
      S(X,Y)=X+SQRT(X**2+Y**2)
      DO 10 I=0,MM
      XP=I+0.5
      XM=I-0.5
      DO 10 J=0,I
      YP=J+0.5
      YM=J-0.5
      A1=S(YP,XP)/S(YM,XP)
      A2=S(XM,YM)/S(XP,YM)
```

```
      A3=S(YM,XM)/S(YP,XM)
      A4=S(XP,YP)/S(XM,YP)
      AK(I,J)=XP*ALOG(A1)+YM*ALOG(A2)+XM*ALOG(A3)+YP*ALOG(A4)
10    AK(J,I)=AK(I,J)
      RETURN
      END
      SUBROUTINE OUTPUT(N,DX,X,Y,H,P)
      DIMENSION X(N),Y(N),H(N,N),P(N,N)
      A=0.0
      WRITE(8,110)A,(Y(I),I=1,N)
      DO I=1,N
      WRITE(8,110)X(I),(H(I,J),J=1,N)
      ENDDO
      WRITE(10,110)A,(Y(I),I=1,N)
      DO I=1,N
      WRITE(10,110)X(I),(P(I,J),J=1,N)
      ENDDO
110   FORMAT(66(E12.6,1X))
      RETURN
      END
      FUNCTION COTH(X)
      COTH=(EXP(X)+EXP(-X))/(EXP(X)-EXP(-X))
      RETURN
      END
```

15.3.2 Example

The pressure and film thickness are obtained according to the given parameters in the above program, that is: AN = 0.56, AL = 0.25. The results are shown in Figure 15.2.

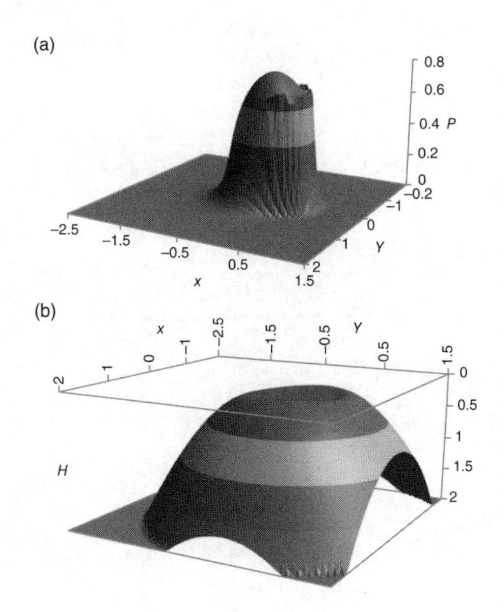

Figure 15.2 Pressure distribution and film thickness of EHL of micropolar fluid in point contact. (a) Pressure distribution and (b) film thickness

References

1. Shizhu Wen, Ping Huang. *Principles of Tribology*, Wiley, 2011.
2. Ping Huang. *Numerical Calculation of Lubrication Methods and Programs*, Wiley, 2013.
3. L G Houpert, Bernard J Hamrock. Fast approach for calculating film thicknesses and pressures in elastohydrodynamically lubricated contacts at heavy loads. *Trans. ASME J. Tribol.*, 1986, 108(3):411–420
4. Wen Shizhu, Ping Huang. *Summary of lubrication calculation methods of elastohydrodynamic lubrication. Fifth National tribology conference*. Wuhan: 1992, 198–204.
5. A A Lubrecht, W E ten Narel, R Bosma. Multigrid, an alternative method for calculating film thickness and pressure profiles in elastohydrodynamic lubricated line contacts. *Trans. ASME J. Tribol.*, 1989, 108(4):551–556, Oct. 21–26, 1992.
6. Huang Ping, Wen Shizhu. Solving elastohydrodynamic lubrication problem with multigrid method in line contact. *J. Tsinghua Univ.*, 1992, 32(5):26–34.
7. Wen Shizhu, Yang Peiran. *Elastohydrodynamic Lubrication*. Beijing: Tsinghua University Press, 1992.
8. Huang Ping, Wen Shizhu. Sectional micro-elastohydrodynamic lubrication. *Trans. ASME J. Tribol.*, 1993, 115(1):148–151.
9. Chandan Singh, Prawal Sinha. The three-dimensional Reynolds equation for micro-polar-fluid-lubricated bearings. *Wear*, 1982, 76(2):199–209.

Numerical Calculation of Elastohydrodynamic Lubrication: Methods and Programs, First Edition.
Ping Huang.

Index

Numerical Calculation of Elastohydrodynamic Lubrication: Methods and Programs, First Edition.
Ping Huang.